T0226019

# Mathematical
# Olympiad
## in China (2011–2014)

Problems and Solutions

# Mathematical Olympiad Series

ISSN: 1793-8570

**Series Editors:** Lee Peng Yee *(Nanyang Technological University, Singapore)*
Xiong Bin *(East China Normal University, China)*

*Published*

Vol. 15 *Mathematical Olympiad in China (2011–2014):*
*Problems and Solutions*
edited by Bin Xiong (East China Normal University, China) &
Peng Yee Lee (Nanyang Technological University, Singapore)

Vol. 14 *Probability and Expectation*
by Zun Shan (Nanjing Normal University, China)
translated by: Shanping Wang (East China Normal University, China)

Vol. 13 *Combinatorial Extremization*
by Yuefeng Feng (Shenzhen Senior High School, China)

Vol. 12 *Geometric Inequalities*
by Gangsong Leng (Shanghai University, China)
translated by: Yongming Liu (East China Normal University, China)

Vol. 11 *Methods and Techniques for Proving Inequalities*
by Yong Su (Stanford University, USA) &
Bin Xiong (East China Normal University, China)

Vol. 10 *Solving Problems in Geometry: Insights and Strategies for*
*Mathematical Olympiad and Competitions*
by Kim Hoo Hang (Nanyang Technological University, Singapore) &
Haibin Wang (NUS High School of Mathematics and Science, Singapore)

Vol. 9 *Mathematical Olympiad in China (2009–2010)*
edited by Bin Xiong (East China Normal University, China) &
Peng Yee Lee (Nanyang Technological University, Singapore)

Vol. 8 *Lecture Notes on Mathematical Olympiad Courses:*
*For Senior Section (In 2 Volumes)*
by Xu Jiagu (Fudan University, China)

The complete list of the published volumes in the series can be found at
http://www.worldscientific.com/series/mos

**Vol. 15** | Mathematical Olympiad Series

# Mathematical Olympiad

## in China (2011–2014)

## Problems and Solutions

Editors

## Xiong Bin
East China Normal University, China

## Lee Peng Yee
Nanyang Technological University, Singapore

 **East China Normal University Press**

 **World Scientific**

*Published by*

East China Normal University Press
3663 North Zhongshan Road
Shanghai 200062
China

and

World Scientific Publishing Co. Pte. Ltd.

5 Toh Tuck Link, Singapore 596224

*USA office:* 27 Warren Street, Suite 401-402, Hackensack, NJ 07601

*UK office:* 57 Shelton Street, Covent Garden, London WC2H 9HE


**Library of Congress Cataloging-in-Publication Data**
Names: Xiong, Bin. | Lee, P. Y. (Peng Yee), 1938–
Title: Mathematical Olympiad in China (2011–2014) : problems and solutions / editors, Bin Xiong,
    East China Normal University, China, Peng Yee Lee, Nanyang Technological University, Singapore.
Description: [Shanghai, China] : East China Normal University Press ; [Hackensack, New Jersey] :
    World Scientific, 2017. | Series: Mathematical olympiad series ; vol. 15
Identifiers: LCCN 2017002034| ISBN 9789813143746 (hardcover : alk. paper) |
    ISBN 9813143746 (hardcover : alk. paper) | ISBN 9789813142930 (pbk. : alk. paper) |
    ISBN 9813142936 (pbk. : alk. paper)
Subjects: LCSH: International Mathematical Olympiad. | Mathematics--Problems, exercises, etc. |
    Mathematics--Competitions--China.
Classification: LCC QA43 .M31453 2017 | DDC 510.76--dc23
LC record available at https://lccn.loc.gov/2017002034


**British Library Cataloguing-in-Publication Data**
A catalogue record for this book is available from the British Library.

For any available supplementary material, please visit
http://www.worldscientific.com/worldscibooks/10.1142/10113#t=suppl


Printed in Singapore

# Preface

The first time China participated in IMO was in 1985, when two students were sent to the 26th IMO. Since 1986, China has a team of six students at every IMO except in 1998 when it was held in Taiwan. So far, up to 2014, China has achieved the number one ranking for 20 times in team effort. A great majority of students received gold medals. The fact that China obtained such encouraging result is due to, on the one hand, Chinese students' hard work and perseverance, and on the other hand, the effort of teachers in schools and the training offered by national coaches. We believe it is also a result of the education system in China, in particular, the emphasis on training of basic skills in science education.

The materials of this book come from a series of four books (in Chinese) on *Forward to IMO: A Collection of Mathematical Olympiad Problems* (2011 - 2014). It is a collection of problems and solutions of the major mathematical competitions in China. It provides a glimpse of how the China national team is selected and formed. First, there is the China Mathematical Competition, a national event. It is held on the second Sunday of October every year. Through the competition, about 300 students are selected to join the China Mathematical Olympiad (commonly known as the winter camp), or in short CMO, in the following January. CMO lasts for five days. Both the type and the difficulty of the problems match those of IMO. Similarly, students are given three problems to solve in 4.5 hours each day. From CMO, about 50 to 60 students are selected to form a national training team. The training takes place for two weeks in the month of March. After four to six tests, plus two qualifying examinations, six students are finally selected to form the national team, taking part in IMO in July of that year.

In view of the differences in education, culture and economy of western part of China in comparison with the coastal part in East China, mathematical competitions in West China did not develop as fast as in the rest of the country. In order to promote the activity of mathematical competition and to enhance the level of mathematical competition, starting from 2001, China Mathematical Olympiad Committee organizes the China Western Mathematical Olympiad. The top two winners will be admitted to the national training team. Through the CWMO, there have been two students entering the national team and receiving Gold Medals for their performance at IMO.

Since 1995, for a quite long period there was no female student in the Chinese national team. In order to encourage more female students participating in the mathematical competition, starting from 2002, China Mathematical Olympiad Committee conducted the China Girls' Mathematical Olympiad. Again, the top two winners will be admitted directly into the national training team.

The authors of this book are coaches of the China national team. They are Xiong Bin, Li Shenghong, Leng Gangsong, Wu Jianping, Chen Yonggao, Li Weigu, Yu Hongbing, Zhu

Huawei, Feng Zhigang, Liu Shixiong, Qu Zhenhua, Wang Weiye, and Zhang Sihui. Those who took part in the translation work are Xiong Bin, Wang Shanping, Liu Yongming. We are grateful to Qiu Zonghu, Wang Jie, Wu Jianping, and Pan Chengbiao for their guidance and assistance to authors. We are grateful to Ni Ming of East China Normal University Press. Their effort has helped make our job easier. We are also grateful to Zhang Ji of World Scientific Publishing for her hard work leading to the final publication of the book.

Xiong Bin
October 2017

# Introduction

Early days

The International Mathematical Olympiad ( IMO ), founded in 1959, is one of the most competitive and highly intellectual activities in the world for high school students.

Even before IMO, there were already many countries which had mathematics competition. They were mainly the countries in Eastern Europe and in Asia. In addition to the popularization of mathematics and the convergence in educational systems among different countries, the success of mathematical competitions at the national level provided a foundation for the setting-up of IMO. The countries that asserted great influence are Hungary, the former Soviet Union and the United States. Here is a brief review of the IMO and mathematical competition in China.

In 1894, the Department of Education in Hungary passed a motion and decided to conduct a mathematical competition for the secondary schools. The well-known scientist, *J. von Etövös*, was the Minister of Education at that time. His support in the event had made it a success and thus it was well publicized. In addition, the success of his son, *R. von Etövös*, who was also a physicist, in proving the principle of equivalence of the general theory of relativity by *A. Einstein* through experiment, had brought Hungary to the world stage in science. Thereafter, the prize for mathematics competition in Hungary was named "*Etövös* prize". This was the first formally organized mathematical competition in the world. In what follows, Hungary had indeed produced a lot of well-known scientists including *L. Fejér*, *G. Szegö*, *T. Radó*, *A. Haar* and *M. Riesz* (in real analysis), *D. König* (in combinatorics), *T. von Kármán* (in aerodynamics), and *J. C. Harsanyi* (in game theory), who had also won the Nobel Prize for Economics in 1994. They all were the winners of Hungary mathematical competition. The top scientific genius of Hungary, *J. von Neumann*, was one of the leading mathematicians in the 20th century. *Neumann* was overseas while the competition took place. Later he did it himself and it took him half an hour to complete. Another mathematician worth mentioning is the highly productive number theorist *P. Erdös*. He was a pupil of *Fejér* and also a winner of the Wolf Prize. *Erdös* was very passionate about mathematical competition and setting competition questions. His contribution to discrete mathematics was unique and greatly significant. The rapid progress and development of discrete mathematics over the subsequent decades had indirectly influenced the types of questions set in IMO. An internationally recognized prize named after Erdös was to honour those who had contributed to the education of mathematical competition. Professor *Qiu Zonghu* from China had won the prize in 1993.

In 1934, a famous mathematician *B. Delone* conducted a mathematical competition for high school students in Leningrad (now St. Petersburg). In 1935, Moscow also started organizing such event. Other than being interrupted during the World War

II, these events had been carried on until today. As for the Russian Mathematical Competition (later renamed as the Soviet Mathematical Competition), it was not started until 1961. Thus, the former Soviet Union and Russia became the leading powers of Mathematical Olympiad. A lot of grandmasters in mathematics including the great *A. N. Kolmogorov* were all very enthusiastic about the mathematical competition. They would personally involve in setting the questions for the competition. The former Soviet Union even called it the Mathematical Olympiad, believing that mathematics is the "gymnastics of thinking". These points of view gave a great impact on the educational community. The winner of the Fields Medal in 1998, *M. Kontsevich*, was once the first runner-up of the Russian Mathematical Competition. *G. Kasparov*, the international chess grandmaster, was once the second runner-up. *Grigori Perelman*, the winner of the Fields Medal in 2006 (but he declined), who solved the Poincaré's Conjecture, was a gold medalist of IMO in 1982.

In the United States of America, due to the active promotion by the renowned mathematician *G. D. Birkhoff* and his son, together with *G. Pólya*, the Putnam mathematics competition was organized in 1938 for junior undergraduates. Many of the questions were within the scope of high school students. The top five contestants of the Putnam mathematical competition would be entitled to the membership of Putnam. Many of these were eventually outstanding mathematicians. There were the famous *R. Feynman* (winner of the Nobel Prize for Physics, 1965), *K. Wilson* (winner of the Nobel Prize for Physics, 1982), *J. Milnor* (winner of the Fields Medal, 1962), *D. Mumford* (winner of the Fields Medal, 1974), and *D.*

*Quillen* (winner of the Fields Medal, 1978).

Since 1972, in order to prepare for the IMO, the United States of America Mathematical Olympiad (USAMO) was organized. The standard of questions posed was very high, parallel to that of the Winter Camp in China. Prior to this, the United States had organized American High School Mathematics Examination (AHSME) for the high school students since 1950. This was at the junior level yet the most popular mathematics competition in America. Originally, it was planned to select about 100 contestants from AHSME to participate in USAMO. However, due to the discrepancy in the level of difficulty between the two competitions and other restrictions, from 1983 onwards, an intermediate level of competition, namely, American Invitational Mathematics Examination (AIME), was introduced. Henceforth both AHSME and AIME became internationally well-known. A few cities in China had participated in the competition and the results were encouraging.

Similarly as in the former Soviet Union, the Mathematical Olympiad education was widely recognized in America. The book "How to Solve it" written by *George Polya* along with many other titles had been translated into many different languages. *George Polya* provided a whole series of general heuristics for solving problems of all kinds. His influence in the educational community in China should not be underestimated.

**International Mathematical Olympiad**

In 1956, the East European countries and the Soviet Union took the initiative to organize the IMO formally. The first International Mathematical Olympiad (IMO) was held in

Brasov, Romania, in 1959. At the time, there were only seven participating countries, namely, Romania, Bulgaria, Poland, Hungary, Czechoslovakia, East Germany and the Soviet Union. Subsequently, the United States of America, United Kingdom, France, Germany and also other countries including those from Asia joined. Today, the IMO had managed to reach almost all the developed and developing countries. Except in the year 1980 due to financial difficulties faced by the host country, Mongolia, there were already 49 Olympiads held and 97 countries participating.

The mathematical topics in the IMO include Algebra, Combinatorics, Geometry, Number theory. These areas had provided guidance for setting questions for the competitions. Other than the first few Olympiads, each IMO is normally held in mid-July every year and the test paper consists of 6 questions in all. The actual competition lasts for 2 days for a total of 9 hours where participants are required to complete 3 questions each day. Each question is 7 points with total up to 42 points. The full score for a team is 252 marks. About half of the participants will be awarded a medal, where 1/12 will be awarded a gold medal. The numbers of gold, silver and bronze medals awarded are in the ratio of 1:2:3 approximately. In the case when a participant provides a better solution than the official answer, a special award is given.

Each participating country will take turn to host the IMO. The cost is borne by the host country. China had successfully hosted the 31st IMO in Beijing. The event had made a great impact on the mathematical community in China. According to the rules and regulations of the IMO, all participating countries are required to send a delegation consisting of a leader, a deputy leader and 6 contestants. The problems are contributed by the participating countries and are later selected carefully by the host country for submission to the international jury set up by the host country. Eventually, only 6 problems will be accepted for use in the competition. The host country does not provide any question. The short-listed problems are subsequently translated, if necessary, in English, French, German, Russian and other working languages. After that, the team leaders will translate the problems into their own languages.

The answer scripts of each participating team will be marked by the team leader and the deputy leader. The team leader will later present the scripts of their contestants to the coordinators for assessment. If there is any dispute, the matter will be settled by the jury. The jury is formed by the various team leaders and an appointed chairman by the host country. The jury is responsible for deciding the final 6 problems for the competition. Their duties also include finalizing the grading standard, ensuring the accuracy of the translation of the problems, standardizing replies to written queries raised by participants during the competition, synchronizing differences in grading between the team leaders and the coordinators and also deciding on the cut-off points for the medals depending on the contestants' results as the difficulties of problems each year are different.

China had participated informally in the 26th IMO in 1985. Only two students were sent. Starting from 1986, except in 1998 when the IMO was held in Taiwan, China had always sent 6 official contestants to the IMO. Today, the Chinese contestants not only performed outstandingly in the IMO, but also in the International Physics, Chemistry, Informatics, and Biology Olympiads. This can be regarded as an indication that China pays great attention to the training of basic skills in mathematics and science education.

### Winners of the IMO

Among all the IMO medalists, there were many of them who eventually became great mathematicians. They were also awarded the Fields Medal, Wolf Prize and Nevanlinna Prize (a prominent mathematics prize for computing and informatics). In what follows, we name some of the winners.

*G. Margulis*, a silver medalist of IMO in 1959, was awarded the Fields Medal in 1978. *L. Lovasz*, who won the Wolf Prize in 1999, was awarded the Special Award in IMO consecutively in 1965 and 1966. *V. Drinfeld*, a gold medalist of IMO in 1969, was awarded the Fields Medal in 1990. *J. -C. Yoccoz* and *T. Gowers*, who were both awarded the Fields Medal in 1998, were gold medalists in IMO in 1974 and 1981 respectively. A silver medalist of IMO in 1985, *L. Lafforgue*, won the Fields Medal in 2002. A gold medalist of IMO in 1982, *Grigori Perelman* from Russia, was awarded the Fields Medal in 2006 for solving the final step of the Poincaré conjecture. In 1986, 1987, and 1988, *Terence Tao* won a bronze, silver, and gold medal respectively. He was the youngest participant to date in the IMO, first competing at the age of ten. He was also awarded the Fields Medal in 2006. Gold medalist of IMO 1988 and 1989, *Ngo Bau Chao*, won the Fields Medal in 2010, together with the bronze medalist of IMO 1988, *E. Lindenstrauss*. Gold medalist of IMO 1994 and 1995, **Maryam Mirzakhani** won the Fields Medal in 2014. A gold medalist of

IMO in 1995, Artur Avila won the Fields Medal in 2014.

A silver medalist of IMO in 1977, *P. Shor*, was awarded the Nevanlinna Prize. A gold medalist of IMO in 1979, *A. Razborov*, was awarded the Nevanlinna Prize. Another gold medalist of IMO in 1986, *S. Smirnov*, was awarded the Clay Research Award. *V. Lafforgue*, a gold medalist of IMO in 1990, was awarded the European Mathematical Society prize. He is *L. Lafforgue*'s younger brother.

Also, a famous mathematician in number theory, *N. Elkies*, who is also a professor at Harvard University, was awarded a gold medal of IMO in 1982. Other winners include *P. Kronheimer* awarded a silver medal in 1981 and *R. Taylor* a contestant of IMO in 1980.

### Mathematical competition in China

Due to various reasons, mathematical competition in China started relatively late but is progressing vigorously.

"We are going to have our own mathematical competition too!" said *Hua Luogeng*. *Hua* is a house-hold name in China. The first mathematical competition was held concurrently in Beijing, Tianjin, Shanghai and Wuhan in 1956. Due to the political situation at the time, this event was interrupted a few times. Until 1962, when the political environment started to improve, Beijing and other cities started organizing the competition though not regularly. In the era of Cultural Revolution, the whole educational system in China was in chaos. The mathematical competition came to a complete halt. In contrast, the mathematical competition in the former Soviet Union was still on-going during the war and at a time under the difficult political situation. The competitions in Moscow were interrupted only 3 times between 1942 and 1944. It was indeed commendable.

In 1978, it was the spring of science. *Hua Luogeng* conducted the Middle School Mathematical Competition for 8 provinces in China. The mathematical competition in China was then making a fresh start and embarked on a road of rapid development. *Hua* passed away in 1985. In commemorating him, a competition named *Hua Luogeng* Gold Cup was set up in 1986 for students in Grade 6 and 7 and it has a great impact.

The mathematical competitions in China before 1980 can be considered as the initial period. The problems were set within the scope of middle school textbooks. After 1980, the competitions were gradually moving towards the senior middle school level. In 1981, the Chinese Mathematical Society decided to conduct the China Mathematical Competition, a national event for high schools.

In 1981, the United States of America, the host country of IMO, issued an invitation to China to participate in the event. Only in 1985, China sent two contestants to participate informally in the IMO. The results were not encouraging. In view of this, another activity called the Winter Camp was conducted after the China Mathematical Competition. The Winter Camp was later renamed as the China Mathematical Olympiad or CMO. The winning team would be awarded the *Chern Shiing-Shen* Cup. Based on the outcome at the Winter Camp, a selection would be made to form the 6-member national team for IMO. From 1986 onwards, other than the year when IMO was organized in Taiwan, China had been sending a 6-member team to IMO. Up to 2011, China had been awarded the overall team champion for 17 times.

In 1990, China had successfully hosted the 31st IMO. It showed that the standard of mathematical competition in China has leveled that of other leading countries. First, the fact that China achieves the highest marks at the 31st IMO for the team is an evidence of the effectiveness of the pyramid approach in selecting the contestants in China. Secondly, the Chinese mathematicians had simplified and modified over 100 problems and submitted them to the team leaders of the 35 countries for their perusal. Eventually, 28 problems were recommended. At the end, 5 problems were chosen (IMO requires 6 problems). This is another evidence to show that China has achieved the highest quality in setting problems. Thirdly, the answer scripts of the participants were marked by the various team leaders and assessed by the coordinators who were nominated by the host countries. China had formed a group 50 mathematicians to serve as coordinators who would ensure the high accuracy and fairness in marking. The marking process was completed half a day earlier than it was scheduled. Fourthly, that was the first ever IMO organized in Asia. The outstanding performance by China had encouraged the other developing countries, especially those in Asia. The organizing and coordinating work of the IMO by the host country was also reasonably good.

In China, the outstanding performance in mathematical competition is a result of many contributions from the all quarters of mathematical community. There are the older generation of mathematicians, middle-aged mathematicians and also the middle and elementary school teachers. There is one person who deserves a special mention and he is *Hua Luogeng*. He initiated and promoted the mathematical competition. He is also the author of the following books: Beyond *Yang hui's*

Triangle, Beyond the *pi* of *Zu Chongzhi*, Beyond the Magic Computation of *Sun-zi*, Mathematical Induction, and Mathematical Problems of Bee Hive. These were his books derived from mathematics competitions. When China resumed mathematical competition in 1978, he participated in setting problems and giving critique to solutions of the problems. Other outstanding books derived from the Chinese mathematics competitions are: Symmetry by *Duan Xuefu*, Lattice and Area by *Min Sihe*, One Stroke Drawing and Postman Problem by *Jiang Boju*.

After 1980, the younger mathematicians in China had taken over from the older generation of mathematicians in running the mathematical competition. They worked and strived hard to bring the level of mathematical competition in China to a new height. *Qiu Zonghu* is one such outstanding representative. From the training of contestants and leading the team 3 times to IMO to the organizing of the 31th IMO in China, he had contributed prominently and was awarded the *P. Erdös* prize.

**Preparation for IMO**

Currently, the selection process of participants for IMO in China is as follows.

First, the China Mathematical Competition, a national competition for high Schools, is organized on the second Sunday in October every year. The objectives are: to increase the interest of students in learning mathematics, to promote the development of co-curricular activities in mathematics, to help improve the teaching of mathematics in high schools, to discover and cultivate the talents and also to prepare for the

IMO. This happens since 1981. Currently there are about 200,000 participants taking part.

Through the China Mathematical Competition, around 350 of students are selected to take part in the China Mathematical Olympiad or CMO, that is, the Winter Camp. The CMO lasts for 5 days and is held in January every year. The types and difficulties of the problems in CMO are similar to the IMO. There are also 3 problems to be completed within 4.5 hours each day. However, the score for each problem is 21 marks which add up to 126 marks in total. Starting from 1990, the Winter Camp instituted the *Chern Shiing-Shen* Cup for team championship. In 1991, the Winter Camp was officially renamed as the China Mathematical Olympiad (CMO). It is similar to the highest national mathematical competition in the former Soviet Union and the United States.

The CMO awards the first, second and third prizes. Among the participants of CMO, about 60 students are selected to participate in the training for IMO. The training takes place in March every year. After 6 to 8 tests and another 2 rounds of qualifying examinations, only 6 contestants are short-listed to form the China IMO national team to take part in the IMO in July.

Besides the China Mathematical Competition (for high schools), the Junior Middle School Mathematical Competition is also developing well. Starting from 1984, the competition is organized in April every year by the Popularization Committee of the Chinese Mathematical Society. The various provinces, cities and autonomous regions would rotate to host the event. Another mathematical competition for the junior middle schools is also conducted in April every year by the Middle School

Mathematics Education Society of the Chinese Educational Society since 1998 till now.

The *Hua Luogeng* Gold Cup, a competition by invitation, had also been successfully conducted since 1986. The participating students comprise elementary six and junior middle one students. The format of the competition consists of a preliminary round, semi-finals in various provinces, cities and autonomous regions, then the finals.

Mathematical competition in China provides a platform for students to showcase their talents in mathematics. It encourages learning of mathematics among students. It helps identify talented students and to provide them with differentiated learning opportunity. It develops co-curricular activities in mathematics. Finally, it brings about changes in the teaching of mathematics.

# Contents

# China Mathematical Competition

*2010* （Fujian）

Commissioned by Chinese Mathematical Society, Fujian Mathematical Society organized the 2010 China Mathematical Competition held on October 17, 2010.

Compared to the competitions in the previous years, while the test time and problem types remain unchanged in this competition, the allocation of marks to each problem is adjusted slightly to make it more reasonable.

The time for the first round test is 80 minutes, and that for the supplementary test is 150 minutes.

**Part I   Short-Answer Questions (Questions 1 – 8, eight marks each)**

**①** The range of $f(x) = \sqrt{x-5} - \sqrt{24-3x}$ is _____ .

**Solution.** It is easy to see that $f(x)$ is increasing on its domain $[5, 8]$. Therefore, its range is $[-3, \sqrt{3}]$.    □

**②** The minimum of $y = (a\cos^2 x - 3)\sin x$ is $-3$. Then the range of real number $a$ is _____ .

**Solution.** Let $\sin x = t$. The expression is then changed to $g(t) = (-at^2 + a - 3)t$, or

$$g(t) = -at^3 + (a-3)t.$$

From $-at^3 + (a-3)t \geqslant -3$, we get

$$-at(t^2 - 1) - 3(t-1) \geqslant 0,$$
$$(t-1)(-at(t+1) - 3) \geqslant 0.$$

Since $t - 1 \leqslant 0$, we have $-at(t+1) - 3 \leqslant 0$, or

$$a(t^2 + t) \geqslant -3. \qquad \qquad ①$$

When $t = 0, -1$, expression ① always holds; when $0 < t \leqslant 1$, we have $0 < t^2 + t \leqslant 2$; and when $-1 < t < 0$, $-\dfrac{1}{4} \leqslant t^2 + t < 0$. Therefore, $-\dfrac{3}{2} \leqslant a \leqslant 12$.    □

**③** The number of integral points (i.e., the points whose $x$- and $y$-coordinates are both integers) within the area (not including the boundary) enclosed by the right branch of hyperbola $x^2 - y^2 = 1$ and line $x = 100$ is _____ .

**Solution.** By symmetry, we only need to consider the part of the area above the $x$-axis. Suppose line $y = k$ intercepts the right branch of the hyperbola and line $x = 100$ at points $A_k$ and $B_k$ ($k = 1, 2, \ldots, 99$), respectively. Then the number of integral points within the segment $A_kB_k$ is $99 - k$. Therefore, the number of integral points within the area above the $x$-axis is

$$\sum_{k=1}^{99} (99 - k) = 99 \times 49 = 4851.$$

Finally, we obtain the total number of integral points within the whole area as $2 \times 4851 + 98 = 9800$.  □

**④** It is known that $\{a_n\}$ is an arithmetic sequence with non-zero common difference and $\{b_n\}$ a geometric sequence, satisfying $a_1 = 3$, $b_1 = 1$, $a_2 = b_2$, $3a_5 = b_3$; furthermore, there are constants $\alpha$ and $\beta$ such that for every positive integer $n$, we have $a_n = \log_a b_n + \beta$. Then $\alpha + \beta = $ _____.

**Solution.** Let the common difference of $\{a_n\}$ be $d$ and the common ratio of $\{b_n\}$ be $q$. Then

$$3 + d = q, \tag{①}$$

$$3(3 + 4d) = q^2. \tag{②}$$

Substituting ① into ②, we have $9 + 12d = d^2 + 6d + 9$. Then we get $d = 6$ and $q = 9$.

Therefore, $3 + 6(n - 1) = \log_a 9^{n-1} + \beta$ or $6n - 3 = (n - 1)\log_a 9 + \beta$ holds for every positive integer $n$. Letting $n = 1$ and $n = 2$ in turn, we find that $\alpha = \sqrt[3]{3}$ and $\beta = 3$.

Consequently, $\alpha + \beta = \sqrt[3]{3} + 3$.  □

**⑤** Function $f(x) = a^{2x} + 3a^x - 2$ ($a > 0$, $a \neq 1$) reaches the maximum value 8 on interval $[-1, 1]$. Then its minimum value on this interval is _____ .

**Solution.** Let $a^x = y$. The original function is then changed to

$g(y) = y^2 + 3y - 2$, which is increasing over $\left(-\dfrac{3}{2}, +\infty\right)$.

When $0 < a < 1$, we have $y \in [a, a^{-1}]$ and

$$g(y)_{\max} = a^{-2} + 3a^{-1} - 2 = 8 \Rightarrow a^{-1} = 2 \Rightarrow a = \dfrac{1}{2}.$$

Then

$$g(y)_{\min} = \left(\dfrac{1}{2}\right)^2 + 3 \times \dfrac{1}{2} - 2 = -\dfrac{1}{4}.$$

When $a > 1$, we have $y \in [a^{-1}, a]$ and

$$g(y)_{\max} = a^2 + 3a - 2 = 8 \Rightarrow a = 2.$$

Then     $g(y)_{\min} = 2^{-2} + 3 \times 2^{-1} - 2 = -\dfrac{1}{4}.$

In summary, the minimum value of $f(x)$ on $x \in [-1, 1]$ is $-\dfrac{1}{4}$.

$\square$

**6**  Two persons roll two dice in turn. Whoever gets the sum number greater than 6 first will win the game. The probability for the person rolling first to win is _____ .

**Solution.** The probability for rolling two dice to get the sum number greater than 6 is $\dfrac{21}{36} = \dfrac{7}{12}$. Therefore, the required probability is

$$\dfrac{7}{12} + \left(\dfrac{5}{12}\right)^2 \times \dfrac{7}{12} + \left(\dfrac{5}{12}\right)^4 \times \dfrac{7}{12} + \cdots = \dfrac{7}{12} \times \dfrac{1}{1 - \dfrac{25}{144}} = \dfrac{12}{17}.$$

$\square$

**7** The lengths of the nine edges of regular triangular prism $ABC - A_1B_1C_1$ are equal, $P$ is the midpoint of $CC_1$, and the dihedral angle $B - A_1P - B_1 = \alpha$. Then $\sin \alpha = $ _____.

**Solution 1.** Let the line through segment $AB$ be $x$-axis with the origin $O$ being the midpoint of $AB$ and let the line through segment $OC$ be $y$-axis to establish a space rectangular coordinate system shown in Fig. 7.1. Assuming the length of an edge is 2, we have $B(1, 0, 0)$, $B_1(1, 0, 2)$, $A_1(-1, 0, 2)$, $P(0, \sqrt{3}, 1)$. Then

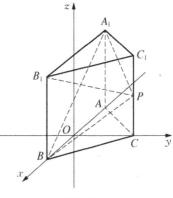

Fig. 7.1

$$\overrightarrow{BA_1} = (-2, 0, 2), \overrightarrow{BP} = (-1, \sqrt{3}, 1),$$
$$\overrightarrow{B_1A_1} = (-2, 0, 0), \overrightarrow{B_1P} = (-1, \sqrt{3}, -1).$$

Let vectors $\vec{m} = (x_1, y_1, z_1)$ and $\vec{n} = (x_2, y_2, z_2)$ be perpendicular to $BA_1P$ and $B_1A_1P$, respectively. We have

$$\begin{cases} \vec{m} \cdot \overrightarrow{BA_1} = -2x_1 + 2z_1 = 0, \\ \vec{m} \cdot \overrightarrow{BP} = -x_1 + \sqrt{3} y_1 + z_1 = 0, \end{cases}$$
$$\begin{cases} \vec{n} \cdot \overrightarrow{B_1A_1} = -2x_2 = 0, \\ \vec{n} \cdot \overrightarrow{B_1P} = -x_2 + \sqrt{3} y_2 - z_2 = 0. \end{cases}$$

We can then assume that $\vec{m} = (1, 0, 1)$, $\vec{n} = (0, 1, \sqrt{3})$. From $| \vec{m} \cdot \vec{n} | = | \vec{m} | \cdot | \vec{n} | | \cos \alpha |$, we have

$$\sqrt{3} = \sqrt{2} \cdot 2 | \cos \alpha | \Rightarrow | \cos \alpha | = \frac{\sqrt{6}}{4}.$$

Therefore, $\sin\alpha = \dfrac{\sqrt{10}}{4}$.

**Solution 2.** As seen in Fig. 7. 2, we have
$PC = PC_1$, $PA_1 = PB$. Suppose $A_1B$ and
$AB_1$ intersect at $O$. Then we get $OA_1 = OB$, $OA = OB_1$, $A_1B \perp AB_1$.

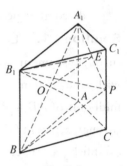

Fig. 7. 2

Since $PA = PB_1$, then $PO \perp AB_1$.
Therefore $AB_1 \perp$ plane $PA_1B$.

On plane $PA_1B$ through $O$ draw line
$OE \perp A_1P$ with foot point $E$.

Connecting $B_1E$, $\angle B_1EO$ is then the plane angle of the
dihedral angle $B-A_1P-B_1$. Assuming $AA_1 = 2$, it is easy to find
that $PB = PA_1 = \sqrt{5}$, $A_1O = B_1O = \sqrt{2}$, $PO = \sqrt{3}$.

In right triangle $\triangle PA_1O$, we have $A_1O \cdot PO = A_1P \cdot OE$,
or $\sqrt{2} \cdot \sqrt{3} = \sqrt{5} \cdot OE$. So $OE = \dfrac{\sqrt{6}}{\sqrt{5}}$.

As $B_1O = \sqrt{2}$, we have

$$B_1E = \sqrt{B_1O^2 + OE^2} = \sqrt{2 + \frac{6}{5}} = \frac{4\sqrt{5}}{5}.$$

Finally,

$$\sin\alpha = \sin\angle B_1EO = \frac{B_1O}{B_1E} = \frac{\sqrt{2}}{\dfrac{4\sqrt{5}}{5}} = \frac{\sqrt{10}}{4}. \qquad \square$$

**8** The number of positive integer solutions of equation $x + y + z = 2010$ with $x \leqslant y \leqslant z$ is _____.

**Solution.** It is easy to find that the number of positive integer
solutions of $x + y + z = 2010$ is $C_{2009}^2 = 2009 \times 1004$.

We now classify these solutions into three categories:

(1) $x = y = z$, the number in this category is obviously 1;

(2) there are exactly two that are equal among $x$, $y$, $z$ — the number in this category is 1003;

(3) $x$, $y$, $z$ are different from each other — suppose the number in this category is $k$.

From

$$1 + 3 \times 1003 + 6k = 2009 \times 1004,$$

we have

$$
\begin{aligned}
6k &= 2009 \times 1004 - 3 \times 1003 - 1 \\
&= 2006 \times 1005 - 2009 + 3 \times 2 - 1 \\
&= 2006 \times 1005 - 2004.
\end{aligned}
$$

We get $\quad k = 1003 \times 335 - 334 = 335\,671.$

Therefore, the number of positive integer solutions satisfying $x \leqslant y \leqslant z$ is

$$1 + 1003 + 335\,671 = 336\,675. \qquad \square$$

**Part II  Word Problems (16 marks for Question 9, 20 marks each for Questions 10 and 11, and then 56 marks in total)**

**9**  It is known that $f(x) = ax^3 + bx^2 + cx + d$ $(a \neq 0)$, and $|f'(x)| \leqslant 1$ for $0 \leqslant x \leqslant 1$. Please find the maximum value of $a$.

**Solution 1.** $f'(x) = 3ax^2 + 2bx + c$. We have

$$
\begin{cases}
f'(0) = c, \\
f'\left(\dfrac{1}{2}\right) = \dfrac{3}{4}a + b + c, \\
f'(1) = 3a + 2b + c.
\end{cases}
$$

Then

$$3a = 2f'(0) + 2f'(1) - 4f'\left(\frac{1}{2}\right).$$

We get

$$3 \mid a \mid = \left| 2f'(0) + 2f'(1) - 4f'\left(\frac{1}{2}\right) \right|$$

$$\leqslant 2 \mid f'(0) \mid + 2 \mid f'(1) \mid + 4 \left| f'\left(\frac{1}{2}\right) \right| \leqslant 8.$$

Therefore, $a \leqslant \dfrac{8}{3}$. Furthermore, it is easy to find that

$f(x) = \dfrac{8}{3}x^3 - 4x^2 + x + m$ (where $m$ is any constant) satisfies

the given condition. Therefore, the maximum value of $a$ is $\dfrac{8}{3}$.

**Solution 2.** Let $g(x) = f'(x) + 1$. Then $0 \leqslant g(x) \leqslant 2$ for $0 \leqslant$

$x \leqslant 1$. Let $z = 2x - 1$. Then $x = \dfrac{z+1}{2}$ and $-1 \leqslant z \leqslant 1$. Let

$$h(z) = g\left(\frac{z+1}{2}\right) = \frac{3a}{4}z^2 + \frac{3a+2b}{2}z + \frac{3a}{4} + b + c + 1.$$

It is easy to check that $0 \leqslant h(z) \leqslant 2$ and $0 \leqslant h(-z) \leqslant 2$ for

$-1 \leqslant z \leqslant 1$.

Therefore, $0 \leqslant \dfrac{h(z) + h(-z)}{2} \leqslant 2$ for $-1 \leqslant z \leqslant 1$. And

that is

$$0 \leqslant \frac{3a}{4}z^2 + \frac{3a}{4} + b + c + 1 \leqslant 2.$$

Then we have $\dfrac{3a}{4} + b + c + 1 \geqslant 0$ and $\dfrac{3a}{4}z^2 \leqslant 2$. From $0 \leqslant$

$z^2 \leqslant 1$ we get $a \leqslant \dfrac{8}{3}$.

As $f(x) = \dfrac{8}{3}x^3 - 4x^2 + x + m$ (where $m$ is any constant) satisfies the given condition. We obtain that the maximum value of $a$ is $\dfrac{8}{3}$. $\qquad\square$

**10** Given two moving points $A(x_1, y_1)$ and $B(x_2, y_2)$ on parabola curve $y^2 = 6x$ with $x_1 + x_2 = 4$ and $x_1 \neq x_2$, and the perpendicular bisector of segment $AB$ intersects $x$-axis at point $C$. Find the maximum area of $\triangle ABC$.

**Solution 1.** Let the midpoint of $AB$ be $M(x_0, y_0)$. Then $x_0 = \dfrac{x_1 + x_2}{2} = 2$ and $y_0 = \dfrac{y_1 + y_2}{2}$. We have

$$k_{AB} = \frac{y_2 - y_1}{x_2 - x_1} = \frac{y_2 - y_1}{\dfrac{y_2^2}{6} - \dfrac{y_1^2}{6}} = \frac{6}{y_2 + y_1} = \frac{3}{y_0}.$$

The equation of the perpendicular bisector of $AB$ is

$$y - y_0 = -\frac{y_0}{3}(x - 2). \qquad \qquad ①$$

It is easy to find that one solution of it is $x = 5$, $y = 0$. Therefore, the intersection $C$ is a fixed point with coordinate $(5, 0)$.

From ①, we know the equation of line $AB$ is $y - y_0 = \dfrac{3}{y_0}(x - 2)$, or

$$x = \frac{y_0}{3}(y - y_0) + 2. \qquad \qquad ②$$

Substituting ② in $y^2 = 6x$, we get $y^2 = 2y_0(y - y_0) + 12$, or

$$y^2 - 2y_0 y + 2y_0^2 - 12 = 0. \qquad \qquad ③$$

As $y_1$ and $y_2$ are two real roots of ③ and $y_1 \neq y_2$, we have

$$\Delta = 4y_0^2 - 4(2y_0^2 - 12) = -4y_0^2 + 48 > 0.$$

Therefore, $-2\sqrt{3} < y_0 < 2\sqrt{3}$. Then we have

$$
\begin{aligned}
|AB| &= \sqrt{(x_1 - x_2)^2 + (y_1 - y_2)^2} \\
&= \sqrt{\left(1 + \left(\frac{y_0}{3}\right)^2\right)(y_1 - y_2)^2} \\
&= \sqrt{\left(1 + \frac{y_0^2}{9}\right)[(y_1 + y_2)^2 - 4y_1 y_2]} \\
&= \sqrt{\left(1 + \frac{y_0^2}{9}\right)(4y_0^2 - 4(2y_0^2 - 12))} \\
&= \frac{2}{3}\sqrt{(9 + y_0^2)(12 - y_0^2)}.
\end{aligned}
$$

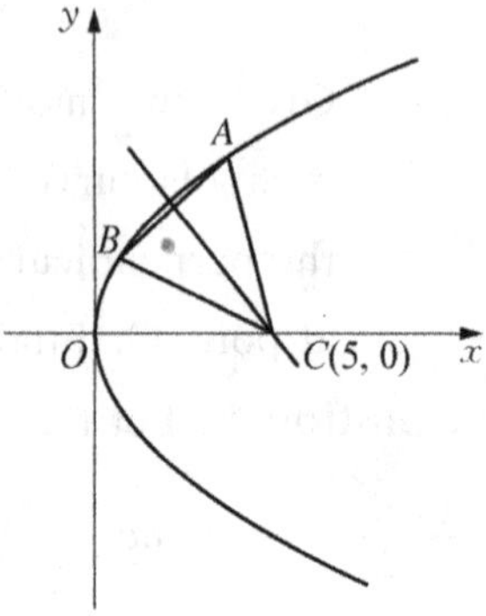

Fig. 10. 1

The distance from point $C(5, 0)$ to segment $AB$ is

$$h = |CM| = \sqrt{(5-2)^2 + (0 - y_0)^2} = \sqrt{9 + y_0^2}.$$

Therefore,

$$
\begin{aligned}
S_{\triangle ABC} &= \frac{1}{2}|AB| \cdot h = \frac{1}{3}\sqrt{(9 + y_0^2)(12 - y_0^2)} \cdot \sqrt{9 + y_0^2} \\
&= \frac{1}{3}\sqrt{\frac{1}{2}(9 + y_0^2)(24 - 2y_0^2)(9 + y_0^2)} \\
&\leqslant \frac{1}{3}\sqrt{\frac{1}{2}\left(\frac{9 + y_0^2 + 24 - 2y_0^2 + 9 + y_0^2}{3}\right)^3} \\
&= \frac{14}{3}\sqrt{7}.
\end{aligned}
$$

The equality holds if and only if $9 + y_0^2 = 24 - 2y_0^2$, i. e. $y_0 = \pm\sqrt{5}$. Then we get

$$A\left(\frac{6 + \sqrt{35}}{3}, \sqrt{5} + \sqrt{7}\right), B\left(\frac{6 - \sqrt{35}}{3}, \sqrt{5} - \sqrt{7}\right)$$

and    $A\left(\frac{6 + \sqrt{35}}{3}, -(\sqrt{5} + \sqrt{7})\right), B\left(\frac{6 - \sqrt{35}}{3}, -\sqrt{5} + \sqrt{7}\right).$

Consequently, the maximum area of $\triangle ABC$ is $\frac{14}{3}\sqrt{7}$.

**Solution 2.** Similar to Solution 1, we get that $C$, the intersection of the perpendicular bisector of $AB$ and the $x$-axis, is a fixed point with coordinate $(5, 0)$.

Let $x_1 = t_1^2$, $x_2 = t_2^2$, $t_1 > t_2$, $t_1^2 + t_2^2 = 4$. Then $S_{\triangle ABC}$ is the absolute value of

$$\frac{1}{2}\begin{vmatrix} 5 & 0 & 1 \\ t_1^2 & \sqrt{6}\,t_1 & 1 \\ t_2^2 & \sqrt{6}\,t_2 & 1 \end{vmatrix},$$

so

$$\begin{aligned} S_{\triangle ABC}^2 &= \left(\frac{1}{2}(5\sqrt{6}\,t_1 + \sqrt{6}\,t_1^2 t_2 - \sqrt{6}\,t_1 t_2^2 - 5\sqrt{6}\,t_2)\right)^2 \\ &= \frac{3}{2}(t_1 - t_2)^2 (t_1 t_2 + 5)^2 \\ &= \frac{3}{2}(4 - 2t_1 t_2)(t_1 t_2 + 5)(t_1 t_2 + 5) \\ &\leqslant \frac{3}{2}\left(\frac{14}{3}\right)^3. \end{aligned}$$

Therefore, $S_{\triangle ABC} \leqslant \frac{14}{3}\sqrt{7}$ and the equality holds if and only if $(t_1 - t_2)^2 = t_1 t_2 + 5$ and $t_1^2 + t_2^2 = 4$. We then get $t_1 = \frac{\sqrt{7} + \sqrt{5}}{\sqrt{6}}$ and $t_2 = -\frac{\sqrt{7} - \sqrt{5}}{\sqrt{6}}$, which implies either

$$A\left(\frac{6 + \sqrt{35}}{3}, \sqrt{5} + \sqrt{7}\right), \quad B\left(\frac{6 - \sqrt{35}}{3}, \sqrt{5} - \sqrt{7}\right)$$

or

$$A\left(\frac{6 + \sqrt{35}}{3}, -(\sqrt{5} + \sqrt{7})\right), \quad B\left(\frac{6 - \sqrt{35}}{3}, -\sqrt{5} + \sqrt{7}\right).$$

Finally, the maximum area of $\triangle ABC$ is $\frac{14}{3}\sqrt{7}$. 　　　□

**11** Prove that equation $2x^3 + 5x - 2 = 0$ has exactly one real root (denoted as $r$), and there is a unique strictly increasing sequence $\{a_n\}$ such that $\frac{2}{5} = r^{a_1} + r^{a_2} + r^{a_3} + \cdots$.

**Solution.** Let $f(x) = 2x^3 + 5x - 2$. Then we have $f'(x) = 6x^2 + 5 > 0$, which means $f(x)$ is strictly increasing. Furthermore, $f(0) = -2 < 0$, $f\left(\frac{1}{2}\right) = \frac{3}{4} > 0$. Therefore, $f(x)$ has a unique real root $r \in \left(0, \frac{1}{2}\right)$. From $2r^3 + 5r - 2 = 0$, we have

$$\frac{2}{5} = \frac{r}{1 - r^3} = r + r^4 + r^7 + r^{10} + \cdots.$$

Therefore, sequence $a_n = 3n - 2$ $(n = 1, 2, \ldots)$ satisfies the required condition.

Assume there are two different positive integer sequences

$$a_1 < a_2 < \cdots < a_n < \cdots \text{ and } b_1 < b_2 < \cdots < b_n < \cdots$$

satisfying

$$r^{a_1} + r^{a_2} + r^{a_3} + \cdots = r^{b_1} + r^{b_2} + r^{b_3} + \cdots = \frac{2}{5}.$$

Deleting the terms that appear at the both sides of the expression, we have

$$r^{s_1} + r^{s_2} + r^{s_3} + \cdots = r^{t_1} + r^{t_2} + r^{t_3} + \cdots,$$

where $s_1 < s_2 < s_3 < \cdots$, $t_1 < t_2 < t_3 < \cdots$ with all the $s_i$ and $t_j$ different from each other.

We may as well assume that $s_1 < t_1$. Then

$$r^{s_1} < r^{s_1} + r^{s_2} + \cdots = r^{t_1} + r^{t_2} + \cdots,$$

$$1 < r^{t_1 - s_1} + r^{t_2 - s_1} + \cdots \leqslant r + r^2 + \cdots$$

$$= \frac{1}{1-r} - 1 < \frac{1}{1 - \frac{1}{2}} - 1 = 1.$$

It is a contradiction. This proves that $\{a_n\}$ is unique.

## 2011 (Hubei)

Commissioned by Chinese Mathematical Society, Hubei Mathematical Society organized the 2011 China Mathematical Competition held on October 16, 2011.

### Part I Short-Answer Questions (Questions 1 – 8, eight marks each)

1 Let $A = \{a_1, a_2, a_3, a_4\}$. Suppose the set of sums of all the elements in every ternary subset of $A$ is $B = \{-1, 3, 5, 8\}$. Then $A = $ _____.

**Solution.** Obviously, every element of $A$ appears three times in all the ternary subsets. Then we have

$$3(a_1 + a_2 + a_3 + a_4) = (-1) + 3 + 5 + 8 = 15,$$

or $a_1 + a_2 + a_3 + a_4 = 5$. Therefore, the four elements of $A$ are $5 - (-1) = 6$, $5 - 3 = 2$, $5 - 5 = 0$, $5 - 8 = -3$, respectively. The answer is $A = \{-3, 0, 2, 6\}$. ☐

2 The value field of $f(x) = \dfrac{\sqrt{x^2 + 1}}{x - 1}$ is _____.

**Solution.** Let $x = \tan\theta$, $-\dfrac{\pi}{2} < \theta < \dfrac{\pi}{2}$ and $\theta \neq \dfrac{\pi}{4}$. We have

$$f(x) = \frac{\dfrac{1}{\cos\theta}}{\tan\theta - 1} = \frac{1}{\sin\theta - \cos\theta} = \frac{1}{\sqrt{2}\sin\left(\theta - \dfrac{\pi}{4}\right)}.$$

Let $u = \sqrt{2}\sin\left(\theta - \dfrac{\pi}{4}\right)$. Then $-\sqrt{2} \leqslant u < 1$ and $u \neq 0$.

Therefore, $f(x) = \dfrac{1}{u} \in \left(-\infty, -\dfrac{\sqrt{2}}{2}\right] \cup (1, +\infty)$.

**3** Suppose $a$ and $b$ are positive real numbers satisfying $\dfrac{1}{a} + \dfrac{1}{b}$ $\leqslant 2\sqrt{2}$ and $(a - b)^2 = 4(ab)^3$. Then $\log_a b = \underline{\qquad}$.

**Solution.** From $\dfrac{1}{a} + \dfrac{1}{b} \leqslant 2\sqrt{2}$, we have $a + b \leqslant 2\sqrt{2}\,ab$. On the other hand,

$$(a + b)^2 = 4ab + (a - b)^2 = 4ab + 4(ab)^3$$
$$\geqslant 4 \cdot 2\sqrt{ab \cdot (ab)^3} = 8(ab)^2,$$

and that means

$$a + b \geqslant 2\sqrt{2}\,ab. \qquad\qquad ①$$

Therefore,

$$a + b = 2\sqrt{2}\,ab. \qquad\qquad ②$$

The equality in ① holds only when $ab = 1$. Associating it with ②, we find

$$\begin{cases} a = \sqrt{2} - 1, \\ b = \sqrt{2} + 1, \end{cases} \text{and} \begin{cases} a = \sqrt{2} + 1, \\ b = \sqrt{2} - 1. \end{cases}$$

So the answer is $\log_a b = -1$.                                $\square$

**4** Suppose $\cos^5\theta - \sin^5\theta < 7(\sin^3\theta - \cos^3\theta)$, $\theta \in [0, 2\pi)$.

Then the range of $\theta$ is _____.

**Solution.** From the inequality

$$\cos^5\theta - \sin^5\theta < 7(\sin^3\theta - \cos^3\theta),$$

we have

$$\sin^3\theta + \frac{1}{7}\sin^5\theta > \cos^3\theta + \frac{1}{7}\cos^5\theta.$$

Since $f(x) = x^3 + \frac{1}{7}x^5$ is increasing over $(-\infty, +\infty)$,

then $\sin\theta > \cos\theta$, and that means

$$2k\pi + \frac{\pi}{4} < \theta < 2k\pi + \frac{5\pi}{4}(k \in \mathbf{Z}).$$

But $\theta \in [0, 2\pi)$, so the range of $\theta$ is $\left(\frac{\pi}{4}, \frac{5\pi}{4}\right)$.   □

**5** Seven students are arranged to attend five sporting events. It is required that students $A$ and $B$ cannot attend the same event, every event is attended by at least one student, and each student must attend one and only one event. Then the number of the arrangement plans meeting the required condition is _____. (the answer should be given in numerical value)

**Solution.** There are two possible cases that satisfy the required conditions:

(1) there is an event attended by three students — this case has $C_7^3 \cdot 5! - C_5^1 \cdot 5! = 3600$ plans;

(2) there are two events each attended by two students — this case has $\frac{1}{2}(C_7^2 \cdot C_5^2) \cdot 5! - C_5^2 \cdot 5! = 11\,400$ plans.

Therefore, there are $3600 + 11\,400 = 15\,000$ plans that meet the required condition. $\qquad\square$

**6**  Given a tetrahedron $ABCD$, it is known that $\angle ADB = \angle BDC = \angle CDA = 60°$, $AD = BD = 3$ and $CD = 2$. Then the radius of the sphere circumscribing $ABCD$ is _____.

**Solution.** Let the center of the sphere circumscribing $ABCD$ be $O$. Then $O$ is on the vertical line of plane $ABD$ through point $N$ the circumcenter of $\triangle ABD$. It is known that $\triangle ABD$ is regular, so $N$ is the center of it. Let $P$ and $M$ be the midpoints of $AB$ and $CD$, respectively. Then $N$ is on $DP$ with $ON \perp DP$ and $OM \perp CD$.

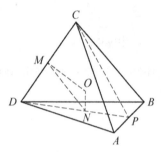

Fig. 6. 1

Let $\theta$ denote the angle between $CD$ and plane $ABD$. From

$$\angle CDA = \angle CDB = \angle ADB = 60°,$$

we find $\cos\theta = \dfrac{1}{\sqrt{3}}$, $\sin\theta = \dfrac{\sqrt{2}}{\sqrt{3}}$.

Since $DM = \dfrac{1}{2}CD = 1$, $DN = \dfrac{2}{3}\cdot DP = \dfrac{2}{3}\cdot\dfrac{\sqrt{3}}{2}\cdot 3 = \sqrt{3}$,

by cosine theorem we have, in $\triangle DMN$,

$$MN^2 = DM^2 + DN^2 - 2\cdot DM\cdot DN\cdot\cos\theta$$

$$= 1^2 + (\sqrt{3})^2 - 2\cdot 1\cdot\sqrt{3}\cdot\dfrac{1}{\sqrt{3}} = 2,$$

that is $MN = \sqrt{2}$. The radius of the sphere circumscribing $ABCD$ is then

$$OD = \frac{MN}{\sin \theta} = \frac{\sqrt{2}}{\frac{\sqrt{2}}{\sqrt{3}}} = \sqrt{3}.$$

The answer is $R = \sqrt{3}$.    $\square$

**7** Line $x - 2y - 1 = 0$ and parabola $y^2 = 4x$ intersect at points $A$, $B$, point $C$ is on the parabola, and $\angle ACB = 90°$. Then the coordinate of $C$ is _____.

**Solution.** Let $A(x_1, y_1)$, $B(x_2, y_2)$, $C(t^2, 2t)$. From

$$\begin{cases} x - 2y - 1 = 0, \\ y^2 = 4x, \end{cases}$$

we get $y^2 - 8y - 4 = 0$, which means $y_1 + y_2 = 8$, $y_1 \cdot y_2 = -4$.

Since $x_1 = 2y_1 + 1$, $x_2 = 2y_2 + 1$, we have

$$x_1 + x_2 = 2(y_1 + y_2) + 2 = 18,$$
$$x_1 \cdot x_2 = 4y_1 \cdot y_2 + 2(y_1 + y_2) + 1 = 1.$$

Furthermore, by $\angle ACB = 90°$, we have $\overrightarrow{CA} \cdot \overrightarrow{CB} = 0$, which means

$$(t^2 - x_1)(t^2 - x_2) + (2t - y_1)(2t - y_2) = 0,$$

that is

$$t^4 - (x_1 + x_2)t^2 + x_1 \cdot x_2 + 4t^2 - 2(y_1 + y_2)t + y_1 \cdot y_2 = 0.$$

Then

$$t^4 - 14t^2 - 16t - 3 = 0,$$

or

$$(t^2 + 4t + 3)(t^2 - 4t - 1) = 0.$$

Obviously, $t^2 - 4t - 1 \neq 0$; otherwise, we have $t^2 - 2 \cdot 2t - 1 = 0$, which means $C$ is on $x - 2y - 1 = 0$, i.e., $C$

coincides with either $A$ or $B$. So $t^2 + 4t + 3 = 0$. Then $t_1 = -1$, $t_2 = -3$.

Therefore, the coordinate of $C$ is $(1, -2)$ or $(9, -6)$. $\square$

**8** Let $a_n = C_{200}^n \cdot (\sqrt[3]{6})^{200-n} \cdot \left(\dfrac{1}{\sqrt{2}}\right)^n$ ($n = 1, 2, \ldots, 95$). Then

the number of terms that are integers in $\{a_n\}$ is _____ .

**Solution.** We have $a_n = C_{200}^n \cdot 3^{\frac{200-n}{3}} \cdot 2^{\frac{400-5n}{6}}$. When $a_n$ ($1 \leqslant n \leqslant 95$) is an integer, $\dfrac{200-n}{3}$ and $\dfrac{400-5n}{6}$ must be integers. Then $6 \mid n + 4$.

When $n = 2, 8, 14, 20, 26, 32, 38, 44, 50, 56, 62, 68, 74, 80$, $\dfrac{200-n}{3}$ and $\dfrac{400-5n}{6}$ are all non-negative integers. So the corresponding $a_n$, totally 14, are integers.

When $n = 86$, we have $a_{86} = C_{200}^{86} \cdot 3^{38} \cdot 2^{-5}$. The number of the factors of 2 in 200! is

$$\left[\frac{200}{2}\right] + \left[\frac{200}{2^2}\right] + \left[\frac{200}{2^3}\right] + \left[\frac{200}{2^4}\right] + \left[\frac{200}{2^5}\right] + \left[\frac{200}{2^6}\right] + \left[\frac{200}{2^7}\right] = 197.$$

By the same reason, the numbers of the factors of 2 in 86! and 114! are 82 and 110, respectively. Therefore, the number of the factors of 2 in $C_{200}^{86} = \dfrac{200!}{86! \cdot 114!}$ is $197 - 82 - 110 = 5$. So $a_{86}$ is an integer.

When $n = 92$, we have $a_{92} = C_{200}^{92} \cdot 3^{36} \cdot 2^{-10}$. In the same way, we find the numbers of the factors of 2 in 92! and 108! are 88 and 105, respectively, which means that in $C_{200}^{86}$ is $197 - 88 - 105 = 4$. Therefore, $a_{92}$ is not an integer.

Overall, the required number is $14 + 1 = 15$. $\square$

**Part II**   **Word Problems (16 marks for Question 9, 20 marks each for Questions 10 and 11, and then 56 marks in total)**

**9**   Suppose $f(x) = |\lg(x+1)|$ and real numbers $a$, $b$ $(a < b)$ satisfy $f(a) = f\left(-\dfrac{b+1}{b+2}\right)$, $f(10a + 6b + 21) = 4\lg 2$.

Find the values of $a$, $b$.

**Solution.** As $f(a) = f\left(-\dfrac{b+1}{b+2}\right)$, we have

$$|\lg(a+1)| = \left|\lg\left(-\frac{b+1}{b+2}+1\right)\right| = \left|\lg\left(\frac{1}{b+2}\right)\right| = |\lg(b+2)|.$$

Then either $a+1 = b+2$ or $(a+1)(b+2) = 1$. Since $a < b$, so $a+1 \neq b+2$. Therefore, $(a+1)(b+2) = 1$.

From $f(a) = |\lg(a+1)|$ we know $0 < a+1$. Then

$$0 < a+1 < b+1 < b+2,$$

which implies

$$0 < a+1 < 1 < b+2.$$

Therefore,

$$(10a + 6b + 21) + 1 = 10(a+1) + 6(b+2)$$
$$= 6(b+2) + \frac{10}{b+2} > 1.$$

Then

$$f(10a + 6b + 21) = \left|\lg\left[6(b+2) + \frac{10}{b+2}\right]\right|$$
$$= \lg\left[6(b+2) + \frac{10}{b+2}\right].$$

On the other hand,

$$f(10a + 6b + 21) = 4\lg 2.$$

So

$$\lg\left[6(b+2)+\frac{10}{b+2}\right] = 4\lg 2,$$

which means $6(b+2)+\dfrac{10}{b+2} = 16$. Then either $b = -\dfrac{1}{3}$ or $b = -1$ (discarded).

Substituting $b = -\dfrac{1}{3}$ into $(a+1)(b+2) = 1$, we find $a = -\dfrac{2}{5}$.

Therefore $a = -\dfrac{2}{5}$, $b = -\dfrac{1}{3}$.                        □

**10** Suppose sequence $\{a_n\}$ satisfies $a_1 = 2t - 3(t \in \mathbf{R}$ and $t \neq \pm 1)$,

$$a_{n+1} = \frac{(2t^{n+1} - 3)a_n + 2(t-1)t^n - 1}{a_n + 2t^n - 1}(n \in \mathbf{N}^*).$$

(1) Find the formula of general term about $\{a_n\}$.

(2) If $t > 0$, find out which is larger between $a_{n+1}$ and $a_n$.

**Solution:** (1) The given expression can be rewritten as

$$a_{n+1} = \frac{2(t^{n+1} - 1)(a_n + 1)}{a_n + 2t^n - 1} - 1.$$

Then

$$\frac{a_{n+1} + 1}{t^{n+1} - 1} = \frac{2(a_n + 1)}{a_n + 2t^n - 1} = \frac{\dfrac{2(a_n + 1)}{t^n - 1}}{\dfrac{a_n + 1}{t^n - 1} + 2}.$$

Let $\dfrac{a_n + 1}{t^n - 1} = b_n$. Then $b_{n+1} = \dfrac{2b_n}{b_n + 2}$, with $b_1 = \dfrac{a_1 + 1}{t - 1} = \dfrac{2t - 2}{t - 1} = 2.$

Furthermore, $\dfrac{1}{b_{n+1}} = \dfrac{1}{b_n} + \dfrac{1}{2}$, $\dfrac{1}{b_1} = \dfrac{1}{2}$. Then

$$\frac{1}{b_n} = \frac{1}{b_1} + (n-1) \cdot \frac{1}{2} = \frac{n}{2}.$$

Therefore, $\dfrac{a_n + 1}{t^n - 1} = \dfrac{2}{n}$, which means $a_n = \dfrac{2(t^n - 1)}{n} - 1$.

(2) We have

$$
\begin{aligned}
a_{n+1} - a_n &= \frac{2(t^{n+1} - 1)}{n+1} - \frac{2(t^n - 1)}{n} \\
&= \frac{2(t-1)}{n(n+1)} [n(1 + t + \cdots + t^{n-1} + t^n) - \\
&\quad (n+1)(1 + t + \cdots + t^{n-1})] \\
&= \frac{2(t-1)}{n(n+1)} [nt^n - (1 + t + \cdots + t^{n-1})] \\
&= \frac{2(t-1)}{n(n+1)} [(t^n - 1) + (t^n - t) + \cdots + (t^n - t^{n-1})] \\
&= \frac{2(t-1)^2}{n(n+1)} [(t^{n-1} + t^{n-2} + \cdots + 1) + \\
&\quad t(t^{n-2} + t^{n-3} + \cdots + 1) + \cdots + t^{n-1}].
\end{aligned}
$$

It is obvious that $a_{n+1} - a_n > 0$ for $t > 0 (t \neq 1)$. Therefore, $a_{n+1} > a_n$. $\qquad \square$

**11** Straight line $l$ with slope $\dfrac{1}{3}$ intercepts ellipse $C : \dfrac{x^2}{36} + \dfrac{y^2}{4} = 1$ at points $A$, $B$, and point $P(3\sqrt{2}, \sqrt{2})$ is in the top-left of $l$ (as shown in Fig. 11. 1).

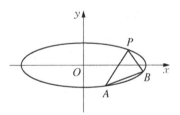

Fig. 11. 1

(1) Prove that the center of the inscribed circle of $\triangle PAB$

is on a given line;

(2) When $\angle APB = 60°$, find the area of $\triangle PAB$.

**Solution.** (1) Let $l$ be a straight line such that $y = \dfrac{1}{3}x + m$,

and $A(x_1, y_1)$, $B(x_2, y_2)$.

Substituting $y = \dfrac{1}{3}x + m$ into $\dfrac{x^2}{36} + \dfrac{y^2}{4} = 1$, and simplifying

it, we have

$$2x^2 + 6mx + 9m^2 - 36 = 0.$$

Then $x_1 + x_2 = -3m$, $x_1 x_2 = \dfrac{9m^2 - 36}{2}$, $k_{PA} = \dfrac{y_1 - \sqrt{2}}{x_1 - 3\sqrt{2}}$,

$k_{PB} = \dfrac{y_2 - \sqrt{2}}{x_2 - 3\sqrt{2}}$. Therefore,

$$\begin{aligned}
k_{PA} + k_{PB} &= \frac{y_1 - \sqrt{2}}{x_1 - 3\sqrt{2}} + \frac{y_2 - \sqrt{2}}{x_2 - 3\sqrt{2}} \\
&= \frac{(y_1 - \sqrt{2})(x_2 - 3\sqrt{2}) + (y_2 - \sqrt{2})(x_1 - 3\sqrt{2})}{(x_1 - 3\sqrt{2})(x_2 - 3\sqrt{2})}.
\end{aligned}$$

In the expression above, the numerator is equal to

$$\begin{aligned}
&\left(\frac{1}{3}x_1 + m - \sqrt{2}\right)(x_2 - 3\sqrt{2}) + \left(\frac{1}{3}x_2 + m - \sqrt{2}\right)(x_1 - 3\sqrt{2}) \\
&= \frac{2}{3}x_1 x_2 + (m - 2\sqrt{2})(x_1 + x_2) - 6\sqrt{2}(m - \sqrt{2}) \\
&= \frac{2}{3} \cdot \frac{9m^2 - 36}{2} + (m - 2\sqrt{2})(-3m) - 6\sqrt{2}(m - \sqrt{2}) \\
&= 3m^2 - 12 - 3m^2 + 6\sqrt{2}m - 6\sqrt{2}m + 12 = 0.
\end{aligned}$$

Therefore, $k_{PA} + k_{PB} = 0$. Since $P$ is in the top-left of $l$, we know that the bisector of $\angle APB$ is parallel to the $y$-axis. Therefore, the center of the inscribed circle of $\triangle PAB$ is on line $x = 3\sqrt{2}$.

(2) When $\angle APB = 60°$, by the result in (1), we have $k_{PA} = \sqrt{3}$, $k_{PB} = -\sqrt{3}$. Then the equation for line $PA$ is $y - \sqrt{2} = \sqrt{3}(x - 3\sqrt{2})$. Substituting it into $\dfrac{x^2}{36} + \dfrac{y^2}{4} = 1$, and eliminating $y$, we get

$$14x^2 + 9\sqrt{6}(1 - 3\sqrt{3})x + 18(13 - 3\sqrt{3}) = 0,$$

which has roots $x_1$ and $3\sqrt{2}$. So $x_1 \cdot 3\sqrt{2} = \dfrac{18(13 - 3\sqrt{3})}{14}$, i.e.

$x_1 = \dfrac{3\sqrt{2}(13 - 3\sqrt{3})}{14}$. Then we find

$$|PA| = \sqrt{1 + (\sqrt{3})^2} \cdot |x_1 - 3\sqrt{2}| = \dfrac{3\sqrt{2}(3\sqrt{3} + 1)}{7}.$$

In the same way, we have $|PB| = \dfrac{3\sqrt{2}(3\sqrt{3} - 1)}{7}$.

Therefore,

$$
\begin{aligned}
S_{\triangle PAB} &= \frac{1}{2} \cdot |PA| \cdot |PB| \cdot \sin 60° \\
&= \frac{1}{2} \cdot \frac{3\sqrt{2}(3\sqrt{3} + 1)}{7} \cdot \frac{3\sqrt{2}(3\sqrt{3} - 1)}{7} \cdot \frac{\sqrt{3}}{2} \\
&= \frac{117\sqrt{3}}{49}.
\end{aligned}
$$

□

## $2012$ (Shaanxi)

Commissioned by Chinese Mathematical Society, Shaanxi Mathematical Society organized the 2012 China Mathematical

Competition held on October 14, 2012.

## Part I   Short-Answer Questions (Questions 1 – 8, eight marks each)

**1** Let $P$ be a point on the image of $y = x + \dfrac{2}{x}(x > 0)$.

Through $P$ draw lines perpendicular to $y = x$ and $y$-axis with foot points $A$, $B$, respectively. Then the value of $\overrightarrow{PA} \cdot \overrightarrow{PB}$ is _____ .

**Solution 1.** Let $P\left(x_0, x_0 + \dfrac{2}{x_0}\right)$. The expression for line $PA$ is then

$$y - \left(x_0 + \frac{2}{x_0}\right) = -(x - x_0),$$

or                 $$y = -x + 2x_0 + \frac{2}{x_0}.$$

From

$$\begin{cases} y = x, \\ y = -x + 2x_0 + \dfrac{2}{x_0}, \end{cases}$$

we get $A\left(x_0 + \dfrac{1}{x_0}, \ x_0 + \dfrac{1}{x_0}\right)$.

On the other hand, we have $B\left(0, \ x_0 + \dfrac{2}{x_0}\right)$. Then $\overrightarrow{PA} = \left(\dfrac{1}{x_0}, \ -\dfrac{1}{x_0}\right)$ and $\overrightarrow{PB} = (-x_0, \ 0)$. Therefore,

$$\overrightarrow{PA} \cdot \overrightarrow{PB} = \frac{1}{x_0} \cdot (-x_0) = -1.$$

The answer is $-1$.

**Solution 2.** As seen in Fig. 1. 1, the distances from $P\left(x_0,\ x_0 + \dfrac{2}{x_0}\right)$ to lines $y = x$ and $y$-axis, respectively, are

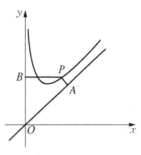

Fig. 1. 1

$$| PA | = \frac{\left| x_0 - \left( x_0 + \dfrac{2}{x_0} \right) \right|}{\sqrt{2}} = \frac{\sqrt{2}}{x_0}$$

and

$$| PB | = x_0.$$

Since $O$, $A$, $P$ and $B$ are concyclic points, then

$$\angle APB = \pi - \angle AOB = \frac{3\pi}{4}.$$

Therefore, $\overrightarrow{PA} \cdot \overrightarrow{PB} = | \overrightarrow{PA} | \cdot | \overrightarrow{PB} | \cdot \cos \dfrac{3\pi}{4} = -1.$  ☐

**2**  Suppose $\triangle ABC$ with angles $A$, $B$ and $C$, and the corresponding sides $a$, $b$ and $c$ satisfies equation $a \cos B - b \cos A = \dfrac{3}{5} c$. Then the value of $\dfrac{\tan A}{\tan B}$ is _____.

**Solution 1.** By the given condition and the Law of Cosines, we have

$$a \cdot \frac{c^2 + a^2 - b^2}{2ca} - b \cdot \frac{b^2 + c^2 - a^2}{2bc} = \frac{3}{5} c,$$

or $a^2 - b^2 = \dfrac{3}{5} c^2$. Therefore,

$$\frac{\tan A}{\tan B} = \frac{\sin A \cos B}{\sin B \cos A} = \frac{a \cdot \dfrac{c^2 + a^2 - b^2}{2ca}}{b \cdot \dfrac{b^2 + c^2 - a^2}{2bc}}$$

$$= \frac{c^2 + a^2 - b^2}{b^2 + c^2 - a^2} = \frac{\frac{8}{5}c^2}{\frac{2}{5}c^2} = 4.$$

The answer is 4.

**Solution 2.** As seen in Fig. 2. 1, through $C$ draw $CD \perp AB$ with foot point $D$. We have $a \cos B = DB$ and $b \cos A = AD$. By the given condition, we have $DB - AD = \frac{3}{5}c$. Combining it with $DB + AD = c$, we get $AD = \frac{1}{5}c$ and $DB = \frac{4}{5}c$. Therefore,

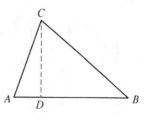

Fig. 2. 1

$$\frac{\tan A}{\tan B} = \frac{\dfrac{CD}{AD}}{\dfrac{CD}{BD}} = \frac{BD}{AD} = 4.$$

**Solution 3.** By the projection theorem, we have $a \cos B + b \cos A = c$. Combining it with $a \cos B - b \cos A = \frac{3}{5}c$, we get $a \cos B = \frac{4}{5}c$ and $b \cos A = \frac{1}{5}c$. Therefore,

$$\frac{\tan A}{\tan B} = \frac{\sin A \cos B}{\sin B \cos A} = \frac{a \cdot \cos B}{b \cdot \cos A} = \frac{\frac{4}{5}c}{\frac{1}{5}c} = 4. \qquad \square$$

**3** Let $x, y, z \in [0, 1]$. Then the maximum value of $M = \sqrt{|x - y|} + \sqrt{|y - z|} + \sqrt{|z - x|}$ is _____.

**Solution.** We may assume $0 \leqslant x \leqslant y \leqslant z \leqslant 1$. Then

$$M = \sqrt{y - x} + \sqrt{z - y} + \sqrt{z - x}.$$

Since

$$\sqrt{y-x} + \sqrt{z-y} \leqslant \sqrt{2[(y-x)+(z-y)]} = \sqrt{2(z-x)},$$

we have

$$M \leqslant \sqrt{2(z-x)} + \sqrt{z-x} = (\sqrt{2}+1)\sqrt{z-x} \leqslant \sqrt{2}+1.$$

The equality holds if and only if $y-x = z-y$, $x = 0$, $z = 1$ (i.e. $x = 0$, $y = \dfrac{1}{2}$, $z = 1$).

Therefore, the answer is $M_{\max} = \sqrt{2}+1$.  □

**4** Let the focus and directrix of parabola $y^2 = 2px\,(p>0)$ be $F$ and $l$, respectively. $A$ and $B$ are moving points on the parabola satisfying $\angle AFB = \dfrac{\pi}{3}$. Let the projection of $M$ the midpoint of segment $AB$ on $l$ be $N$. Then the maximum value of $\dfrac{|MN|}{|AB|}$ is _____.

**Solution 1.** Suppose $\angle ABF = \theta\left(0 < \theta < \dfrac{2\pi}{3}\right)$. Then by the Law of Sine, we have

$$\frac{|AF|}{\sin\theta} = \frac{|BF|}{\sin\left(\dfrac{2\pi}{3}-\theta\right)} = \frac{|AB|}{\sin\dfrac{\pi}{3}}.$$

And then

$$\frac{|AF|+|BF|}{\sin\theta + \sin\left(\dfrac{2\pi}{3}-\theta\right)} = \frac{|AB|}{\sin\dfrac{\pi}{3}}.$$

So

$$\frac{|AF|+|BF|}{|AB|} = \frac{\sin\theta + \sin\left(\dfrac{2\pi}{3}-\theta\right)}{\sin\dfrac{\pi}{3}} = 2\cos\left(\theta - \dfrac{\pi}{3}\right).$$

As seen in Fig. 4. 1, by using the definition of a parabola and the property of a trapezoid, we have $|MN| = \dfrac{|AF| + |BF|}{2}$. Then

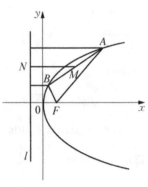

$$\frac{|MN|}{|AB|} = \cos\left(\theta - \frac{\pi}{3}\right).$$

Therefore, $\dfrac{|MN|}{|AB|}$ reaches the

Fig. 4. 1

maximum value 1 when $\theta = \dfrac{\pi}{3}$.

The answer is 1.

**Solution 2.** By using the definition of a parabola and the property of a trapezoid, we have $|MN| = \dfrac{|AF| + |BF|}{2}$. In $\triangle AFB$, by using the Law of Cosines we have

$$
\begin{aligned}
|AB|^2 &= |AF|^2 + |BF|^2 - 2|AF| \cdot |BF| \cos\frac{\pi}{3} \\
&= (|AF| + |BF|)^2 - 3|AF| \cdot |BF| \\
&\geqslant (|AF| + |BF|)^2 - 3\left(\frac{|AF| + |BF|}{2}\right)^2 \\
&= \left(\frac{|AF| + |BF|}{2}\right)^2 = |MN|^2.
\end{aligned}
$$

The equality holds if and only if $|AF| = |BF|$. Therefore, the maximum value of $\dfrac{|MN|}{|AB|}$ is 1. $\qquad\square$

**5** Suppose two regular triangular pyramids $P - ABC$ and $Q - ABC$ sharing the same base are inscribed in the same sphere. If the angle between the side-face and the base of $P - ABC$ is $45°$, then the tangent value of the angle between the side-face and the base of $Q - ABC$ is _____ .

**Solution.** As seen in Fig. 5.1, connecting $PQ$, then $PQ$ is perpendicular to plane $ABC$ with the foot point $H$ being the center of $\triangle ABC$. The center of the sphere $O$ is also on $PQ$. Connect and extend $CH$ to let it intersect with $AB$ at point $M$. $M$ is then the midpoint of $AB$, and $CM \perp AB$. It is easy to see that $\angle PMH$ and $\angle QMH$ are the plane angles formed by the sides-faces and

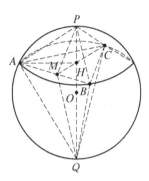

Fig. 5.1

the bases of the regular triangular pyramids $P - ABC$ and $Q - ABC$,

respectively. Then $\angle PMH = 45°$, so $PH = MH = \dfrac{1}{2}AH$.

Since $\angle PAQ = 90°$, $AH \perp PQ$, then $AH^2 = PH \cdot QH$. We

then have $AH^2 = \dfrac{1}{2}AH \cdot QH$.

Therefore, $QH = 2AH = 4MH$.

Finally, $\tan\angle QMH = \dfrac{QH}{MH} = 4$.

The answer is 4. $\qquad\qquad\qquad\qquad\square$

**6** Let $f(x)$ be an odd function on **R**, and $f(x) = x^2$ for $x \geqslant$ 0. Suppose for any $x \in [a, a + 2]$, $f(x + a) \geqslant 2f(x)$. Then the range of real number $a$ is _____ .

**Solution.** According to the given condition, we have

$$f(x) = \begin{cases} x^2 & (x \geqslant 0), \\ -x^2 & (x < 0). \end{cases}$$

So $2f(x) = f(\sqrt{2}x)$. Therefore, the original inequality is equivalent to $f(x + a) \geqslant f(\sqrt{2}x)$.

As $f(x)$ is increasing over **R**, then $x + a \geqslant \sqrt{2}x$, i.e.,

$$a \geqslant (\sqrt{2} - 1)x.$$

Furthermore, since $x \in [a, a + 2]$, $(\sqrt{2} - 1)x$ reaches $(\sqrt{2} - 1)(a + 2)$ the maximum value when $x = a + 2$.

Therefore, $a \geqslant (\sqrt{2} - 1)(a + 2)$, from which we obtain $a \geqslant \sqrt{2}$, i.e., $a \in [\sqrt{2}, +\infty)$.

The answer is then $[\sqrt{2}, +\infty)$.    □

---

**7** The sum of all the positive integers $n$ satisfying $\dfrac{1}{4} < \sin \dfrac{\pi}{n} < \dfrac{1}{3}$ is _____.

**Solution.** As $\sin x$ is a convex function for $x \in \left(0, \dfrac{\pi}{6}\right)$, we have $\dfrac{3}{\pi}x < \sin x < x$. Then

$$\sin \frac{\pi}{13} < \frac{\pi}{13} < \frac{1}{4}, \ \sin \frac{\pi}{12} > \frac{3}{\pi} \times \frac{\pi}{12} = \frac{1}{4},$$

$$\sin \frac{\pi}{10} < \frac{\pi}{10} < \frac{1}{3}, \ \sin \frac{\pi}{9} > \frac{3}{\pi} \times \frac{\pi}{9} = \frac{1}{3},$$

that is

$$\sin \frac{\pi}{13} < \frac{1}{4} < \sin \frac{\pi}{12} < \sin \frac{\pi}{11} < \sin \frac{\pi}{10} < \frac{1}{3} < \sin \frac{\pi}{9}.$$

Therefore, all the possible values of positive integers $n$ are 10, 11, 12, and their sum is 33.

The answer is 33.    □

---

**8** An information station employs four different codes, $A$, $B$, $C$ and $D$, for communication, but each week uses only one of them. The code used in a definite week is randomly selected with equal chance among the three ones that have not been used in the last week. Suppose the code used in the first week is $A$. Then the probability that $A$ is also used in the seventh week is _____ . (expressed as an irreducible fraction)

**Solution.** Let $P_k$ denote the probability that code $A$ is used in the $k$th week. Then the probability that $A$ is not used in the $k$th week is $1 - P_k$. Therefore, we have

$$P_{k+1} = \frac{1}{3}(1 - P_k).$$

Or

$$P_{k+1} - \frac{1}{4} = -\frac{1}{3}\left(P_k - \frac{1}{4}\right).$$

As $P_1 = 1$, $\left\{P_k - \frac{1}{4}\right\}$ is then a geometric sequence with $\frac{3}{4}$ as the first term and $-\frac{1}{3}$ as the common ratio. So we have

$$P_k - \frac{1}{4} = \frac{3}{4}\left(-\frac{1}{3}\right)^{k-1}.$$

Or

$$P_k = \frac{3}{4}\left(-\frac{1}{3}\right)^{k-1} + \frac{1}{4}.$$

Therefore, $P_7 = \dfrac{61}{243}$.

The answer is $\dfrac{61}{243}$. ◻

**Part II   Word Problems (56 marks in total for three questions)**

**9**  (16 marks) Suppose $f(x) = a\sin x - \dfrac{1}{2}\cos 2x + a - \dfrac{3}{a} + \dfrac{1}{2}$, $a \in \mathbf{R}$, $a \neq 0$.

(1) If $f(x) \leqslant 0$ for any $x \in \mathbf{R}$, find the range of $a$.

(2) If $a \geqslant 2$ and there exists $x \in \mathbf{R}$ such that $f(x) \leqslant 0$, find the range of $a$.

**Solution.** (1) We have $f(x) = \sin^2 x + a\sin x + a - \dfrac{3}{a}$. Let $t = \sin x (-1 \leqslant t \leqslant 1)$. Then

$$g(t) = t^2 + at + a - \dfrac{3}{a}.$$

The sufficient and necessary condition for $f(x) \leqslant 0$, $\forall x \in \mathbf{R}$ is

$$\begin{cases} g(-1) = 1 - \dfrac{3}{a} \leqslant 0, \\ g(1) = 1 + 2a - \dfrac{3}{a} \leqslant 0. \end{cases}$$

Therefore, we obtain the range $a$ of is $(0, 1]$.

(2) As $a \geqslant 2$, then $-\dfrac{a}{2} \leqslant -1$. We have

$$g(t)_{\min} = g(-1) = 1 - \dfrac{3}{a}.$$

Then $f(x)_{\min} = 1 - \dfrac{3}{a}$. Therefore, the sufficient and necessary condition for $f(x) \leqslant 0$, $\exists x \in \mathbf{R}$ is $1 - \dfrac{3}{a} \leqslant 0$, or $0 < a \leqslant 3$.

Finally, we obtain that the range of $a$ is $[2, 3]$.  □

**10** (20 marks) It is known that each term of sequence $\{a_n\}$ is a non-zero real number, and for any positive integer $n$ holds the equation

$$(a_1 + a_2 + \cdots + a_n)^2 = a_1^3 + a_2^3 + \cdots + a_n^3.$$

(1) When $n = 3$, find out all the sequences consisting of three terms $a_1$, $a_2$, $a_3$.

(2) Does there exist an infinite sequence $\{a_n\}$ such that $a_{2013} = -2012$? If it exists, write out the formula of general term; if not, give your reason.

**Solution.** (1) When $n = 1$, we have $a_1^2 = a_1^3$. Since $a_1 \neq 0$, we get $a_1 = 1$.

When $n = 2$, we have $(1 + a_2)^2 = 1 + a_2^3$. Since $a_2 \neq 0$, we get $a_2 = 2$ or $a_2 = -1$.

When $n = 3$, we have $(1 + a_2 + a_3)^2 = 1 + a_2^3 + a_3^3$. For $a_2 = 2$, we get $a_3 = 3$ or $a_3 = -2$; for $a_2 = -1$, we get $a_3 = 1$.

In summary, we get three sequences consisting of three terms that satisfy the given condition:

$$\{1, 2, 3\}, \{1, 2, -2\}, \text{ and } \{1, -1, 1\}.$$

(2) Let $S_n = a_1 + a_2 + \cdots + a_n$. Then we have

$$S_n^2 = a_1^3 + a_2^3 + \cdots + a_n^3 (n \in \mathbf{N}),$$
$$(S_n + a_{n+1})^2 = a_1^3 + a_2^3 + \cdots + a_n^3 + a_{n+1}^3.$$

Finding out the difference of the two expressions above and by $a_{n+1} \neq 0$, we have $2S_n = a_{n+1}^2 - a_{n+1}$.

When $n = 1$, we know from (1) that $a_1 = 1$.

When $n \geq 2$, we have

$$2a_n = 2(S_n - S_{n-1}) = (a_{n+1}^2 - a_{n+1}) - (a_n^2 - a_n).$$

And that is

$$(a_{n+1} + a_n)(a_{n+1} - a_n - 1) = 0.$$

Then we get $a_{n+1} = -a_n$ or $a_{n+1} = a_n + 1$.

Finally, from $a_1 = 1$ and $a_{2013} = -2012$, we find the formula of general term for a required sequence as

$$a_n = \begin{cases} n, & 1 \leqslant n \leqslant 2012, \\ 2012(-1)^n, & n \geqslant 2013. \end{cases}$$

$\square$

**11** (20 marks) As seen in Fig. 11.1, in the rectangular coordinate system $XOY$, the side of the rhombus $ABCD$ is 4, and $|OB| = |OD| = 6$.

(1) Prove that $|OA| \cdot |OC|$ is a constant.

(2) When point $A$ is moving on the half circle $(x - 2)^2 + y^2 = 4(2 \leqslant x \leqslant 4)$, find the trace of $C$.

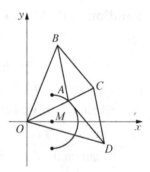

Fig. 11.1

**Solution.** (1) Since $|OB| = |OD|$ and $|AB| = |AD| = |BC| = |CD|$, then $O$, $A$, $C$ are collinear.

As seen in Fig. 11.2, connecting $BD$, then $BD$ is perpendicular to $AC$ and through its midpoint $K$. So we have

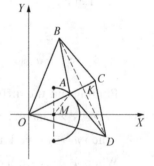

Fig. 11.2

$$|OA| \cdot |OC| = (|OK| - |AK|)(|OK| + |AK|)$$
$$= |OK|^2 - |AK|^2$$
$$= (|OB|^2 - |BK|^2) - (|AB|^2 - |BK|^2)$$
$$= |OB|^2 - |AB|^2 = 6^2 - 4^2 = 20 \text{ (a constant)}.$$

(2) Let $C(x, y)$, $A(2 + 2\cos \alpha, 2\sin \alpha)$, where

$$\alpha = \angle XMA \left( -\frac{\pi}{2} \leqslant \alpha \leqslant \frac{\pi}{2} \right).$$

Then $\angle XOC = \frac{\alpha}{2}$. As

$$|OA|^2 = (2 + 2\cos \alpha)^2 + (2\sin \alpha)^2 = 8(1 + \cos \alpha) = 16\cos^2 \frac{\alpha}{2},$$

then $|OA| = 4\cos \frac{\alpha}{2}$. Combining it with the result in (1), we

get $|OC| \cos \frac{\alpha}{2} = 5$.

Then we have $x = |OC| \cos \frac{\alpha}{2} = 5$, and $y = |OC| \sin \frac{\alpha}{2} = $

$5\tan \frac{\alpha}{2} \in [-5, 5]$.

Therefore, the trace of point $C$ is a segment with the ends $(5, 5)$ and $(5, -5)$. $\qquad \square$

## *2013* (Jilin)

Commissioned by Chinese Mathematical Society, Jilin Mathematical Society organized the 2013 China Mathematical Competition held on October 13, 2013.

**Part I   Short-Answer Questions (Questions 1 – 8, eight marks each)**

**1**   Given $A = \{2, 0, 1, 3\}$, let $B = \{x \mid -x \in A, 2 - x^2 \notin A\}$. Then the sum of elements in $B$ is _____.

**Solution.** It is easy to find that $B \subseteq \{-2, 0, -1, -3\}$. We have $2 - x^2 \notin A$ when $x = -2, -3$, and $2 - x^2 \in A$ when $x = 0, -1$. Therefore, $B = \{-2, -3\}$, the sum of whose elements is $-5$.

The answer is $-5$. $\qquad\qquad$ □

**2**   In a plane rectangular coordinate system $xOy$, points $A$, $B$ are on the parabola $y^2 = 4x$, satisfying $\overrightarrow{OA} \cdot \overrightarrow{OB} = -4$, and point $F$ is the focus of the parabola. Then $S_{\triangle OFA} \cdot S_{\triangle OFB} = $ _____.

**Solution.** Let $F(1, 0)$, $A(x_1, y_1)$, $B(x_2, y_2)$. Then $x_1 = \dfrac{y_1^2}{4}$, $x_2 = \dfrac{y_2^2}{4}$, and

$$-4 = \overrightarrow{OA} \cdot \overrightarrow{OB} = x_1 x_2 + y_1 y_2 = \frac{1}{16}(y_1 y_2)^2 + y_1 y_2,$$

from which we have $\dfrac{1}{16}(y_1 y_2 + 8)^2 = 0$, or $y_1 y_2 = -8$. Therefore,

$$S_{\triangle OFA} \cdot S_{\triangle OFB} = \left(\frac{1}{2} \mid OF \mid \cdot \mid y_1 \mid\right) \cdot \left(\frac{1}{2} \mid OF \mid \cdot \mid y_2 \mid\right)$$

$$= \frac{1}{4} \cdot \mid OF \mid^2 \cdot \mid y_1 y_2 \mid = 2.$$

The answer is 2. $\qquad\qquad$ □

**3**   Suppose in $\triangle ABC$ we have $\sin A = 10 \sin B \sin C$, $\cos A = 10 \cos B \cos C$. Then $\tan A = $ _____.

**Solution.** As $\sin A - \cos A = 10(\sin B \sin C - \cos B \cos C) = -10\cos(B + C) = 10\cos A$, we have $\sin A = 11\cos A$. Therefore $\tan A = 11$.

The answer is 11. ☐

④ Suppose the side of the base and the height of regular triangular pyramid $P - ABC$ are 1 and $\sqrt{2}$, respectively. Then the radius of the inscribed sphere of the pyramid is

_____.

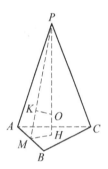

Fig. 4. 1

**Solution.** As seen in Fig. 4. 1, suppose the projections of the inscribed sphere's center $O$ on faces $ABC$ and $ABP$ are $H$, $K$, respectively, the midpoint of $AB$ is $M$, and the radius of the sphere is $r$. Then $P$, $K$, $M$ are collinear, $\angle PHM = \angle PKO = \dfrac{\pi}{2}$, and

$$OH = OK = r, \quad PO = PH - OH = \sqrt{2} - r,$$

$$MH = \frac{\sqrt{3}}{6}AB = \frac{\sqrt{3}}{6}, \quad PM = \sqrt{MH^2 + PH^2} = \sqrt{\frac{1}{12} + 2} = \frac{5\sqrt{3}}{6}.$$

Then we have

$$\frac{r}{\sqrt{2} - r} = \frac{OK}{PO} = \sin\angle KPO = \frac{MH}{PM} = \frac{1}{5}.$$

Therefore, $r = \dfrac{\sqrt{2}}{6}$.

The answer is $\dfrac{\sqrt{2}}{6}$. ☐

⑤ Let $a$, $b$ be real numbers, and $f(x) = ax + b$ satisfies $|f(x)| \leqslant 1$ for any $x \in [0, 1]$. Then the maximum of $ab$

is _____.

**Solution.** It is easy to find that $a = f(1) - f(0)$, $b = f(0)$. Then

$$ab = f(0) \cdot (f(1) - f(0)) = -\left(f(0) - \frac{1}{2}f(1)\right)^2 + \frac{1}{4}(f(1))^2$$

$$\leqslant \frac{1}{4}(f(1))^2 \leqslant \frac{1}{4}.$$

When $2f(0) = f(1) = \pm 1$, i.e., $a = b = \pm\frac{1}{2}$, we get $ab = \frac{1}{4}$.

Therefore, the maximum of $ab$ is $\frac{1}{4}$.

The answer is $\frac{1}{4}$. ▢

**6** Take randomly five different numbers from 1, 2, ..., 20. Then the probability that there are at least two adjacent numbers among them is _____.

**Solution.** Suppose $a_1 < a_2 < a_3 < a_4 < a_5$ taken from 1, 2, ..., 20. If $a_1$, $a_2$, $a_3$, $a_4$, $a_5$ are not adjacent to each other, then we have

$$1 \leqslant a_1 < a_2 - 1 < a_3 - 2 < a_4 - 3 < a_5 - 4 \leqslant 16,$$

from which we know that the number of ways to select five numbers not adjacent to each other from 1, 2, ..., 20 is the same as selecting five different numbers from 1, 2, ..., 16, i.e., $C_{16}^5$. Therefore, the required probability is

$$\frac{C_{20}^5 - C_{16}^5}{C_{20}^5} = 1 - \frac{C_{16}^5}{C_{20}^5} = \frac{232}{323}.$$

The answer is $\frac{232}{323}$. ▢

**7** Suppose real numbers $x$, $y$ satisfy $x - 4\sqrt{y} = 2\sqrt{x-y}$.
Then the range of $x$ is _____.

**Solution.** Let $\sqrt{y} = a$, $\sqrt{x-y} = b(a, b \geqslant 0)$. Then $x = y + (x-y) = a^2 + b^2$. The equation in the question becomes $a^2 + b^2 - 4a = 2b$, which is equivalent to

$$(a - 2)^2 + (b - 1)^2 = 5 \ (a, b \geqslant 0).$$

As seen in Fig. 7.1, the trace of point $(a, b)$ in plane $aOb$ is the part of the circle with center $(2, 1)$ and radius $\sqrt{5}$ satisfying $a, b \geqslant 0$, i.e., the union of point $O$ and arc $\overset{\frown}{ACB}$. Then

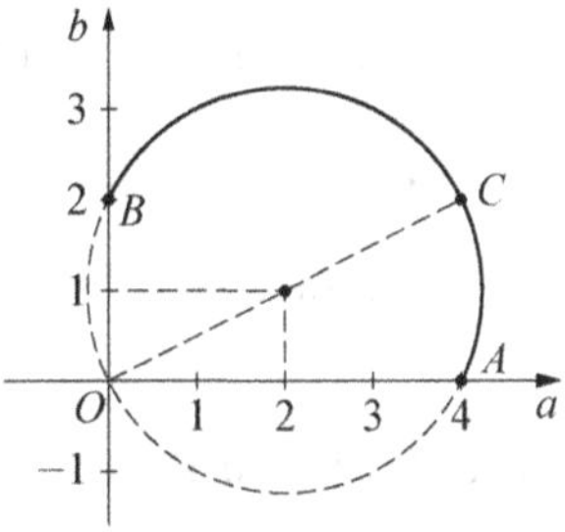

Fig. 7.1

$$\sqrt{a^2 + b^2} \in \{0\} \cup [2, 2\sqrt{5}].$$

Therefore, $x = a^2 + b^2 \in \{0\} \cup [4, 20]$.

The answer is $\{0\} \cup [4, 20]$. ☐

**8** Suppose sequence $\{a_n\}$ consists of nine terms, which satisfy $a_1 = a_9 = 1$ and $\dfrac{a_{i+1}}{a_i} \in \left\{2, 1, -\dfrac{1}{2}\right\}$ for any $i \in \{1, 2, \ldots, 8\}$. Then the number of sequences like this is _____.

**Solution.** Let $b_i = \dfrac{a_{i+1}}{a_i} (1 \leqslant i \leqslant 8)$. Then for each $\{a_n\}$ satisfying the given condition, we have

$$\prod_{i=1}^{8} b_i = \prod_{i=1}^{8} \frac{a_{i+1}}{a_i} = \frac{a_9}{a_1} = 1, \text{ with } b_i \in \left\{2, 1, -\frac{1}{2}\right\}(1 \leqslant i \leqslant 8).$$

$$\text{①}$$

Conversely, a sequence of eight terms $\{b_n\}$ satisfying ① can uniquely determine a sequence $\{a_n\}$ in the question.

In each $\{b_n\}$, there are obviously even number of $-\dfrac{1}{2}$ and the same number of 2, with the remainder being 1. Or, in other words, the numbers of $-\dfrac{1}{2}$ and 2 are both $2k$, while the number of 1 is $8 - 4k$. Here, it is easy to check that $k$ can only be 0, 1, 2. Once $k$ is given, there are $C_8^{2k} C_{8-2k}^{2k}$ ways to construct $\{b_n\}$. Therefore, the total number of $\{b_n\}$ satisfying ① is

$$N = 1 + C_8^2 C_6^2 + C_8^4 C_4^4 = 1 + 28 \times 15 + 70 \times 1 = 491.$$

The answer is 491.                                           □

**Part II   Word Problems (56 marks in total for three questions)**

**9**   ( 16   marks )   Suppose   positive   number   sequence $\{x_n\}$ satisfies $S_n \geqslant 2S_{n-1}$, $n = 2, 3, \ldots$, where $S_n = x_1 + \cdots + x_n$. Prove that there exists a constant $C > 0$, such that

$$x_n \geqslant C \cdot 2^n, \quad n = 1, 2, \ldots.$$

**Solution.** When $n \geqslant 2$, $S_n \geqslant 2S_{n-1}$ is equivalent to

$$x_n \geqslant x_1 + \cdots + x_{n-1}. \tag{①}$$

Let $C = \dfrac{1}{4}x_1$. We will prove

$$x_n \geqslant C \cdot 2^n, \quad n = 1, 2, \ldots. \tag{②}$$

by induction.

When $n = 1$, it is obviously true. When $n = 2$, we have $x_2 \geqslant x_1 = C \cdot 2^2$.

When $n \geqslant 3$, assume $x_k \geqslant C \cdot 2^k$, $k = 1, 2, \ldots, n - 1$. Then from ①, we have

$$x_n \geqslant x_1 + (x_2 + \cdots + x_{n-1})$$
$$\geqslant x_1 + (C \cdot 2^2 + \cdots + C \cdot 2^{n-1})$$
$$= C(2^2 + 2^2 + 2^3 + \cdots + 2^{n-1}) = C \cdot 2^n.$$

Therefore, ② holds for every $n$. $\qquad\square$

**10** (20 marks) Given an ellipse equation $\dfrac{x^2}{a^2} + \dfrac{y^2}{b^2} = 1 (a > b > 0)$ in a plane rectangular coordinate system $xOy$, let $A_1$, $A_2$, $F_1$, $F_2$ be its left and right end-points, left and right focuses, respectively, and $P$ be any point on the ellipse different from $A_1$, $A_2$. Suppose there are points $Q$, $R$ satisfying $QA_1 \perp PA_1$, $QA_2 \perp PA_2$, $RF_1 \perp PF_1$, $RF_2 \perp PF_2$. Find and prove the relationship between the length of segment $QR$ and $b$.

**Solution.** Let $c = \sqrt{a^2 - b^2}$. Then $A_1(-a, 0)$, $A_2(a, 0)$, $F_1(-c, 0)$, $F_2(c, 0)$.

Denote $P(x_0, y_0)$, $Q(x_1, y_1)$, $R(x_2, y_2)$, where $\dfrac{x_0^2}{a^2} + \dfrac{y_0^2}{b^2} = 1$, $y_0 \neq 0$.

From $QA_1 \perp PA_1$, $QA_2 \perp PA_2$, we have

$$\overrightarrow{A_1Q} \cdot \overrightarrow{A_1P} = (x_1 + a)(x_0 + a) + y_1 y_0 = 0, \qquad ①$$

$$\overrightarrow{A_2Q} \cdot \overrightarrow{A_2P} = (x_1 - a)(x_0 - a) + y_1 y_0 = 0. \qquad ②$$

Subtracting ① and ②, we have $2a(x_1 + x_0) = 0$, i.e. $x_1 = -x_0$. Substituting it into ①, we get $-x_0^2 + a^2 + y_1 y_0 = 0$, or $y_1 = \dfrac{x_0^2 - a^2}{y_0}$. Then $Q\left(-x_0, \dfrac{x_0^2 - a^2}{y_0}\right)$.

From $RF_1 \perp PF_1$, $RF_2 \perp PF_2$, in the same way we obtain $R\left(-x_0, \dfrac{x_0^2 - c^2}{y_0}\right)$. Therefore,

$$| QR | = \left| \frac{x_0^2 - a^2}{y_0} - \frac{x_0^2 - c^2}{y_0} \right| = \frac{b^2}{| y_0 |}.$$

Since $| y_0 | \in (0, b]$, then $| QR | \geqslant b$, where the equality holds if and only if $| y_0 | = b$ (i.e., $P(0, \pm b)$). ⬜

**11** (20 marks) Find all the positive real number pairs $(a, b)$, such that $f(x) = ax^2 + b$ satisfies

$f(xy) + f(x + y) \geqslant f(x)f(y)$ (for any real numbers $x$, $y$).

**Solution.** The given condition is equivalent to

$$(ax^2y^2 + b) + (a(x + y)^2 + b) \geqslant (ax^2 + b)(ay^2 + b). \quad ①$$

In ①, let $y = 0$. We have $b + (ax^2 + b) \geqslant (ax^2 + b) \cdot b$, or

$$(1 - b)ax^2 + b(2 - b) \geqslant 0.$$

As $a > 0$ and $ax^2$ can be sufficiently large, then $1 - b \geqslant 0$, i.e., $0 < b \leqslant 1$.

In ①, let $y = -x$. We have $(ax^4 + b) + b \geqslant (ax^2 + b)^2$, or

$$(a - a^2)x^4 - 2abx^2 + (2b - b^2) \geqslant 0. \quad ②$$

Denote the left-hand side of ② as $g(x)$. It is obvious that $a - a^2 \neq 0$ (otherwise, from $a > 0$ we know $a = 1$. Then $g(x) = -2bx^2 + (2b - b^2)$ with $b > 0$, which means $g(x)$ can be negative. A contradiction). Then

$$g(x) = (a - a^2)\left(x^2 - \frac{ab}{a - a^2}\right)^2 - \frac{(ab)^2}{a - a^2} + (2b - b^2)$$

$$= (a - a^2)\left(x^2 - \frac{b}{1 - a}\right)^2 + \frac{b}{1 - a}(2 - 2a - b)$$

$$\geqslant 0$$

holds for any real number $x$. So we have $a - a^2 > 0$, i.e., $0 < a < 1$.

Furthermore, from $\dfrac{b}{1-a} > 0$ and

$$g\left(\sqrt{\dfrac{b}{1-a}}\right) = \dfrac{b}{1-a}(2 - 2a - b) \geqslant 0,$$

we have $2a + b \leqslant 2$.

So far, we get the necessary condition that $a$, $b$ must satisfy as follows:

$$0 < b \leqslant 1,\ 0 < a < 1,\ 2a + b \leqslant 2. \qquad \text{③}$$

We are going to prove that for any pair $(a, b)$ satisfying ③ and any real numbers $x$, $y$, ① holds, or equivalently,

$$h(x, y) = (a - a^2)x^2 y^2 + a(1 - b)(x^2 + y^2) +$$
$$2axy + (2b - b^2) \geqslant 0.$$

As a matter of fact, when ③ holds, we then have

$$a(1 - b) \geqslant 0,\ a - a^2 > 0 \text{ and } \dfrac{b}{1-a}(2 - 2a - b) \geqslant 0.$$

Combining it with $x^2 + y^2 \geqslant -2xy$, we get

$$h(x, y) \geqslant (a - a^2)x^2 y^2 + a(1 - b)(-2xy) + 2axy + (2b - b^2)$$
$$= (a - a^2)x^2 y^2 + 2abxy + (2b - b^2)$$
$$= (a - a^2)\left(xy + \dfrac{b}{1-a}\right)^2 + \dfrac{b}{1-a}(2 - 2a - b) \geqslant 0.$$

Therefore, the set of all the pairs $(a, b)$ meeting the given condition is

$$\{(a, b) \mid 0 < b \leqslant 1,\ 0 < a < 1,\ 2a + b \leqslant 2\}.$$

□

# China Mathematical Competition

# (Complementary Test)

**1** (40 marks) As seen in Fig. 1.1, the circumcenter of acute triangle $ABC$ is $O$, $K$ is a point (not the midpoint) on the side $BC$, $D$ is a point on the extended line of segment $AK$, lines $BD$ and $AC$ intersect at point $N$, and lines $CD$ and $AB$ intersect at point $M$. Prove if $OK \perp MN$, then $A$, $B$,

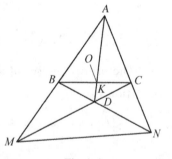

Fig. 1. 1

*D*, *C* are concyclic.

**Solution.** By reduction to absurdity, assume that *A*, *B*, *D*, *C* are not concyclic. Let the circumcircle (with radius *r*) of *ABC* intersect *AD* at point *E*. Join *BE* and extend it to intersect line *AN* at point *Q*; join *CE* and extend it to intersect *AM* at *P*. Join *PQ*, as seen in Fig. 1. 2.

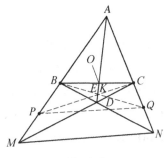

Fig. 1. 2

We have

$$PK^2 = \text{the power of } P \text{ with respect to } \odot O +$$
$$\text{the power of } Q \text{ with respect to } \odot O$$
$$= (PO^2 - r^2) + (KO^2 - r^2).$$

(We will prove it in the appendix)

In the same way,

$$QK^2 = (QO^2 - r^2) + (KO^2 - r^2).$$

Then we have

$$PO^2 - PK^2 = QO^2 - QK^2.$$

Therefore, $OK \perp PQ$. By the given condition $OK \perp MN$, we get that $PQ \parallel MN$. Then we have

$$\frac{AQ}{QN} = \frac{AP}{PM}. \tag{①}$$

By Menelaus' Theorem, we obtain

$$\frac{NB}{BD} \cdot \frac{DE}{EA} \cdot \frac{AQ}{QN} = 1, \tag{②}$$

$$\frac{MC}{CD} \cdot \frac{DE}{EA} \cdot \frac{AP}{PM} = 1. \tag{③}$$

From ①, ②, ③, we get $\dfrac{NB}{BD} = \dfrac{MC}{CD}$, or $\dfrac{ND}{BD} = \dfrac{MD}{DC}$. Then $\triangle DMN \backsim \triangle DCB$, which implies $\angle DMN = \angle DCB$. Then $BC \parallel MN$. Therefore, $OK \perp BC$, and that means $K$ is the midpoint of $BC$, which is a contradiction. This completes the proof that $A$, $B$, $D$, $C$ are concyclic.

**Appendix.** We are going to prove

$$PK^2 = \text{the power of } P \text{ with respect to } \odot O +$$
$$\text{the power of } Q \text{ with respect to } \odot O.$$

Extend $PK$ to point $F$, such that

$$PK \cdot KF = AK \cdot KE \qquad\qquad ④$$

(see Fig. 1.3). Then $P$, $E$, $F$, $A$ are concyclic, and we have

$$\angle PFE = \angle PAE = \angle BCE.$$

Then $E$, $C$, $F$, $K$ are concyclic, and we have

$$PK \cdot PF = PE \cdot PC. \qquad ⑤$$

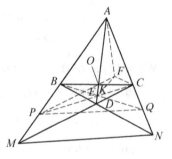

Fig. 1.3

From ⑤ and ④, we get $PK^2 = PE \cdot PC - AK \cdot KE =$ the power of $P$ with respect to $\odot O +$ the power of $Q$ with respect to $\odot O$.

$\square$

**Remark.** If $E$ is on the extended line of $AD$, then the proof is similar.

**2** (40 marks) Given positive integer $k$, let $r = k + \dfrac{1}{2}$. Define

$$f^{(1)}(r) = f(r) = r\lceil r \rceil, \ f^{(l)}(r) = f(f^{(l-1)}(r)), \ x \in \mathbf{R}^+, \ l \geqslant 2.$$

(Here $\lceil x \rceil$ denotes the minimum integer not less than $x$; e. g. , $\left\lceil \dfrac{1}{2} \right\rceil = 1$, $\lceil 1 \rceil = 1$. ) Prove that there exists a positive integer $m$ such that $f^{(m)}(r)$ is an integer.

**Solution.** Define $v_2(n)$ as the exponent of 2 in positive integer $n$. We will prove that $f^{(m)}(r)$ is an integer for $m = v_2(k) + 1$. We use mathematical induction on $v_2(k) = v$.

When $v = 0$, $k$ is odd and $k + 1$ is even. Then

$$f(r) = f^{(1)}(r) = \left( k + \frac{1}{2} \right)\left\lceil k + \frac{1}{2} \right\rceil = \left( k + \frac{1}{2} \right)(k + 1)$$

is an integer.

Assume the proposition is true for $v - 1 (v \geqslant 1)$. Then for $v \geqslant 1$, let

$$k = 2^v + \alpha_{v+1} \cdot 2^{v+1} + \alpha_{v+2} \cdot 2^{v+2} + \cdots,$$

where $\alpha_i \in \{0, 1\}$ for $i = v + 1, v + 2, \ldots$. We have

$$f(r) = \left( k + \frac{1}{2} \right)\left\lceil k + \frac{1}{2} \right\rceil = \left( k + \frac{1}{2} \right)(k + 1)$$

$$= \frac{1}{2} + \frac{k}{2} + k^2 + k$$

$$= \frac{1}{2} + 2^{v-1} + (\alpha_{v+1} + 1) \cdot 2^v + (\alpha_{v+1} + \alpha_{v+2}) \cdot$$

$$2^{v+1} + \cdots + 2^{2v} + \cdots$$

$$= k' + \frac{1}{2}, \tag{①}$$

with

$$k' = 2^{v-1} + (\alpha_{v+1} + 1) \cdot 2^v + (\alpha_{v+1} + \alpha_{v+2}) \cdot 2^{v+1} + \cdots + 2^{2v} + \cdots.$$

Obviously, $v_2(k') = v - 1$. Let $r' = k' + \dfrac{1}{2}$. By assumption, we know $f^{(v)}(r')$ is an integer, which is equal to

$f^{(v+1)}(r)$ by ①. The proof is then completed.          □

**3** (50 marks) Given integer $n > 2$, suppose positive real numbers $a_1, a_2, \ldots, a_n$ satisfy $a_k \leqslant 1$, $k = 1, 2, \ldots, n$.

Let $A_k = \dfrac{a_1 + a_2 + \cdots + a_k}{k}$, $k = 1, 2, \ldots, n$.

Prove $\left| \displaystyle\sum_{k=1}^{n} a_k - \sum_{k=1}^{n} A_k \right| < \dfrac{n-1}{2}$.

**Solution.** For $1 \leqslant k \leqslant n-1$, we have $0 < \sum_{i=1}^{k} a_i \leqslant k$ and $0 < \sum_{i=k+1}^{n} a_i \leqslant n-k$. By using the fact that $|x - y| < \max\{x, y\}$ for $x, y > 0$, we get

$$
\begin{aligned}
| A_n - A_k | &= \left| \left( \frac{1}{n} - \frac{1}{k} \right) \sum_{i=1}^{k} a_i + \frac{1}{n} \sum_{i=k+1}^{n} a_i \right| \\
&= \left| \frac{1}{n} \sum_{i=k+1}^{n} a_i - \left( \frac{1}{k} - \frac{1}{n} \right) \sum_{i=1}^{k} a_i \right| \\
&< \max \left\{ \frac{1}{n} \sum_{i=k+1}^{n} a_i, \left( \frac{1}{k} - \frac{1}{n} \right) \sum_{i=1}^{k} a_i \right\} \\
&\leqslant \max \left\{ \frac{1}{n} (n-k), \left( \frac{1}{k} - \frac{1}{n} \right) k \right\} \\
&= 1 - \frac{k}{n}.
\end{aligned}
$$

Therefore,

$$
\begin{aligned}
\left| \sum_{k=1}^{n} a_k - \sum_{k=1}^{n} A_k \right| &= \left| nA_n - \sum_{k=1}^{n} A_k \right| \\
&= \left| \sum_{k=1}^{n-1} (A_n - A_k) \right| \leqslant \sum_{k=1}^{n-1} | A_n - A_k | \\
&< \sum_{k=1}^{n-1} \left( 1 - \frac{k}{n} \right) = \frac{n-1}{2}.
\end{aligned}
$$

This completes the proof.          □

**4** (50 marks) The code setting of a cipher lock is established on an $n$-regular-polygon with vertices $A_1$, $A_2$, ..., $A_n$: each vertex is assigned a number (0 or 1) and a color (red or blue), such that either the numbers or the colors on each pair of adjacent vertices are the same. We ask: How many code-sets can be realized for this lock?

**Solution.** Given an arbitrary code-set for the lock, if two adjacent vertices have different numbers, we label the sides linking them by letter $a$; if they have different colors, we label it by $b$; if both the numbers and colors are the same, we label it by $c$. Once the number and color on vertex $A_1$ are set (there are four different sets for it), we can then set $A_2$, $A_3$, ..., $A_n$ one by one according to the letters labelled on each side. In order to let it return to the initial set of $A_1$ finally, the numbers of sides labelled $a$ and $b$ must be both even. So the number of code-sets for the lock is four times of the number of labelled-side sequences which satisfy the condition that the numbers of sides labelled by $a$ and $b$ are both even.

Suppose there are $2i\left(0 \leqslant i \leqslant \left[\dfrac{n}{2}\right]\right)$ sides labelled by $a$, and $2j\left(0 \leqslant j \leqslant \left[\dfrac{n-2i}{2}\right]\right)$ sides labelled by $b$. Then there are $C_n^{2i}$ ways to label $2i$ sides by $a$ from $n$ ones, $C_{n-2i}^{2j}$ ways to label $2j$ sides by $b$ from $n-2i$ ones, and the remaining sides are labelled by $c$. Therefore, by the Multiplication Principle, there are $C_n^{2i}C_{n-2i}^{2j}$ ways to label all the sides. So there are totally

$$4\sum_{i=0}^{\left[\frac{n}{2}\right]}\left(C_n^{2i}\sum_{j=0}^{\left[\frac{n-2i}{2}\right]}C_{n-2i}^{2j}\right) \qquad ①$$

code-sets for the lock. Here we stipulate $C_0^0 = 1$.

When $n$ is odd, we have $n - 2i > 0$, and then

$$\sum_{j=0}^{\left[\frac{n-2i}{2}\right]} C_{n-2i}^{2j} = 2^{n-2i-1}.$$  ②

Substituting it into ①, we get

$$4 \sum_{i=0}^{\left[\frac{n}{2}\right]} \left( C_n^{2i} \sum_{j=0}^{\left[\frac{n-2i}{2}\right]} C_{n-2i}^{2j} \right) = 4 \sum_{i=0}^{\left[\frac{n}{2}\right]} (C_n^{2i} 2^{n-2i-1}) = 2 \sum_{i=0}^{\left[\frac{n}{2}\right]} (C_n^{2i} 2^{n-2i})$$

$$= \sum_{k=0}^{n} C_n^k 2^{n-k} + \sum_{k=0}^{n} C_n^k 2^{n-k} (-1)^k$$

$$= (2 + 1)^n + (2 - 1)^n$$

$$= 3^n + 1.$$

When $n$ is even, if $i < \frac{n}{2}$, then ② remains true; if $i = \frac{n}{2}$, then all the sides of the polygon are labelled by $a$, and that means there is only one way to label the sides. Therefore, there are totally

$$4 \sum_{i=0}^{\left[\frac{n}{2}\right]} \left( C_n^{2i} \sum_{j=0}^{\left[\frac{n-2i}{2}\right]} C_{n-2i}^{2j} \right) = 4 \times \left( 1 + \sum_{i=0}^{\left[\frac{n}{2}\right]-1} (C_n^{2i} 2^{n-2i-1}) \right)$$

$$= 2 + 4 \sum_{i=0}^{\left[\frac{n}{2}\right]} (C_n^{2i} 2^{n-2i-1}) = 3^n + 3$$

code-sets for the lock.

In summary,

the number of code-sets for the lock $= \begin{cases} 3^n + 1 \text{ when } n \text{ is odd,} \\ 3^n + 3 \text{ when } n \text{ is even.} \end{cases}$

$\square$

# *2011* (Hubei)

1. (40 marks) As seen in Fig. 1.1, points $P$, $Q$ are, respectively, the midpoints of $AC$, $BD$ — the two diagonals of cyclic quadrilateral $ABCD$. Let $\angle BPA = \angle DPA$. Prove $\angle AQB = \angle CQB$.

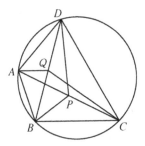

Fig. 1. 1

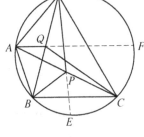

Fig. 1. 2

**Solution.** As shown in Fig. 1. 2, we extend segment $DP$ to intercept with the circle at point $E$. Then $\angle CPE = \angle DPA = \angle BPA$. Since $P$ is the midpoint of $AC$, we get $\overset{\frown}{AB} = \overset{\frown}{CE}$, which means $\angle CDP = \angle BDA$. Furthermore, $\angle ABD = \angle PCD$. Therefore, $\triangle ABD \backsim \triangle PCB$. Then $\dfrac{AB}{BD} = \dfrac{PC}{CD}$, i. e., $AB \cdot CD = PC \cdot BD$.

Then we have

$$AB \cdot CD = \frac{1}{2}AC \cdot BD = AC \cdot \left(\frac{1}{2}BD\right) = AC \cdot BQ,$$

or $\dfrac{AB}{AC} = \dfrac{BQ}{CD}$. Combining it with $\angle ABQ = \angle ACD$, we derive that $\triangle ABQ \backsim \triangle ACD$. So $\angle QAB = \angle DAC$.

Extending segment $AQ$ to intercept with the circle at point $F$, we then have

$$\angle CAB = \angle QAB - \angle QAC = \angle DAC - \angle QAC = \angle DAF,$$

which means $\overset{\frown}{BC} = \overset{\frown}{DF}$. Furthermore, as $Q$ is the midpoint of $BD$, then $\angle CQB = \angle DQF$.

Since $\angle AQB = \angle DQF$, we then have $\angle AQB = \angle CQB$. The proof is completed. $\qquad\qquad\qquad\qquad\qquad\qquad\square$

**2** (40 marks) Prove for any integer $n \geqslant 4$, there exists a polynomial of degree $n$,

$$f(x) = x^n + a_{n-1}x^{n-1} + \cdots + a_1 x + a_0$$

with the following properties.

(1) $a_0, a_1, \ldots, a_{n-1}$ are all positive integers;

(2) For any positive integer $m$ and arbitrary $k\,(k \geqslant 2)$ positive integers $r_1, r_2, \ldots, r_k$ that are different from each other, we have

$$f(m) \neq f(r_1)f(r_2)\cdots f(r_k).$$

**Solution.** Let

$$f(x) = (x+1)(x+2)\cdots(x+n) + 2. \qquad\qquad ①$$

Obviously, $f(x)$ is a monic polynomial of degree $n$ with positive integer coefficients. We are going to prove that $f(x)$ has property (2).

For any integer $t$, since $n \geqslant 4$, we know that there exists definitely a multiple of 4 in any $n$ consecutive numbers $t+1$, $t+2, \ldots, t+n$. Then from ①, we have $f(t) \equiv 2(\bmod 4)$.

Then for any $k$ $(k \geqslant 2)$ positive integers $r_1, r_2, \ldots, r_k$, we have

$$f(r_1)f(r_2)\cdots f(r_k) \equiv 2^k \equiv 0(\bmod 4).$$

On the other hand, for any positive integer $m$, we have

$f(m) \equiv 2 \pmod 4$. Therefore,

$$f(m) \not\equiv f(r_1)f(r_2)\cdots f(r_k) \pmod 4,$$

which implies that $f(m) \neq f(r_1)f(r_2)\cdots f(r_k)$. We then find the required $f(x)$ and complete the proof.　　□

**③** (50 marks) Let $a_1$, $a_2$, $\ldots$, $a_n$ $(n \geqslant 4)$ be positive real numbers with $a_1 < a_2 < \cdots < a_n$. For any positive real number $r$, the number of ternary groups $(i, j, k)$ satisfying $\dfrac{a_j - a_i}{a_k - a_j} = r$ $(1 \leqslant i < j < k \leqslant n)$ is denoted as $f_n(r)$. Prove $f_n(r) < \dfrac{n^2}{4}$.

**Solution.** Given $j (1 < j < n)$, the number of ternary groups $(i, j, k)$ satisfying $1 \leqslant i < j < k \leqslant n$ and

$$\frac{a_j - a_i}{a_k - a_j} = r \qquad\qquad ①$$

is denoted as $g_j(r)$. For fixed $i$, $j$ with $i < j$, there is at most one $k$ satisfying ①; so there are $j - 1$ ways to choose $i$, which means $g_j(r) \leqslant j - 1$. In a similar way, for fixed $j$, $k$ with $k > j$, there is at most one $i$ satisfying ①; so there are $n - j$ ways to choose $k$, which means $g_j(r) \leqslant n - j$. Therefore,

$$g_j(r) \leqslant \min\{j - 1, n - j\}.$$

Then, when $n$ is even (i.e., $n = 2m$), we have

$$f_n(r) = \sum_{j=2}^{n-1} g_j(r) = \sum_{j=2}^{m-1} g_j(r) + \sum_{j=m}^{2m-1} g_j(r)$$

$$\leqslant \sum_{j=2}^{m}(j - 1) + \sum_{j=m+1}^{2m-1}(2m - j) = \frac{m(m-1)}{2} + \frac{m(m-1)}{2}$$

$$= m^2 - m < m^2 = \frac{n^2}{4}.$$

When $n$ is odd (i.e. $n = 2m + 1$), we have

$$f_n(r) = \sum_{j=2}^{n-1} g_j(r) = \sum_{j=2}^{m} g_j(r) + \sum_{j=m+1}^{2m} g_j(r)$$

$$\leqslant \sum_{j=2}^{m} (j - 1) + \sum_{j=m+1}^{2m} (2m + 1 - j)$$

$$= m^2 < \frac{n^2}{4}.$$

The proof is completed. □

**4** (50 marks) Given a $3 \times 9$ array $A$ with each cell containing a positive integer, we say a $m \times n$ ($1 \leqslant m \leqslant 3$, $1 \leqslant n \leqslant 9$) subarray of $A$ is a "good rectangle" if the sum of the numbers in its cells is a multiple of 10, and call a $1 \times 1$ cell of $A$ "bad" if it is not contained in any "good rectangle". Find the maximum number of "bad cells" in $A$.

**Solution.** We first claim that the number of "bad cells" in $A$ is no more than 25. Otherwise, there will be at most one cell in $A$ that is not "bad". Without loss of generality, we assume the cells in the first row of $A$ are all "bad". Then let the numbers from top to bottom in the $i$th column be $a_i$, $b_i$, $c_i$ ($i = 1$, 2, ..., 9) in turn, and define

$$S_k = \sum_{i=1}^{k} a_i, \quad T_k = \sum_{i=1}^{k} (b_i + c_i), \quad k = 1, 2, \ldots, 9,$$

with $S_0 = T_0 = 0$. We are going to prove that three number groups $S_0$, $S_1$, ..., $S_9$, $T_0$, $T_1$, ..., $T_9$, and $S_0 + T_0$, $S_1 + T_1$, ..., $S_9 + T_9$ each form a complete set of residues modulo 10:

If there exist $m$, $n$, $0 \leqslant m < n \leqslant 9$ such that $S_m \equiv S_n$ (mod 10), then

$$\sum_{i=m+1}^{n} a_i = S_n - S_m \equiv 0 \pmod{10},$$

which means that the cells in the first row and from columns $m+1$ to $n$ form a "good rectangle". But it is a contradiction to the assumption that the cells in the first row are all "bad".

If there exist $m$, $n$, $0 \leqslant m < n \leqslant 9$ such that $T_m \equiv T_n \pmod{10}$, then

$$\sum_{i=m+1}^{n} (b_i + c_i) = T_n - T_m \equiv 0 \pmod{10}.$$

So the cells ranging from rows 2 to 3 and columns $m+1$ to $n$ form a "good rectangle", which means there are at least two cells that are not "bad". But it is also a contradiction.

In a similar way, we can also prove that there are no $m$, $n$, $0 \leqslant m < n \leqslant 9$ such that

$$S_m + T_m \equiv S_n + T_n \pmod{10}.$$

Therefore, we have

$$\sum_{k=0}^{9} S_k \equiv \sum_{k=0}^{9} T_k \equiv \sum_{k=0}^{9} (S_k + T_k) \equiv 0 + 1 + 2 + \cdots + 9$$
$$\equiv 5 \pmod{10}.$$

Then

$$\sum_{k=0}^{9} (S_k + T_k) \equiv \sum_{k=0}^{9} S_k + \sum_{k=0}^{9} T_k \equiv 5 + 5 \equiv 0 \pmod{10}.$$

It is again a contradiction! Therefore, the number of "bad cells" in $A$ is no more than 25.

On the other hand, we can construct a $3 \times 9$ array in the following and check that each cell in it that does not contain number 10 is "bad".

| 1 | 1 | 1 | 2 | 1 | 1 | 1 | 1 | 10 |
|---|---|---|---|---|---|---|---|---|
| 1 | 1 | 1 | 1 | 1 | 1 | 1 | 1 | 1 |
| 1 | 1 | 1 | 10 | 1 | 1 | 1 | 1 | 2 |

Therefore, we find out that the maximum number of "bad cells" in $A$ is 25. □

# 2012 (Shaanxi)

**1** (40 marks) As seen in Fig. 1.1, in acute triangle $\triangle ABC$, $AB > AC$, $M$, $N$ are two different points on $BC$ such that $\angle BAM = \angle CAN$, and $O_1$, $O_2$ are the circumcenters of $\triangle ABC$, $\triangle AMN$, respectively. Prove $O_1$, $O_2$, $A$ are collinear.

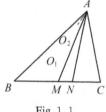
Fig. 1.1

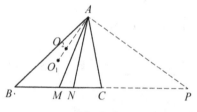
Fig. 1.2

**Solution.** As shown in Fig. 1.2, we connect $AO_1$, $AO_2$, and through $A$ draw a line perpendicular to $AO_1$ and intersect with the extended line of $BC$ at point $P$. Then $AP$ is a tangent line of $\odot O_1$, which means $\angle B = \angle PAC$.

As $\angle BAM = \angle CAN$, we have

$$\angle AMP = \angle B + \angle BAM = \angle PAC + \angle CAN = \angle PAN.$$

Then $AP$ is also a tangent line of $\odot O_2$ the circumcircle of

$\triangle AMN$. Therefore, $AP \perp AO_2$, and that means $O_1$, $O_2$, $A$ are collinear. The proof is complete.    □

**2** (40 marks) Let $A = \{2, 2^2, \ldots, 2^n, \ldots\}$. Prove:

(1) For any $a \in A$, $b \in \mathbf{N}^*$, if $b < 2a - 1$, then $b(b+1)$ will not be a multiple of $2a$.

(2) For any $a \in \bar{A}(= \mathbf{N}^* - A)$ satisfying $a \neq 1$, there exists $b \in \mathbf{N}^*$ satisfying $b < 2a - 1$, such that $b(b+1)$ is a multiple of $2a$.

**Solution.** (1) For any $a \in A$, $a = 2^k$ ($k \in \mathbf{N}^*$). Then $2a = 2^{k+1}$. Let $b$ be any positive integer strictly less than $2a - 1$. Then $(b+1) \leqslant 2a - 1$.

Between $b$ and $b + 1$, one is an odd number that contains no prime factor 2, and the other is an even number that contains at most the $k$th power of 2. Therefore, $b(b+1)$ is definitely not a multiple of $2a$.

(2) For $a \in \bar{A}$ and $a \neq 1$, suppose $a = 2^k m$ where $k$ is a non-negative integer and $m$ is an odd number greater than 1. Then $2a = 2^{k+1}m$. We will present three different proofs in the following.

Proof 1. Let $b = mx$, $b + 1 = 2^{k+1}y$. Eliminating $b$, we have $2^{k+1}y - mx = 1$. Since $(2^{k+1}, m) = 1$, the equation has integral solutions that can be expressed as

$$\begin{cases} x = x_0 + 2^{k+1}t, \\ y = y_0 + mt \end{cases}$$ (where $t \in \mathbf{Z}$, and $(x_0, y_0)$ is a special

solution of the equation).

Denote the smallest solution among them as $(x^*, y^*)$. Then $x^* < 2^{k+1}$.

Therefore, $b = mx^* < 2a - 1$ and $b(b+1)$ is a multiple of $2a$.

Proof 2. Since $(2^{k+1}, m) = 1$, by the Chinese Remainder

Theorem, the congruence equation

$$\begin{cases} x \equiv 0 (\bmod\ 2^{k+1}), \\ x \equiv m - 1 (\bmod\ m) \end{cases}$$

has a solution $x = b$ with $b \in (0, 2^{k+1}m)$. It is easy to see that $b < 2a - 1$ and $b(b + 1)$ is a multiple of $2a$.

**Proof 3.** Since $(2^{k+1}, m) = 1$, then there exists $r \in \mathbf{N}^*$, $r \leqslant m - 1$, such that $2^r \equiv 1 (\bmod\ m)$.

Take $t \in \mathbf{N}^*$ such that $tr > k + 1$. Then $2^{tr} \equiv 1 (\bmod\ m)$. It is easy to see that there exists

$$b = (2^{tr} - 1) - q \cdot 2^{k+1}m > 0 (q \in \mathbf{N}),$$

such that $0 < b < 2a - 1$. Then we have $m \mid b$, $2^{k+1} \mid b + 1$. Therefore, $b(b + 1)$ is a multiple of $2a$.  $\square$

**3** (50 marks) Let $P_0, P_1, P_2, \ldots, P_n$ be $n + 1$ points on a plane, and the minimum distance between each two points of them is $d(d > 0)$. Prove

$$\mid P_0P_1 \mid \cdot \mid P_0P_2 \mid \cdot \cdots \cdot \mid P_0P_n \mid > \left(\frac{d}{3}\right)^n \sqrt{(n+1)!}.$$

**Solution 1.** We may assume that $\mid P_0P_1 \mid \leqslant \mid P_0P_2 \mid \leqslant \cdots \leqslant \mid P_0P_n \mid$.

At first, we will prove that $\mid P_0P_k \mid > \frac{d}{3}\sqrt{k+1}$ for any positive integer $n$.

Obviously, $\mid P_0P_k \mid \geqslant d \geqslant \frac{d}{3}\sqrt{k+1}$ for $k = 1, 2, \ldots, 8$, and the second equality holds only when $k = 8$. Then we only need to prove that $\mid P_0P_k \mid \geqslant d \geqslant \frac{d}{3}\sqrt{k+1}$ for $k \geqslant 9$.

Take each $P_i (i = 0, 1, 2, \ldots, k)$ as the center to draw a circle with radius $\frac{d}{2}$. Then these circles are either externally tangent to or apart from each other. Take $P_0$ as the center to draw a circle with radius $|P_0P_k| + \dfrac{d}{2}$. Then the previous $k+1$ smaller circles are all located in this larger one.

Then $\pi\left(|P_0P_k| + \dfrac{d}{2}\right)^2 > (k+1)\pi\left(\dfrac{d}{2}\right)^2$, from which we have $|P_0P_k| > \dfrac{d}{2}(\sqrt{k+1} - 1)$.

It is easy to check that $\dfrac{\sqrt{k+1} - 1}{2} > \dfrac{\sqrt{k+1}}{3}$ for $k \geqslant 9$.

Then $|P_0P_k| > \dfrac{d}{3}\sqrt{k+1}$ for $k \geqslant 9$.

Over all, we have $|P_0P_k| > \dfrac{d}{3}\sqrt{k+1}$ for $k \geqslant 9$.

Therefore,

$$|P_0P_1| \cdot |P_0P_2| \cdot \cdots \cdot |P_0P_n| > \left(\dfrac{d}{3}\right)^n \sqrt{(n+1)!}.$$

**Solution 2.** We may assume $|P_0P_1| \leqslant |P_0P_2| \leqslant \cdots \leqslant |P_0P_n|$.

Take each $P_i\,(i = 0, 1, 2, \ldots, k)$ as the center to draw a circle with radius $\dfrac{d}{2}$. Then these circles are either externally tangent to or apart from each other.

Let $Q$ be any point on $\odot P_i$. Since

$$|P_0Q| \leqslant |P_0P_i| + |P_iQ| = |P_0P_i| + \dfrac{d}{2}$$

$$\leqslant |P_0P_k| + \dfrac{1}{2}|P_0P_k| = \dfrac{3}{2}|P_0P_k|,$$

we get that the circle with center $P_0$ and radius $\dfrac{3}{2}|P_0P_k|$ cover the previous $k+1$ smaller circles. Then we have $\pi\left(\dfrac{3}{2}|P_0P_k|\right)^2 > (k+1)\pi\left(\dfrac{d}{2}\right)^2$, and that is

$$| P_0 P_k | > \frac{d}{3} \sqrt{k+1} (i = 0, 1, 2, \ldots, k).$$

Therefore,

$$| P_0 P_1 | \cdot | P_0 P_2 | \cdot \cdots \cdot | P_0 P_n | > \left( \frac{d}{3} \right)^n \sqrt{(n+1)!}. \quad \square$$

**4** (50 marks) Let $S_n = 1 + \frac{1}{2} + \cdots + \frac{1}{n}$, where $n$ is a positive integer. Prove that for any real numbers $a$, $b$ with $0 \leqslant a < b \leqslant 1$, there are infinite many terms in the sequence $\{S_n - [S_n]\}$ that are within $(a, b)$. (Here $[x]$ denotes the largest integer not greater than real number $x$.)

**Solution 1.** For any $n \in \mathbf{N}^*$, we have

$$S_{2^n} = 1 + \frac{1}{2} + \frac{1}{3} + \cdots + \frac{1}{2^n} = 1 + \frac{1}{2} + \left( \frac{1}{2^1+1} + \frac{1}{2^2} \right) + \\ \left( \frac{1}{2^{n-1}+1} + \cdots + \frac{1}{2^n} \right)$$

$$> 1 + \frac{1}{2} + \left( \frac{1}{2^2} + \frac{1}{2^2} \right) + \cdots + \left( \frac{1}{2^n} + \cdots + \frac{1}{2^n} \right)$$

$$= 1 + \frac{1}{2} + \frac{1}{2} + \cdots + \frac{1}{2} > \frac{1}{2} n.$$

Let $N_0 = \left[ \frac{1}{b-a} \right] + 1$, $m = [S_{N_0}] + 1$. Then $\frac{1}{b-a} < N_0$, $\frac{1}{N_0} < b - a$, and $S_{N_0} < m \leqslant m + a$.

Let $N_1 = 2^{2(m+1)}$. Then $S_{N_1} = S_{2^{2(m+1)}} > m + 1 \geqslant m + b$.

We claim that there exist $n \in \mathbf{N}^*$ with $N_0 < n < N_1$ such that $m + a < S_n < m + b$ (or, in other words, $S_n - [S_n] \in (a, b)$).

Otherwise, assuming the claim is false, then there must exist $k > N_0$ such that $S_{k-1} \leqslant m + a$ and $S_k \geqslant m + b$.

Then $S_k - S_{k-1} \geqslant b - a$. But it contradicts the fact that

$S_k - S_{k-1} = \dfrac{1}{k} < \dfrac{1}{N_0} < b - a$. Therefore, the claim is true.

Furthermore, assume there are only a finite number of positive integers $n_1, \ldots, n_k$ satisfying

$$S_{n_j} - [\,S_{n_j}\,] \in (a, b) \ (1 \leqslant j \leqslant k).$$

Define $c = \min\limits_{1 \leqslant j \leqslant k} \{S_{n_j} - [\,S_{n_j}\,]\}$. Then there exists no $n \in \mathbf{N}^*$ such that $S_n - [S_n] \in (a, c)$. It contradicts the above claim.

Therefore, there are infinite terms in the sequence $\{S_n - [S_n]\}$ that are within $(a, b)$.

The proof is complete.

**Solution 2.** For any $n \in \mathbf{N}^*$, we have

$$S_{2^n} = 1 + \frac{1}{2} + \frac{1}{3} + \cdots + \frac{1}{2^n} = 1 + \frac{1}{2} + \left(\frac{1}{2^1 + 1} + \frac{1}{2^2}\right) +$$
$$\left(\frac{1}{2^{n-1} + 1} + \cdots + \frac{1}{2^n}\right)$$
$$> 1 + \frac{1}{2} + \left(\frac{1}{2^2} + \frac{1}{2^2}\right) + \cdots + \left(\frac{1}{2^n} + \cdots + \frac{1}{2^n}\right)$$
$$= 1 + \frac{1}{2} + \frac{1}{2} + \cdots + \frac{1}{2} > \frac{1}{2}n.$$

Therefore, $S_n$ can be larger than any positive number as long as $n$ becomes sufficiently large.

Let $N_0 = \left[\dfrac{1}{b-a}\right] + 1$. Then $\dfrac{1}{N_0} < b - a$, and when $k > N_0$, we have

$$S_k - S_{k-1} = \frac{1}{k} < \frac{1}{N_0} < b - a.$$

So for any positive integer $m > S_{N_0}$, there exists $n > N_0$ such that $S_n - m \in (a, b)$, or, in other words, $m + a < S_n < m + b$. Otherwise, there must be $k > N_0$ such that $S_{k-1} \leqslant m + a$ and $S_k \geqslant m + b$, i.e., $S_k - S_{k-1} \geqslant b - a$. But it contradicts the fact that $S_k - S_{k-1} = \dfrac{1}{k} < \dfrac{1}{N_0} < b - a$.

Now let $m_i = [S_{N_0}] + i$ $(i = 1, 2, 3, \ldots)$. Then there exists $n_i > N_0$ such that $m_i + a < S_{n_i} < m_i + b$, i.e., $S_{n_i} - [S_{n_i}] \in (a, b)$.

Therefore, there are infinite terms in the sequence $\{S_n - [S_n]\}$ that are within $(a, b)$. □

# · 2013 (Jilin)

**1** (40 marks) As seen in Fig. 1. 1, $AB$ is a chord of circle $\omega$, $P$ is a point on arc $AB$, and $E$, $F$ are 2 points on $AB$ satisfying $AE = EF = FB$. Connect $PE$, $PF$ and extend them to intersect with $\omega$ at $C$, $D$, respectively. Prove

$$EF \cdot CD = AC \cdot BD.$$

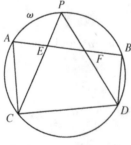

Fig. 1. 1

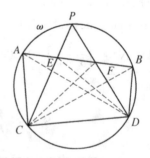

Fig. 1. 2

**Solution.** As shown in Fig. 1. 2, we connect $AD$, $BC$, $CF$, $DE$. Since $AE = EF = FB$, we have

$$\frac{BC \cdot \sin\angle BCE}{AC \cdot \sin\angle ACE} = \frac{\text{distance between } B \text{ and } CP}{\text{distance between } A \text{ and } CP} = \frac{BE}{AE} = 2.$$

①

In the same way,

$$\frac{AD \cdot \sin\angle ADF}{BD \cdot \sin\angle BDF} = \frac{\text{distance between } A \text{ and } PD}{\text{distance between } B \text{ and } PD} = \frac{AF}{BF} = 2.$$

②

On the other hand, since

$$\angle BCE = \angle BCP = \angle BDP = \angle BDF,$$

$$\angle ACE = \angle ACP = \angle ADP = \angle ADF,$$

multiplying ① by ②, we have $\dfrac{BC \cdot AD}{AC \cdot BD} = 4$, or

$$BC \cdot AD = 4AC \cdot BD. \qquad\qquad ③$$

By Ptolemy's Theorem, we have

$$AD \cdot BC = AC \cdot BD + AB \cdot CD. \qquad ④$$

Combining ③ and ④, we get $AB \cdot CD = 3AC \cdot BD$, and that is

$$EF \cdot CD = AC \cdot BD.$$

The proof is complete. □

**2** (40 marks) Given positive integers $u$, $v$, the sequence $\{a_n\}$ is defined as: $a_1 = u + v$, and for $m \geqslant 1$,

$$\begin{cases} a_{2m} = a_m + u, \\ a_{2m+1} = a_m + v. \end{cases}$$

Denote $S_m = a_1 + a_2 + \cdots + a_m$ ($m = 1, 2, \ldots$). Prove that there are infinite terms in sequence $\{S_n\}$ that are square numbers.

**Solution.** For positive integer $n$, we have

$$S_{2^{n+1}-1} = a_1 + (a_2 + a_3) + (a_4 + a_5) + \cdots + (a_{2^{n+1}-2} + a_{2^{n+1}-1})$$
$$= u + v + (a_1 + u + a_1 + v) + (a_2 + u + a_2 + v) + \cdots +$$
$$(a_{2^n-1} + u + a_{2^n-1} + v)$$
$$= 2^n(u + v) + 2S_{2^n-1}.$$

Then

$$S_{2^n-1} = 2^{n-1}(u + v) + 2S_{2^{n-1}-1}$$
$$= 2^{n-1}(u + v) + 2(2^{n-2}(u + v) + 2S_{2^{n-2}-1})$$
$$= 2 \cdot 2^{n-1}(u + v) + 2^2 S_{2^{n-2}-1}$$
$$= \cdots = (n - 1) \cdot 2^{n-1}(u + v) + 2^{n-1}(u + v)$$
$$= (u + v) n \cdot 2^{n-1}.$$

Suppose $u + v = 2^k \cdot q$, where $k$ is a non-negative integer, and $q$ is an odd number. Take $n = q \cdot l^2$, where $l$ is any positive integer satisfying $l \equiv k - 1 \pmod 2$. Then $S_{2^n-1} = q^2 l^2 \cdot 2^{k-1+q \cdot l^2}$, and

$$k - 1 + q \cdot l^2 \equiv k - 1 + l^2 \equiv k - 1 + (k - 1)^2$$
$$= k(k - 1) \equiv 0 \pmod 2.$$

Therefore, $S_{2^n-1}$ is a square number. Since there are infinite $l$'s, there are infinite terms in $\{S_n\}$ that are square numbers. The proof is complete. □

**3** (50 marks) Suppose there are $m$ questions in an examination attended by $n$ students, where $m$, $n \geqslant 2$ are given natural numbers. The marking rule for each question is as follows: if there are exactly $x$ students failing to answer the question correctly, then they will each get 0 marks, and those who answer it correctly will each get $x$ marks. The total marks of a student are the sum of marks he/she gets from the $m$ questions. Now rank the total marks of the $n$ students as $p_1 \geqslant p_2 \geqslant \cdots \geqslant p_n$. Find the maximum possible value of $p_1 + p_n$.

**Solution.** For any $k = 1, 2, \ldots, m$, assuming there are $x_k$ students failing to answer the $k$th question correctly, then there are $n - x_k$ ones who answer it correctly and each gets $x_k$ marks from it accordingly. Suppose the sum of the $n$ students' total marks is S. Then we have

$$\sum_{i=1}^{n} p_i = S = \sum_{k=1}^{m} x_k (n - x_k) = n \sum_{k=1}^{m} x_k - \sum_{k=1}^{m} x_k^2.$$

As each student gets at most $x_k$ marks from the $k$th question, we have

$$p_1 \leqslant \sum_{k=1}^{m} x_k.$$

Since $p_2 \geqslant \cdots \geqslant p_n$, then $p_n \leqslant \dfrac{p_2 + p_3 + \cdots + p_n}{n-1} = \dfrac{S - p_1}{n-1}$.

Therefore,

$$p_1 + p_n \leqslant p_1 + \frac{S - p_1}{n-1} = \frac{n-2}{n-1} p_1 + \frac{S}{n-1}$$

$$\leqslant \frac{n-2}{n-1} \cdot \sum_{k=1}^{m} x_k + \frac{1}{n-1} \cdot \left( n \sum_{k=1}^{m} x_k - \sum_{k=1}^{m} x_k^2 \right)$$

$$= 2 \sum_{k=1}^{m} x_k - \frac{1}{n-1} \cdot \sum_{k=1}^{m} x_k^2.$$

By the Cauchy Inequality, we have

$$\sum_{k=1}^{m} x_k^2 \geqslant \frac{1}{m} \left( \sum_{k=1}^{m} x_k \right)^2.$$

Then

$$p_1 + p_n \leqslant 2 \sum_{k=1}^{m} x_k - \frac{1}{m(n-1)} \cdot \left( \sum_{k=1}^{m} x_k \right)^2$$

$$= -\frac{1}{m(n-1)} \cdot \left( \sum_{k=1}^{m} x_k - m(n-1) \right)^2 + m(n-1)$$

$$\leqslant m(n-1).$$

On the other hand, if there is a student who answers all the questions correctly, while the other $n - 1$ students fail to answer any questions, then we have

$$p_1 + p_n = p_1 = \sum_{k=1}^{m} (n-1) = m(n-1).$$

Therefore, the maximum possible value of $p_1 + p_n$ is $m$ $(n-1)$.                                                                    □

**4** (50 marks) Let $n$, $k$ be integers greater than 1 and satisfy $n < 2^k$. Prove that there are $2k$ integers not divisible by $n$, such that if we divide them into two groups, then there must exist a group in which the sum of some integers can be divided by $n$.

**Solution.** At first, we consider the case that $n = 2^r$, $r \geqslant 1$. Obviously, at this time $r < k$. We take three "$2^{r-1}$"s and $2k - 3$ "1"s — each of them cannot be divided by $n$. If these $2k$ numbers are divided into two groups, then there must exist a group that contains two "$2^{r-1}$"s, whose sum is $2^r$ — divisible by $n$.

Next, we consider the case that $n$ is not a power of 2. At this time, the $2k$ integers we take are

$$-1, -1, -2, -2^2, \ldots, -2^{k-2}, 1, 2, 2^2, \ldots, 2^{k-1}.$$

Then they each cannot be divided by $n$.

Assume these numbers can be divided into two groups, such that any partial sum of numbers in one group is not divisible by

$n$. We may say that "1" is in the first group. Since $(-1) + 1 = 0$ is divisible by $n$, the two "$-1$"s must be in the second group; since $(-1) + (-1) + 2 = 0$, the "2" is in the first group; then the "$-2$" is in the second group.

Now by induction, assuming $1, 2, \ldots, 2^l$ are in the first group and $-1, -1, -2, \ldots, -2^l$ in the second one ($1 \leqslant l < k - 2$), since

$$(-1) + (-1) + (-2) + \cdots + (-2^l) + 2^{l+1} = 0$$

is divisible by $n$, we get that $2^{l+1}$ is in the first group, and then $-2^{l+1}$ in the second.

Therefore, $1, 2, 2^2, \ldots, 2^{k-2}$ is in the first group and $-1$, $-1, -2, -2^2, \ldots, -2^{k-2}$ in the second. Finally, since

$$(-1) + (-1) + (-2) + \cdots + (-2^{k-2}) + 2^{k-1} = 0,$$

then $2^{k-1}$ is in the first group. Therefore, $1, 2, 2^2, \ldots, 2^{k-1}$ are all in the first group.

On the other hand, the knowledge about the binary number system tells us that every positive integer which is not greater than $2^k - 1$ can be represented as the partial sum of $1, 2, 2^2, \ldots, 2^{k-1}$. Since $n \leqslant 2^k - 1$, then it is the partial sum of $1, 2, 2^2, \ldots, 2^{k-1}$ that is of course divisible by $n$ itself. This is a contradiction to the assumption.

Therefore, we have found out $2k$ integers that meet the requirement in the question. The proof is then complete.  □

# China Mathematical Olympiad

The China Mathematical Olympiad, organized by the China Mathematical Olympiad Committee, is held in January every year. About 150 winners of the China Mathematical Competition take part in it. The competition lasts for two days, and there are three problems to be completed within 4. 5 hours each day.

## *2011* (Changchun, Jilin)

### First Day
8:00~12:30 January 15, 2011

1　Let $a_1$, $a_2$, ..., $a_n$ $(n \geqslant 3)$ be real numbers. Prove that

$$\sum_{i=1}^{n} a_i^2 - \sum_{i=1}^{n} a_i a_{i+1} \leqslant \left[\frac{n}{2}\right](M - m)^2,$$

where $a_{n+1} = a_1$, $M = \max_{1 \leqslant i \leqslant n} a_i$, $m = \min_{1 \leqslant i \leqslant n} a_i$. $[x]$ is the largest integer not exceeding $x$.

**Solution.** If $n = 2k$ ($k$ is a positive integer), then

$$2\left(\sum_{i=1}^{n} a_i^2 - \sum_{i=1}^{n} a_i a_{i+1}\right) = \sum_{i=1}^{n} (a_i - a_{i+1})^2 \leqslant n(M - m)^2,$$

therefore,

$$\sum_{i=1}^{n} a_i^2 - \sum_{i=1}^{n} a_i a_{i+1} \leqslant \frac{n}{2}(M - m)^2 = \left[\frac{n}{2}\right](M - m)^2.$$

If $n = 2k + 1$ ($k$ is a positive integer), then for $2k + 1$ numbers arranged in a cyclic way, one can always find three consecutive increasing or decreasing terms (as $\prod_{i=1}^{2k+1}(a_i - a_{i-1})(a_{i+1} - a_i) = \prod_{i=1}^{2k+1}(a_i - a_{i-1})^2 \geqslant 0$, so it is not possible that for every $i$, $a_i - a_{i-1}$ and $a_{i+1} - a_i$ having opposite signs). Without loss of generality, we assume that $a_1$, $a_2$, $a_3$ are monotonic, then

$$(a_1 - a_2)^2 + (a_2 - a_3)^2 \leqslant (a_1 - a_3)^2.$$

Hence,

$$2\left(\sum_{i=1}^{n} a_i^2 - \sum_{i=1}^{n} a_i a_{i+1}\right) = \sum_{i=1}^{n} (a_i - a_{i+1})^2$$

$$\leqslant (a_1 - a_3)^2 + \sum_{i=3}^{n} (a_i - a_{i+1})^2,$$

which transformed the question into the case of $2k$ numbers. We have

$$2\left(\sum_{i=1}^{n} a_i^2 - \sum_{i=1}^{n} a_i a_{i+1}\right) \leqslant (a_1 - a_3)^2 + \sum_{i=3}^{n} (a_i - a_{i+1})^2$$

$$\leqslant 2k(M - m)^2,$$

i.e.,

$$\left(\sum_{i=1}^{n} a_i^2 - \sum_{i=1}^{n} a_i a_{i+1}\right) \leqslant k(M-m)^2 = \left[\frac{n}{2}\right](M-m)^2.$$

$\square$

**2** As shown in Fig. 2.1, $D$ is the midpoint of arc $BC$ of the circumcircle of triangle $ABC$, $X$ lies on arc $BD$, $E$ is the midpoint of arc $AX$, $S$ lies on arc $AC$, $SD$ intersects $BC$ at $R$, $SE$ intersects $AX$ at $T$. Prove that if $RT \parallel DE$, then the incenter of triangle $ABC$ lies on line $RT$.

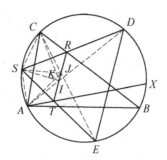

Fig. 2.1

**Proof.** Connect $AD$, denote the intersection of $AD$ and $RT$ by $I$, then $AI$ is the bisector of $\angle BAC$. Connect $AS$, $SI$, then by $RT \parallel DE$, we have

$$\angle STI = \angle SED = \angle SAI,$$

so $A$, $T$, $I$ and $S$ are concyclic, and we denote this circle by $\omega_1$.

Connect $CE$, denote the intersection of $CE$ and $RT$ by $J$, connect $SC$, then

$$\angle SRJ = \angle SDE = \angle SCE,$$

so $S$, $J$, $R$ and $C$ are concyclic, and we denote this circle by $\omega_2$.

Denote by $K$ the intersection point of $\omega_1$ and $\omega_2$ other than $S$, we prove next that $K$ is the intersection of $AJ$ and $CI$.

Denote by $K_1$ the intersection of $\omega_1$ and $AJ$ other than $A$, then

$$\angle SK_1 A = \angle STA = \frac{1}{2}(SA + XE) = \frac{1}{2}(SA + AE)$$

$$= \angle SDE = \angle SRT = \angle SRJ,$$

so $S$, $K_1$, $J$ and $R$ are concyclic, i. e., $K_1$ belongs to $\omega_2$. Similarly, denote by $K_2$ the intersection of $\omega_2$ and $CI$ other than $C$, then $K_2$ belongs to $\omega_1$. Hence, $K_1$ and $K_2$ coincide, and $K$ is the intersection of $AJ$ and $CI$.

As $\angle CAD = \angle CAI$, and $\angle TJE = \angle CJR = \angle CED = \angle CAD$, so $A$, $I$, $J$ and $C$ are concyclic, therefore, $\angle ACI = \angle AJI$.

On the other hand, by the concyclicity of $C$, $K$, $J$ and $R$, we have $\angle BCI = \angle ICR = \angle AJI$, and $\angle ACI = \angle BCI$, so $I$ is the incenter of the triangle $ABC$. $\qquad\square$

**3** Let $A_1$, $A_2$, $\ldots$, $A_n$ be $n$ non-empty subsets of a finite set $A$ of real numbers satisfying the following conditions:

(1) The sum of elements of $A$ is equal to $0$;

(2) Pick arbitrarily a number from each $A_i$, and their sum is strictly positive.

Prove that there exist sets $A_{i_1}$, $A_{i_2}$, $\ldots$, $A_{i_k}$, $1 \leqslant i_1 < i_2 < \cdots < i_k \leqslant n$, such that

$$|A_{i_1} \cup A_{i_2} \cup \cdots \cup A_{i_k}| < \frac{k}{n}|A|.$$

$|X|$ denotes the number of elements of a finite set $X$.

**Solution.** Let $A = \{a_1, \ldots, a_m\}$ with $a_1 > \cdots > a_m$. By (1) we have $a_1 + \cdots + a_m = 0$. Consider the smallest element of each $A_i$, the sum of these numbers is greater than $0$. Assume that there are exactly $k_i$ sets among $A_1$, $\ldots$, $A_n$ whose minimal element is $a_i$, $i = 1, 2, \ldots, m$. Then, one has

$$k_1 + \cdots + k_m = n.$$

By (2), we have

$$k_1 a_1 + \cdots + k_m a_m > 0.$$

For $s = 1, 2, \ldots, m-1$, there are in total $k_1 + \cdots + k_s$ sets, whose minimal elements are greater than or equal to $a_s$. Therefore, the union of these sets is contained in $\{a_1, \ldots, a_s\}$, whence the number of elements does not exceed $s$.

Next, we prove that there exists $s \in \{1, 2, \ldots, m-1\}$ such that $k = k_1 + \cdots + k_s > \dfrac{sn}{m}$. We prove this claim by contradiction. Suppose that

$$k_1 + \cdots + k_s \leqslant \frac{sn}{m}, \quad s = 1, 2, \ldots, m-1.$$

With the help of the Abel transform and the fact that $a_s - a_{s+1} > 0$, $1 \leqslant s \leqslant m-1$, we know that

$$
\begin{aligned}
0 &< \sum_{j=1}^{m} k_j a_j \\
&= \sum_{s=1}^{m-1} (a_s - a_{s+1})(k_1 + \cdots + k_s) + a_m(k_1 + \cdots + k_m) \\
&\leqslant \sum_{s=1}^{m-1} (a_s - a_{s+1}) \frac{sn}{m} + a_m n \\
&= \frac{n}{m} \sum_{j=1}^{m} a_j = 0.
\end{aligned}
$$

We then get a contradiction. For such an $s$, we take the sets among $A_1, \ldots, A_n$, whose minimal elements are greater than $a_s$, say $A_{i_1}, A_{i_2}, \ldots, A_{i_k}$. Then, by the above results, we know that the total number of such sets is $k = k_1 + \cdots + k_s > \dfrac{sn}{m}$, and the number of elements of their union does not exceed $s$, i.e., $| A_{i_1} \cup A_{i_2} \cup \cdots \cup A_{i_k} | \leqslant s < \dfrac{km}{n} = \dfrac{k}{n} | A |.$   $\square$

## Second Day

### 8:00~12:30, January 16, 2011

**④** Given positive integer $n$, let $S = \{1, 2, \ldots, n\}$. Find the minimum of $| A \triangle S | + | B \triangle S | + | C \triangle S |$ for nonempty finite sets $A$ and $B$ of real numbers, where $C = \{a + b \mid a \in A, b \in B\}$, $X \triangle Y = \{x \mid x$ belongs to exactly one of $X$ and $Y\}$, $| X |$ denotes the number of elements of a finite set $X$.

**Solution.** The minimum is $n + 1$.

First, by taking $A = B = S$, we have

$$| A \triangle S | + | B \triangle S | + | C \triangle S | = n + 1.$$

Second, we can prove that $l = | A \triangle S | + | B \triangle S | + | C \triangle S | \geqslant n + 1$. Let $X \backslash Y = \{x \mid x \in X, x \notin Y\}$. We have

$$l = | A \backslash S | + | B \backslash S | + | C \backslash S | + | S \backslash A | + | S \backslash B | + | S \backslash C |.$$

All we need to prove are the following:

(i) $| A \backslash S | + | B \backslash S | + | S \backslash C | \geqslant 1$,

(ii) $| C \backslash S | + | S \backslash A | + | S \backslash B | \geqslant n$.

For (i). In fact, if $| A \backslash S | = | B \backslash S | = 0$, then $A, B \subseteq S$. So 1 cannot be an element of $C$, hence $| S \backslash C | \geqslant 1$, therefore (i) is valid.

For (ii). If $A \cap S = \varnothing$, then $| S \backslash A | \geqslant n$, the claim is already valid. If $A \cap S \neq \varnothing$, we assume that the maximal element of $A \cap S$ is $n - k$, $0 \leqslant k \leqslant n - 1$, then

$$| S \backslash A | \geqslant k. \qquad ①$$

On the other hand, for $i = k + 1, k + 2, \ldots, n$, either $i \notin$

$B$ (then $i \in S \backslash B$ ) or $i \in B$ (then $n - k + i \in C$, i.e. , $n - k + i \in C \backslash S$ ), hence

$$| C \backslash S | + | S \backslash B | \geqslant n - k. \qquad ②$$

From ① and ②, we obtain (ii).

In conclusion, (i) and (ii) are valid, so $l \geqslant n + 1$. Hence, the minimum is $n + 1$. $\qquad \square$

**5**     Given integer $n \geqslant 4$. Find the maximum of

$$\frac{\sum_{i=1}^{n} a_i (a_i + b_i)}{\sum_{i=1}^{n} b_i (a_i + b_i)} \text{ for non-negative real numbers } a_1,$$

$a_2, \ldots, a_n, b_1, b_2, \ldots, b_n$ satisfying

$$a_1 + a_2 + \cdots + a_n = b_1 + b_2 + \cdots + b_n > 0.$$

**Solution.** The maximum is $n - 1$. By homogeneity, we can assume without loss of generality that $\sum_{i=1}^{n} a_i = \sum_{i=1}^{n} b_i = 1$.

First, it is clear that if $a_1 = 1$, $a_2 = a_3 = \cdots = a_n = 0$ and $b_1 = 0$, $b_2 = b_3 = \cdots = b_n = \dfrac{1}{n - 1}$, then $\sum_{i=1}^{n} a_i (a_i + b_i) = 1$,

$\sum_{i=1}^{n} b_i (a_i + b_i) = \dfrac{1}{n - 1}$, hence

$$\frac{\sum\limits_{i=1}^{n} a_i (a_i + b_i)}{\sum\limits_{i=1}^{n} b_i (a_i + b_i)} = n - 1.$$

Now we prove that for any real numbers $a_1, a_2, \ldots, a_n$, $b_1, b_2, \ldots, b_n$ satisfying $\sum_{i=1}^{n} a_i = \sum_{i=1}^{n} b_i = 1$, we have

$$\frac{\sum\limits_{i=1}^{n} a_i (a_i + b_i)}{\sum\limits_{i=1}^{n} b_i (a_i + b_i)} \leqslant n - 1.$$

Note that the denominator is positive, it is equivalent to show that

$$\sum_{i=1}^{n} a_i(a_i + b_i) \leqslant (n-1) \sum_{i=1}^{n} b_i(a_i + b_i),$$

i.e.,

$$(n-1) \sum_{i=1}^{n} b_i^2 + (n-2) \sum_{i=1}^{n} a_i b_i \geqslant \sum_{i=1}^{n} a_i^2.$$

By symmetry, we can assume that $b_1$ is the smallest one among $b_1, b_2, \ldots, b_n$. Then

$$(n-1) \sum_{i=1}^{n} b_i^2 + (n-2) \sum_{i=1}^{n} a_i b_i$$

$$\geqslant (n-1)b_1^2 + (n-1) \sum_{i=2}^{n} b_i^2 + (n-2) \sum_{i=1}^{n} a_i b_1$$

$$\geqslant (n-1)b_1^2 + \left( \sum_{i=2}^{n} b_i \right)^2 + (n-2)b_1$$

$$= (n-1)b_1^2 + (1-b_1)^2 + (n-2)b_1$$

$$= nb_1^2 + (n-4)b_1 + 1$$

$$\geqslant 1 = \sum_{i=1}^{n} a_i \geqslant \sum_{i=1}^{n} a_i^2.$$

$\square$

**6** Prove that for any given positive integers $m$, $n$, there exist infinitely many pairs of coprime positive integers $a$, $b$, such that $a + b \mid am^a + bn^b$.

**Solution.** If $mn = 1$, then the claim is valid. For $mn \geqslant 2$, since

$$n^a(am^a + bn^b) = (a+b)n^{a+b} + a((mn)^a - n^{a+b}),$$

it is sufficient to prove the existence of infinitely many coprime number pairs $a$, $b$, such that

$$a + b \mid (mn)^a - n^{a+b}, (a+b, n) = 1.$$

Let $p = a + b$, we only need to prove that there are infinitely many prime numbers $p$ and positive integer $1 \leqslant a \leqslant p - 1$ such that

$$p \mid (mn)^a - n^p.$$

By Fermat's theorem, i.e., when $a_1 \equiv a_2 \pmod{p-1}$, $a_1 \geqslant 1$, $a_2 \geqslant 1$, $(mn)^{a_1} \equiv (mn)^{a_2} \pmod{p}$.

So we only need to prove that there are infinitely many prime numbers $p$ and positive integer $a$ such that

$$p \mid (mn)^a - n. \tag{①}$$

If there are only finitely many such primes, say $p_1$, $p_2, \ldots, p_r$ (as $mn \geqslant 2$, the existence of such primes is obvious). Suppose that

$$(mn)^2 - n = p_1^{a_1} p_2^{a_2} \ldots p_r^{a_r}, \quad a_i\text{'s are non-negative integers}$$
$$(1 \leqslant i \leqslant r). \tag{②}$$

Let $a = p_1^{a_1} p_2^{a_2} \cdots p_r^{a_r} (p_1 - 1) \cdots (p_r - 1) + 2$, and suppose that

$$(mn)^a - n = p_1^{\beta_1} p_2^{\beta_2} \cdots p_r^{\beta_r}, \quad \beta_i\text{'s are non-negative integers}$$
$$(1 \leqslant i \leqslant r). \tag{③}$$

If $p_i \mid n$, then, by ③ and $a \geqslant 2$, we know that $p_i^{\beta_i} \mid n$, hence $p_i^{\beta_i} \mid (mn)^2 - n$, and by ② we have $\beta_i \leqslant a_i$.

If $p_i \nmid n$, then $p_i \nmid m$, so $(p_i^{a_i+1}, mn) = 1$. By Euler's Theorem (as $\varphi(p_i^{a_i+1}) = p_i^{a_i}(p_i - 1)$ is a factor of $a - 2$)

$$(mn)^a - n \equiv (mn)^2 - n \pmod{p_i^{a_i+1}}.$$

Because $p_i^{a_i+1} \nmid (mn)^2 - n$, the congruence relation above implies that $p_i^{a_i+1} \nmid (mn)^a - n$. So $\beta_i \leqslant a_i$. Hence,

$$(mn)^a - n = p_1^{\beta_1} p_2^{\beta_2} \cdots p_r^{\beta_r} \leqslant p_1^{a_1} p_2^{a_2} \cdots p_r^{a_r} = (mn)^2 - n,$$

which is in contradiction with $a > 2$. So there are infinitely many primes $p$ and positive integers $a$ such that $p \mid (mn)^a - n$.

$\square$

# $2012$   (Xian, Shaanxi)

## First Day

8:00~12:30 January 15, 2012

**1**   As shown in Fig. 1.1, $\angle A$ is the biggest angle in triangle $ABC$. On the circumcircle of $\triangle ABC$, the points $D$ and $E$ are the midpoints of $ABC$ and $ACB$, respectively. Denote by $\odot O_1$ the circle passing through $A$ and $B$, and tangent to line $AC$, by $\odot O_2$ the circle passing through $A$ and $E$, and tangent to line $AD$. $\odot O_1$ intersects $\odot O_2$ at points $A$ and $P$. Prove that $AP$ is the bisector of $\angle BAC$.

**Solution.** Join respectively the pairs of points $EP$, $AE$, $BE$, $BP$, $CD$. For the sake of convenience, we denote by $A$, $B$, $C$ the angles $\angle BAC$, $\angle ABC$, $\angle ACB$, then $A + B + C = 180°$. Take an arbitrary point $X$ on the extension of $CA$, a point $Y$ on the extension of $DA$. It is easy to see that $AD = DC$, $AE = EB$. By the fact that $A$, $B$, $C$, $D$ and $E$ are

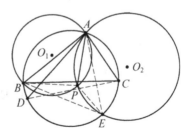

Fig. 1. 1

concyclic, we get

$$\angle BAE = 90° - \frac{1}{2}\angle AEB = 90° - \frac{C}{2}, \ \angle CAD$$

$$= 90° - \frac{1}{2}\angle ADC = 90° - \frac{B}{2}.$$

As line $AC$ and $\odot O_1$ are tangent at point $A$, line $AD$ and $\odot O_2$ are tangent at point $A$, we get

$$\angle APB = \angle BAX = 180° - A, \ \angle ABP = \angle CAP,$$

and

$$\angle APE = \angle EAY = 180° - \angle DAE = 180° - (\angle BAE + \angle CAD - A)$$

$$= 180° - \left(90° - \frac{C}{2}\right) - \left(90° - \frac{B}{2}\right) + A = 90° + \frac{A}{2}.$$

By computation, we obtain

$$\angle BPE = 360° - \angle APB - \angle APE = 90° + \frac{A}{2} = \angle APE.$$

In $\triangle APE$ and $\triangle BPE$, we apply Law of Sine, and take into account that $AE = BE$, we obtain

$$\frac{\sin\angle PAE}{\sin\angle APE} = \frac{PE}{AE} = \frac{PE}{BE} = \frac{\sin\angle PBE}{\sin\angle BPE}.$$

Therefore, $\sin\angle PAE = \sin\angle PBE$. On the other hand, $\angle APE$ and $\angle BPE$ are both obtuse, so $\angle PAE$ and $\angle PBE$ are both acute, thus $\angle PAE = \angle PBE$.

Hence, $\angle BAP = \angle BAE - \angle PAE = \angle ABE - \angle PBE = \angle ABP = \angle CAP$. $\qquad\square$

**2** Given a prime number $p$, let $A$ be a $p \times p$ matrix such that its entries are exactly $1, 2, \ldots, p^2$ in some order. The following operation is allowed for a matrix: add one to each number in a row or a column, or subtract one from each number in a row or a column. The matrix $A$ is called "good" if one can take a finite series of such operations resulting in a matrix with all entries zero. Find the number of good matrices $A$.

**Solution.** We may combine the operations on the same row or column, thus the final result of a series of operations can be realized as substracting integer $x_i$ from each number of $i$-th row and substracting integer $y_j$ from each number of $j$-th column. Thus, the matrix $A$ is good if and only if there exist integers $x_i$, $y_j$, such that $a_{ij} = x_i + y_j$ for all $1 \leqslant i, j \leqslant p$.

Since the entries of $A$ are distinct, $x_1, x_2, \ldots, x_p$ are pairwise distinct, and so are $y_1, y_2, \ldots, y_p$. We may consider only the case that $x_1 < x_2 < \cdots < x_p$ since swapping the value of $x_i$ and $x_j$ results in swapping the $i$-th row and $j$-th row, which is again a good matrix. Similarly, we may consider only the case that $y_1 < y_2 < \cdots < y_p$, thus the matrix is increasing from left to right, also from top to bottom.

From the assumptions above, we have $a_{11} = 1$, $a_{12}$ or $a_{21}$ equals 2. We may consider only the case that $a_{12} = 2$ since the transpose of the matrix is again good. Now we argue by contradiction that the first row is $1, 2, \ldots, p$. Assume on the contrary that $1, 2, \ldots, k$ is on the first row, but $k+1$ is not, $2 \leqslant k < p$, therefore $a_{21} = k+1$. We call $k$ consecutive integers a "block", and we shall prove that the first row consists of several blocks, that is, the first $k$ numbers is a block, the next $k$ numbers is again a block, and so on.

If it is not so, assume the first $n$ groups of $k$ numbers are "blocks", but the next $k$ numbers is not a "block"(or there are no $k$ numbers remaining). It follows that for $j = 1, 2, \ldots, n$,

$y_{(j-1)k+1}$, $y_{(j-1)k+2}$, ..., $y_{jk}$ is a "block", the first $nk$ columns of the matrix can be divided into $pn$ $1 \times k$ submatrices $a_{i,(j-1)k+1}$, $a_{i,(j-1)k+2}$, ..., $a_{i,jk}$, $i = 1, 2, ..., p$, $j = 1, 2, ..., n$, each submatrix is a "block". Now assume $a_{1,nk+1} = a$, let $b$ be the smallest positive integers such that $a + b$ is not on the first row, then $b \leqslant k - 1$. Since $a_{2,nk+1} - a_{1,nk+1} = x_2 - x_1 = a_{21} - a_{11} = k$, we have $a_{2,nk+1} = a + k$, therefore $a + b$ lies in the first $nk$ columns. Therefore, $a + b$ is contained in one of the $1 \times k$ submatrices mentioned above, which is a "block", however $a$, $a + k$ are not in this "block", which is a contradiction.

We showed that the first row is formed by blocks, in particular $k \mid p$, however, $1 < k < p$, and $p$ is a prime, which is impossible. So we conclude that the first row is $1, 2, ..., p$, the $k$-th row must be $(k - 1)p + 1$, $(k - 1)p + 2$, ..., $kp$. Thus up to interchanging rows, columns and transpose, the good matrix is unique, the answer is therefore $2(p!)^2$. $\qquad\square$

**3** Prove that for any real number $M > 2$, there exists a strictly increasing infinite sequence of positive integers $a_1$, $a_2$, ... satisfying both the following two conditions:

(1) $a_i > M^i$ for any positive integer $i$.

(2) An integer $n$ is non-zero if and only if there exists a positive integer $m$ and $b_1$, $b_2$, ..., $b_m \in \{-1, 1\}$, with $n = b_1 a_1 + b_2 a_2 + \cdots + b_m a_m$.

**Solution.** For given $M > 2$, we construct by induction a sequence $\{a_n\}$ that satisfies the requirements. Take $a_1$, $a_2$ that satisfy $a_2 - a_1 = 1$ and $a_1 > M^2$. Now suppose $a_1$, $a_2$, ..., $a_{2k}$ are already chosen, such that $a_i > M^i$, $i = 1, 2, ..., 2k$ and such that the set $A_k = \{b_1 a_1 + \cdots + b_m a_m \mid b_1, ..., b_m = \pm 1, 1 \leqslant m \leqslant 2k\}$ does not contain 0. It is obvious that $A_k$ is symmetric, i.e., $A_k = -A_k$. $A_1 = \{a_1, -a_1, 1, -1\}$. Let $n$ be the smallest positive integer not in $A_k$, $N = \sum_{i=1}^{2k} a_i$, now choose positive integers $a_{2k+1}$, $a_{2k+2}$ satisfying $a_{2k+2} - a_{2k+1} = N + n$, $a_{2k+1} > M^{2k+2}$, $a_{2k+1} > \sum_{i=1}^{2k} a_i$. We now show that $A_{k+1}$ does not contain 0 and $n \in A_{k+1}$. First, $n = -\sum_{i=1}^{2k} a_i - a_{2k+1} + a_{2k+2}$.

On the other hand, if $\sum_{i=1}^{m} b_i a_i = 0$, $m \leqslant 2k + 2$, as $0 \notin A_k$, we must have $m = 2k + 1$ or $2k + 2$.

If $m = 2k + 1$, then $\left| \sum_{i=1}^{2k+1} b_i a_i \right| \geqslant a_{2k+1} - \sum_{i=1}^{2k} a_i > 0$.

If $m = 2k + 2$ and $b_{2k+1}$ and $b_{2k+2}$ are of the same sign, then $\left| \sum_{i=1}^{2k+2} b_i a_i \right| \geqslant a_{2k+1} + a_{2k+2} - \sum_{i=1}^{2k} a_i > 0$; if $b_{2k+1}$ and $b_{2k+2}$ are of different signs, then

$$\left| \sum_{i=1}^{2k+2} b_i a_i \right| = \left| \sum_{i=1}^{2k} b_i a_i \pm (a_{2k+1} - a_{2k+2}) \right| \geqslant |a_{2k+1} - a_{2k+2}| - \sum_{i=1}^{2k} a_i$$

$$= N + n - N = n > 0.$$

The $\{a_n\}$ thus constructed satisfies the requirements since 0 is not contained in any $A_k$, and any non-zero integer between $-k$ and $k$ is contained in $A_k$. $\qquad\square$

## Second Day

8:00~12:30 January 16, 2012

**4** Let $f(x) = (x + a)(x + b)$ where $a$, $b$ are given positive real numbers, $n \geqslant 2$ be a given integer. For non-negative real numbers $x_1$, $x_2$, ..., $x_n$ that satisfy $x_1 + x_2 + \cdots + x_n = 1$, find the maximum of $F = \sum_{1 \leqslant i < j \leqslant n} \min\{f(x_i), f(x_j)\}$.

**Solution 1.** As

$$\min\{f(x_i), f(x_j)\} = \min\{(x_i + a)(x_i + b), (x_j + a)(x_j + b)\}$$

$$\leqslant \sqrt{(x_i + a)(x_i + b)(x_j + a)(x_j + b)}$$

$$\leqslant \frac{1}{2}((x_i + a)(x_j + b) + (x_i + b)(x_j + a))$$

$$= x_i x_j + \frac{1}{2}(x_i + x_j)(a + b) + ab,$$

so

$$F \leqslant \sum_{1 \leqslant i < j \leqslant n} x_i x_j + \frac{a + b}{2} \sum_{1 \leqslant i < j \leqslant n} (x_i + x_j) + C_n^2 \cdot ab$$

$$= \frac{1}{2}\left[\left(\sum_{i=1}^{n} x_i\right)^2 - \sum_{i=1}^{n} x_i^2\right] + \frac{a + b}{2}(n - 1)\sum_{i=1}^{n} x_i + C_n^2 \cdot ab$$

$$= \frac{1}{2}\left(1 - \sum_{i=1}^{n} x_i^2\right) + \frac{n - 1}{2}(a + b) + C_n^2 \cdot ab$$

$$\leqslant \frac{1}{2}\left(1 - \frac{1}{n}\left(\sum_{i=1}^{n} x_i\right)^2\right) + \frac{n - 1}{2}(a + b) + C_n^2 \cdot ab$$

$$= \frac{1}{2}\left(1 - \frac{1}{n}\right) + \frac{n - 1}{2}(a + b) + \frac{n(n - 1)}{2}ab$$

$$= \frac{n - 1}{2}\left(\frac{1}{n} + a + b + nab\right).$$

The equality holds when $x_1 = x_2 = \cdots = x_n = \dfrac{1}{n}$. So the maximum of $F$ is $\dfrac{n - 1}{2}\left(\dfrac{1}{n} + a + b + nab\right)$.

**Solution 2.** We show that the maximum value is attained when $x_1 = x_2 = \cdots = x_n = \dfrac{1}{n}$, and $F_{\max} = \dfrac{n - 1}{2}\left(\dfrac{1}{n} + a + b + nab\right)$.

We induct on $n$ to show a more general statement: for non-negative real numbers $x_1, x_2, \ldots, x_n$ satisfying $x_1 + x_2 + \cdots + x_n = s$ (where $s$ is a fixed non-negative real number), the maximum value of $F = \sum_{1 \leqslant i < j \leqslant n} \min\{f(x_i), f(x_j)\}$ is attained when $x_1 = x_2 = \cdots = x_n = \dfrac{s}{n}$.

Since $F$ is symmetric, we may assume $x_1 \leqslant x_2 \leqslant \cdots \leqslant x_n$. Note that $f(x)$ is strictly increasing on non-negative real numbers, we have

$$F = (n-1)f(x_1) + (n-2)f(x_2) + \cdots + f(x_{n-1}).$$

When $n = 2$, $F = f(x_1) \leqslant f\left(\dfrac{s}{2}\right)$, equality holds when $x_1 = x_2$. Assume that the statement holds for $n$, consider the case of $n+1$. Applying inductive hypothesis on $x_2 + x_3 + \cdots + x_{n+1} = s - x_1$, we have

$$F \leqslant nf(x_1) + \frac{1}{2}n(n-1)f\left(\frac{s-x_1}{n}\right) = g(x_1),$$

where $g(x)$ is a quadratic function of $x$, the leading coefficient is $1 + \dfrac{n-1}{2n^2}$, and the coefficient of $x$ is $a + b - \dfrac{n-1}{2n}\left(a+b+\dfrac{s}{2n}\right)$, therefore, the axis of symmetry is

$$\frac{\dfrac{n-1}{2n}\left(a+b+\dfrac{s}{2n}\right) - a - b}{2 + \dfrac{n-1}{n^2}} \leqslant \frac{s}{2(n+1)}.$$

(The above inequality is equivalent to $[(n-1)s - 2n(n+1)(a+b)](n+1) \leqslant 2s(2n^2 + n - 1)$; obviously, left-hand side $< (n^2-1)s <$ right-hand side.) Therefore, $g\left(\dfrac{s}{n+1}\right)$ is the maximum of $g(x)$ on $\left[0, \dfrac{s}{n+1}\right]$. Thus, $F$ attains its maximum when $x_2 = x_3 = \cdots = x_{n+1} = \dfrac{s-x_1}{n} = \dfrac{s}{n+1} = x_1$, completing the solution. $\qquad\square$

**5** Let $n$ be a square-free positive even number, $k$ be an integer, $p$ be a prime number, satisfying $p < 2\sqrt{n}$, $p \nmid n$, $p \mid n + k^2$. Prove that $n$ can be written as $n = ab + bc + ca$, where $a$, $b$, $c$ are distinctive positive integers.

**Solution.** Since $n$ is even, we have $p \neq 2$. As $p \nmid n$, we have $p \nmid k$. We may assume without loss of generality $0 < k < p$. Set

$$a = k, \ b = p - k, \text{ then } c = \frac{n - k(p - k)}{p} = \frac{n + k^2}{p} - k.$$

By assumption, $c$ is an integer, and $a$, $b$ are distinct positive integers. It remains to be shown that $c > 0$, and $c \neq a$, $b$. By the AM – GM inequality, we have $\frac{n}{k} + k \geq 2\sqrt{n} > p$, thus $n + k^2 > pk$, hence $c > 0$. If $c = a$, then $\frac{n + k^2}{p} - k = k$, thus $n = k(2p - k)$. Since $n$ is even, $k$ is also even, as a consequence $n$ is divisible by 4, which contradicts the fact that $n$ is square-free. If $c = b$, then $n = p^2 - k^2$. Since $n$ is even, $k$ is odd, implying that $n$ is again divisible by 4, which is a contradiction.

We conclude that $a$, $b$, $c$ satisfy all the requirements, completing the proof. $\qquad\square$

**6** Find the smallest positive integer $k$ with the following property: for any $k$ element subset $A$ of the set $S = \{1, 2, \ldots, 2012\}$, there exist three pairwise distinct elements $a$, $b$, $c$ of $S$ such that $a + b$, $b + c$, $c + a$ all belong to $A$.

**Solution.** Without loss of generality, we may assume $a < b < c$. Write $x = a + b$, $z = b + c$, $y = a + c$, then $x < y < z$, $x + y > z$, and $x + y + z$ is even. On the other hand, if there exist $x$, $y$, $z \in A$ such that $x < y < z$, $x + y > z$, and that $x + y + z$ is even, set $a = \frac{x + y - z}{2}$, $b = \frac{x + z - y}{2}$, $c =$

$\dfrac{y + z - x}{2}$, it is clear that $a$, $b$, $c$ are pairwise distinct elements

of $S$, and $x = a + b$, $y = a + c$, $z = b + c$.

The required property is equivalent to the following: for any $k$-element subset $A$ of $S$, there exist three elements $x$, $y$, $z \in A$ such that

$$x < y < z, x + y > z, \text{ and } x + y + z \text{ is even.} \quad (*)$$

If $A = \{1, 2, 3, 5, 7, \ldots, 2011\}$, $|A| = 1007$, and $A$ does not contain three elements satisfying property $(*)$. Therefore, $k \geqslant 1008$.

We next prove that any 1008-element subset of $S$ contains three elements satisfying property $(*)$.

We prove a general statement: For any integer $n \geqslant 4$, any $(n + 2)$-element subset of $\{1, 2, \ldots, 2n\}$ contains three elements satisfying $(*)$. We induct on $n$.

When $n = 4$, let $A$ be a 6-element subset of $\{1, 2, \ldots, 8\}$, then $A \cap \{3, 4, 5, 6, 7, 8\}$ contains at least four elements. If $A \cap \{3, 4, 5, 6, 7, 8\}$ contains three even numbers, then $4$, $6$, $8 \in A$ satisfying $(*)$. If $A \cap \{3, 4, 5, 6, 7, 8\}$ contains exactly two even numbers, then it contains two odd numbers. For any two odd numbers $x$, $y$ of $\{3, 5, 7\}$, two of $(4, x, y)$, $(6, x, y)$, $(8, x, y)$ would satisfy property $(*)$, thus one of them is contained in $A$. If $A \cap \{3, 4, 5, 6, 7, 8\}$ contains exactly one even number $x$, then it contains all three odd numbers, then $(x, 5, 7)$ satisfies $(*)$. The result holds for $n = 4$.

Assuming the result holds for $n$ ($n \geqslant 4$), consider the case of $n + 1$. Let $A$ be $(n + 3)$-elements of $\{1, 2, \ldots, 2n + 2\}$, if $|A \cap \{1, 2, \ldots, 2n\}| \geqslant n + 2$. By inductive hypothesis, the result follows. It remains to consider $|A \cap \{1, 2, \ldots, 2n\}| = n + 1$, and $2n + 1$, $2n + 2 \in A$. If $A$ contains an odd number $x$ in $\{1,$ $2, \ldots, 2n\}$, then $x$, $2n + 1$, $2n + 2$ satisfy ( $*$ ); if no odd number of $\{1, 2, \ldots, 2n\}$ greater than 1 is contained in $A$, then $A = \{1, 2, 4, 6, \ldots, 2n, 2n + 1, 2n + 2\}$, and $4, 6, 8 \in A$ satisfy ( $*$ ).

Hence, the smallest $k$ with the required property is 1008.

$\square$

# $2013$   (Shenyang, Liaoning)

## First Day 8:00 – 12:30
### January 12, 2013

**1** Two circles $K_1$ and $K_2$ of different radii intersect at two points $A$ and $B$, let $C$ and $D$ be two points on $K_1$ and $K_2$, respectively, such that $A$ is the midpoint of the segment $CD$. The extension of $DB$ meets $K_1$ at another point $E$, and the extension of $CB$ meets $K_2$ at another point $F$. Let $l_1$ and $l_2$ be the perpendicular bisectors of $CD$ and $EF$, respectively.

(1) Show that $l_1$ and $l_2$ have a unique common point (denoted by $P$).

(2) Prove that the lengths of $CA$, $AP$ and $PE$ are the side lengths of a right triangle.

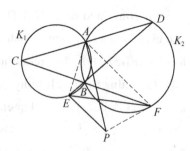

Fig. 1. 1

**Solution** (1) Since $C$, $A$, $B$, $E$ are concyclic, and $D$, $A$, $B$, $F$ are concyclic, $CA = AD$, and by the theorem of power of a point, we have

$$CB \cdot CF = CA \cdot CD = DA \cdot DC = DB \cdot DE. \qquad \textcircled{1}$$

Suppose on the contrary that $l_1$ and $l_2$ do not intersect, then $CD \parallel EF$, hence $\dfrac{CF}{CB} = \dfrac{DE}{DB}$. Plugging into $\textcircled{1}$, we get $CB^2 = DB^2$, thus $CB = DB$, hence $BA \perp CD$. It follows that $CB$ and $DB$ are the diameters of $K_1$ and $K_2$, respectively, hence $K_1$ and $K_2$ have same radii, which contradicts with assumption. Thus, $l_1$ and $l_2$ have a unique common point.

(2) Join $AE$, $AF$ and $PF$, we have

$$\angle CAE = \angle CBE = \angle DBF = \angle DAF.$$

Since $AP \perp CD$, $AP$ is the bisector of $\angle EAF$. Since $P$ is on the perpendicular bisector of the segment $EF$, $P$ is on the circumcircle of $\triangle AEF$. We have

$$\angle EPF = 180° - \angle EAF = \angle CAE + \angle DAF$$
$$= 2\angle CAE = 2\angle CBE.$$

Hence, $B$ is on the circle with center $P$ and radius $PE$, denoting this circle by $\Gamma$. Let $R$ be the radius of $\Gamma$. By the theorem of power of a point, we have

$$2CA^2 = CA \cdot CD = CB \cdot CF = CP^2 - R^2,$$

thus

$$AP^2 = CP^2 - CA^2 = (2CA^2 + R^2) - CA^2 = CA^2 + PE^2.$$

It follows that $CA$, $AP$, $PE$ form the side lengths of a right triangle. $\qquad \square$

**2** Find all non-empty sets $S$ of integers such that $3m - 2n \in$ $S$ for all (not necessarily distinct) $m$, $n \in S$.

**Solution** Call a set $S$ "good" if it satisfies the property as stated in the problem.

(1) If $S$ has only one element, $S$ is "good".

(2) Now we can assume that $S$ contains at least two elements. Let

$$d = \min\{|m - n| : m, n \in S, m \neq n\}.$$

Then there is an integer $a$, such that $a + d$, $a + 2d \in S$. Note that

$$a + 4d = 3(a + 2d) - 2(a + d) \in S,$$
$$a - d = 3(a + d) - 2(a + 2d) \in S,$$
$$a + 5d = 3(a + d) - 2(a - d) \in S,$$
$$a - 2d = 3(a + 2d) - 2(a + 4d) \in S.$$

So we have proved that if

$$a + d, a + 2d \in S,$$

then

$$a - 2d, a - d, a + 4d, a + 5d \in S.$$

Continuing this procedure, we can deduce that

$$\{a + kd \mid k \in \mathbf{Z}, 3 \nmid k\} \subseteq S.$$

Let $S_0 = \{a + kd \mid k \in \mathbf{Z}, 3 \nmid k\}$, it is easy to verify that $S_0$ is "good".

(3) Now we have proved that $S_0 \subseteq S$. If $S \neq S_0$, pick a number $b \in S \backslash S_0$, then there exists an integer $l$, such that $a + ld \leqslant b < a + (l + 1)d$. Since at least one of $l$ and $l + 1$ is indivisible by 3, we know that at least one of $a + ld$, $a + (l + 1)d$ is contained in $S_0$. If $a + ld \in S_0$, note that $0 \leqslant |b - (a +$

$ld) \mid < d$, by the definition of $d$, we must have $b = a + ld$. If $a + (l + 1)d \in S_0$, note that

$$0 < \mid a + (l + 1)d - b \mid \leqslant d,$$

by the definition of $d$, we also have $b = a + ld$.

In both cases, we have proved that there is a number $b \in S \backslash S_0$ of the form $b = a + ld$. This $l$ must be divisible by 3, hence

$$a + ld, a + (l + 1)d, a + (l + 2)d \in S,$$
$$a + (l - 2)d, a + (l - 1)d \in S.$$

So

$$a + (l + 3)d = 3(a + (l + 1)d) - 2(a + ld) \in S,$$
$$a + (l - 3)d = 3(a + (l - 1)d) - 2(a + ld) \in S.$$

Continuing this procedure, we have $a + (l + 3j)d \in S$, $j \in \mathbf{Z}$, which implies that $\{a + kd \mid d \in \mathbf{Z}\} \subseteq S$. We claim that $S = \{a + kd \mid k \in \mathbf{Z}\}$, since for any number $x \notin \{a + kd \mid d \in \mathbf{Z}\}$, there is an element in $y \in \{a + kd \mid k \in \mathbf{Z}\}$ such that $0 < \mid x - y \mid < d$. By the definition of $d$, we must have $x \notin S$. So $S = \{a + kd \mid k \in \mathbf{Z}\}$, and it is easy to verify that such $S$ is "good".

We conclude that there are three classes of "good" sets:

(1) $S = \{a\}$;

(2) $S = \{a + kd \mid k \in \mathbf{Z}, 3 \nmid k\}$;

(3) $S = \{a + kd \mid k \in \mathbf{Z}\}$, here $a, d \in \mathbf{Z}, d > 0$. □

**3** Find all positive real numbers $t$ with the following property: there exists an infinite set $X$ of real numbers such that the inequality

$$\max\{\mid x - (a - d) \mid, \mid y - a \mid, \mid z - (a + d) \mid\} > td$$

holds for all (not necessarily distinct) $x, y, z \in X$, all real numbers $a$ and all positive real numbers $d$.

**Solution** The answer is $0 < t < \dfrac{1}{2}$.

Firstly, for $0 < t < \dfrac{1}{2}$, choose $\lambda \in \left(0, \dfrac{1-2t}{2(1+t)}\right)$, let $x_i = \lambda^i$, $X = \{x_1, x_2, \ldots\}$. We claim that for all (not necessarily distinct) $x$, $y$, $z \in X$, all real numbers $a$ and all positive real numbers $d$, we have the following inequality:

$$\max\{\,|\,x - (a - d)\,|\,,\,|\,y - a\,|\,,\,|\,z - (a + d)\,|\,\} > td.$$

Suppose on the contrary that there exist $a \in \mathbf{R}$, $d \in \mathbf{R}^+$ and $x_i$, $x_j$, $x_k$, such that

$$\max\{\,|\,x_i - (a - d)\,|\,,\,|\,x_j - a\,|\,,\,|\,x_k - (a + d)\,|\,\} \leqslant td.$$

Hence,

$$\begin{cases} -td \leqslant x_i - (a - d) \leqslant td, \\ -td \leqslant x_j - a \leqslant td, \\ -td \leqslant x_k - (a + d) \leqslant td, \end{cases}$$

i.e.,

$$\begin{cases} x_i + (1 - t)d \leqslant a \leqslant x_i + (1 + t)d, \\ x_j - td \leqslant a \leqslant x_j + td, \\ x_k - (1 + t)d \leqslant a \leqslant x_k - (1 - t)d, \end{cases} \qquad (*)$$

which implies that

$$\begin{cases} x_k - (1 + t)d \leqslant a \leqslant x_i + (1 + t)d, \\ x_i + (1 - t)d \leqslant a \leqslant x_j + td, \\ x_j - td \leqslant a \leqslant x_k - (1 - t)d, \end{cases}$$

note that $0 < t < \dfrac{1}{2}$. It follows that

$$
\begin{cases}
d \geqslant \dfrac{x_k - x_i}{2(1+t)}, & \text{①} \\[3mm]
d \leqslant \dfrac{x_j - x_i}{1 - 2t}, & \text{②} \\[3mm]
d \leqslant \dfrac{x_k - x_j}{1 - 2t}. & \text{③}
\end{cases}
$$

By ②, ③ and $d > 0$, we get $x_i < x_j < x_k$, hence $i > j > k$, $\lambda^j + \lambda^{i+1} \leqslant \lambda^{k+1} + \lambda^i$, we get

$$
\frac{x_j - x_i}{x_k - x_i} = \frac{\lambda^j - \lambda^i}{\lambda^k - \lambda^i} \leqslant \lambda. \qquad \text{④}
$$

By ① and ②, we get $\dfrac{x_j - x_i}{1 - 2t} \geqslant \dfrac{x_k - x_i}{2(1+t)}$, hence

$$
\frac{x_j - x_i}{x_k - x_i} \geqslant \frac{1 - 2t}{2(1+t)} > \lambda,
$$

which contradicts ④! Thus our earlier claim about $X$ is proved.

Secondly, for $t \geqslant \dfrac{1}{2}$, we show that for any infinite set $X$, for any $x < y < z$ in $X$, we can choose $a \in \mathbf{R}$ and $d \in \mathbf{R}^+$ such that

$$
\max\{\mid x - (a - d) \mid, \mid y - a \mid, \mid z - (a + d) \mid\} \leqslant td.
$$

In fact, let $d = \dfrac{z - x}{2}$, hence $x + (1 - t)d = z - (1 + t)d$.

Let $a = \max\{x + (1 - t)d, y - td\}$. Since $t \geqslant \dfrac{1}{2}$, we obtain

$$
\begin{cases}
y - x < 2d \leqslant (1 + 2t)d, \\
x - y < 0 \leqslant (2t - 1)d,
\end{cases}
$$

i.e.,

$$
\begin{cases}
y - td \leqslant x + (1 + t)d, \\
x + (1 - t)d \leqslant y + td,
\end{cases}
$$

hence

$$\begin{cases} x + (1-t)d \leqslant a \leqslant x + (1+t)d, \\ y - td \leqslant a \leqslant y + td, \\ z - (1+t)d \leqslant a \leqslant z - (1-t)d, \end{cases}$$

from which we conclude that

$$\max\{\mid x - (a-d) \mid, \mid y - a \mid, \mid z - (a+d) \mid\} \leqslant td.$$

So every $t \geqslant \dfrac{1}{2}$ does not satisfy the requirement of the problem.

In conclusion, the set of all required $t$ is $\left(0, \dfrac{1}{2}\right)$.  $\square$

## Second Day 8:00 – 12:30
### January 13, 2013

**4** Given an integer $n \geqslant 2$, suppose $A_1$, $A_2$, ..., $A_n$ are $n$ nonempty finite sets satisfying $\mid A_i \Delta A_j \mid = \mid i - j \mid$ for all $i, j \in \{1, 2, \ldots, n\}$.

Find the minimum value of $\mid A_1 \mid + \mid A_2 \mid + \cdots + \mid A_n \mid$.

(Here $\mid X \mid$ denotes the number of elements of a finite set $X$ and $X \Delta Y = \{a \mid a \in X, a \notin Y\} \cup \{a \mid a \in Y, a \notin X\}$ for any sets $X$ and $Y$.)

**Solution** For each positive integer $k$, we prove that the minimum value of $S_{2k}$ is $k^2 + 2$; the minimum value of $S_{2k+1}$ is $k(k+1) + 2$.

Firstly, define the sets $A_1$, $A_2$, ..., $A_{2k}$, $A_{2k+1}$ as follows:

$A_i = \{i, i+1, \ldots, k\}$, $i = 1, 2, \ldots, k$; $A_{k+1} = \{k, k+1\}$; $A_{k+j} = \{k+1, k+2, \ldots, k+j-1\}$, $j = 2, 3, \ldots, k+1$.

For this family of sets, it is easy to verify that $|A_i \Delta A_j| = j - i = |i - j|$ holds in the following cases:

(1) $1 \leqslant i < j \leqslant k$;

(2) $1 \leqslant i < j = k + 1$;

(3) $1 \leqslant i < k + 1 < j \leqslant 2k + 1$;

(4) $k + 1 = i < j \leqslant 2k + 1$;

(5) $k + 2 \leqslant i < j \leqslant 2k + 1$.

Moreover, the case of $i = j$ is trivial, and the case of $i > j$ can be reduced to the case of $i < j$. Thus, for all $i, j \in \{1, 2, \ldots, 2k + 1\}$, we have verified that

$$|A_i \Delta A_j| = j - i = |i - j|.$$

For the above $(2k + 1)$ sets, we can easily calculate that

$$S_{2k+1} = \frac{k(k+1)}{2} + 2 + \frac{k(k+1)}{2} = k(k+1) + 2;$$

if we choose the first $2k$ sets, we get that

$$S_{2k} = S_{2k+1} - k = k^2 + 2.$$

Secondly, we show that $S_{2k} \geqslant k^2 + 2$ and $S_{2k+1} \geqslant k(k+1) + 2$. Note the following facts:

*Fact* 1. For any two finite sets $X$, $Y$, we have $|X| + |Y| \geqslant |X \Delta Y|$.

*Fact* 2. For any two non-empty finite sets $X$, $Y$, if $|X \Delta Y| = 1$, then $|X| + |Y| \geqslant 3$.

When $n = 2k$, it follows from Fact 1 that

$$|A_i| + |A_{2k+1-i}| \geqslant |A_i \Delta A_{2k+1-i}| = 2k + 1 - 2i, \, i$$
$$= 1, 2, \ldots, k - 1.$$

By $|A_k \Delta A_{k+1}| = 1$ and Fact 2, we have $|A_k| + |A_{k+1}| \geqslant 3$. So

$$S_{2k} = |A_k| + |A_{k+1}| + \sum_{i=1}^{k-1} (|A_i| + |A_{2k+1-i}|)$$

$$\geqslant 3 + \sum_{i=1}^{k-1} (2k + 1 - 2i) = k^2 + 2.$$

Similarly, when $n = 2k + 1$, we get that

$$|A_i| + |A_{2k+2-i}| \geqslant |A_i \triangle A_{2k+2-i}| = 2k + 2 - 2i, \, i$$

$$= 1, 2, \ldots, k - 1.$$

Since $|A_k \triangle A_{k+1}| = 1$, we have

$$(|A_k| + |A_{k+1}|) + |A_{k+2}| \geqslant 3 + 1 = 4,$$

so

$$S_{2k+1} = |A_k| + |A_{k+1}| + |A_{k+2}| + \sum_{i=1}^{k-1} (|A_i| + |A_{2k+2-i}|)$$

$$\geqslant 4 + \sum_{i=1}^{k-1} (2k + 2 - 2i) = k(k + 1) + 2.$$

In conclusion, the minimum value of $S_{2k}$ is $k^2 + 2$, and the minimum value of $S_{2k+1}$ is $k(k + 1) + 2$. Equivalently, for any $n \geqslant 2$, the minimum value of $S_n$ is $\left[\dfrac{n^2}{4}\right] + 2$.                    □

**⑤** For each positive integer $n$ and each integer $i (0 \leqslant i \leqslant n)$, let $C_n^i \equiv c(n, i) \pmod 2$, where $c(n, i) \in \{0, 1\}$, and define

$$f(n, q) = \sum_{i=0}^{n} c(n, i) q^i.$$

Let $m, n$ and $q$ be positive integers with $q + 1$ not a power of 2. Suppose that $f(m, q) \mid f(n, q)$. Prove that $f(m, r) \mid f(n, r)$ for every positive integer $r$.

**Solution** For each positive integer $n$, we write $n$ in binary representation as $n = 2^{a_1} + 2^{a_2} + \cdots + 2^{a_k}$, where $0 \leqslant a_1 < a_2 < \cdots < a_k$. Define a set $T(n) = \{2^{a_1}, \ldots, 2^{a_k}\}$, $T(0)$ is considered an empty set.

By Lucas' Theorem, $C_n^i$ is odd if and only if $T(i) \subseteq T(n)$, hence

$$f(n, q) = \sum_{A \subseteq T(n)} q^{\sigma(A)} = \prod_{a \in T(n)} (1 + q^a),$$

where $\sigma(A)$ denotes the sum of all elements of $A$.

For $m$, $n$ and $q$ as given by assumption, we show that if

$$f(m, q) = \prod_{a \in T(m)} (1 + q^a) \Big| \prod_{a \in T(n)} (1 + q^a) = f(n, q),$$

then $T(m) \subseteq T(n)$, and consequently, $f(m, r) \mid f(n, r)$ for every $r$.

For any integers $i$, $j$, $0 \leqslant i < j$, we have the following factorization:

$$q^{2^j} - 1 = (q^{2^{j-1}} + 1) \cdots (q^{2^i} + 1)(q^{2^i} - 1),$$

therefore,

$$(q^{2^j} + 1, q^{2^i} + 1) = (q^{2^i} + 1, 2) \mid 2.$$

Let $s(k)$ be the largest odd divisor of a positive integer $k$, then $s(q^{2^i} + 1)$ and $s(q^{2^j} + 1)$ are coprime. Clearly, $q > 1$. If $i > 0$, $q^{2^i} + 1 \equiv 1$ or $2 \pmod 4$, and $q^{2^i} + 1 > 2$, thus $s(q^{2^i} + 1) > 1$. If $i = 0$, since $q + 1$ is not a power of 2, we have $s(q + 1) > 1$. For any $a \in T(m)$, $s(q^a + 1) \mid \prod_{b \in T(n)} s(q^b + 1)$. Since $s(1 + q^a) > 1$, we have $a \in T(n)$, hence $T(m) \subseteq T(n)$, which completes the proof. $\quad\square$

**6** Given positive integers $m$ and $n$, find the smallest integer $N (\geqslant m)$ with the following property: if an $N$-element set of integers contains a complete residue system modulo $m$, then it has a non-empty subset such that the sum of its elements is divisible by $n$.

**Solution** The answer is

$$N = \max\left\{m, m + n - \frac{1}{2}m[(m, n) + 1]\right\}.$$

First, we show that $N \geqslant \max\left\{m, m + n - \frac{1}{2}m[(m, n) + 1]\right\}$.

Let $d = (m, n)$, and write $m = dm_1$, $n = dn_1$. If $n > \frac{1}{2}m(d + 1)$, there exists a complete residue system modulo $m$, $x_1$, $x_2, \ldots, x_m$, such that their residue modulo $n$ consists exactly of $m_1$ groups of $1, 2, \ldots, d$. For example, the following $m$ numbers have the required property:

$$i + dn_1 j, \ i = 1, 2, \ldots, d, \ j = 1, 2, \ldots, m_1.$$

Finding another set of $k = n - \frac{1}{2}m(d + 1) - 1$ numbers $y_1$, $y_2, \ldots, y_k$ that are congruent to 1 modulo $n$, the set

$$A = \{x_1, x_2, \ldots, x_m, y_1, \ldots, y_k\}$$

contains a complete residue system modulo $m$, however, none of its non-empty subsets has its sum of elements divisible by $n$. In fact, the sum of the (smallest non-negative) residue modulo $n$ of all elements of $A$ is greater than zero and less than or equal to $m_1(1 + 2 + \cdots + d) + k = n - 1$. Thus,

$$N \geqslant m + n - \frac{1}{2}m(d + 1),$$

i.e.,

$$N \geqslant \max\left\{m, m + n - \frac{1}{2}m[(m, n) + 1]\right\}.$$

Next, we show that $N = \max\left\{m, \; m + n - \dfrac{1}{2}m[(m, n) + 1]\right\}$ has the required property.

The following key fact is frequently used in the proof: among any $k$ integers, one can find a (non-empty) subset whose sum is divisible by $k$. Let $a_1, a_2, \ldots, a_k$ be integers, $S_i = a_1 + a_2 + \cdots + a_i$. If some $S_i$ is divisible by $k$, then the result is true. Otherwise, there exist $1 \leqslant i < j \leqslant k$, such that $S_i \equiv S_j \pmod{k}$, then $S_j - S_i = a_{i+1} + \cdots + a_j$ is divisible by $k$, the result is again true. The following fact is an easy corollary of the previous result: among any $k$ integers, each of which is a multiple of $a$, one can find a (non-empty) subset whose sum is divisible by $ka$.

Returning to the problem, we shall discuss two cases.

Case 1: $n \leqslant \dfrac{1}{2}m(d+1)$, and $N = m$.

We call a finite set of integers a $k$-set if the sum of all its elements is divisible by $k$. Let $x_1, x_2, \ldots, x_m$ be a complete residue system modulo $m$. Clearly, we can divide these numbers into $m_1$ groups, each group consisting of a complete residue system modulo $d$. Let $y_1, y_2, \ldots, y_d$ be a complete residue system modulo $d$, and $y_i \equiv i \pmod{d}$. If $d$ is odd, we can divide each group into $\dfrac{d+1}{2}$ $d$-set, for example, $\{y_1, y_{d-1}\}, \ldots, \{y_{\frac{d-1}{2}}, y_{\frac{d+1}{2}}\}, \{y_d\}$. We get $\dfrac{1}{2}m_1(d+1)$ $d$-sets. Since $n_1 \leqslant \dfrac{1}{2}m_1(d+1)$, we can choose some of these $d$-sets such that the sum of their elements is divisible by $n_1 d \,(= n)$. If $d$ is even, similarly, a complete residue system modulo $d$ can be divided into $\dfrac{d}{2}$ $d$-sets, with $y_{\frac{d}{2}}$ remaining. Two remaining numbers can form another $d$-set. In the end, we divide $x_1, x_2, \ldots, x_m$ into $\dfrac{1}{2}m_1 d +$

$\left[\dfrac{m_1}{2}\right]$ $d$-sets (possibly with a number left if $m_1$ is odd).

Since $n_1 \leqslant \dfrac{1}{2}m_1(d+1) = \dfrac{1}{2}m_1 d + \dfrac{m_1}{2}$, we have $n_1 \leqslant \dfrac{1}{2}$ $m_1 d + \left[\dfrac{m_1}{2}\right]$, again we can find some of these $d$-sets such that the sum of all their elements is divisible by $n_1 d = n$.

Case 2: $n > \dfrac{1}{2}m(d+1)$, $N = m + n - \dfrac{1}{2}m(d+1)$.

Let $A$ be an $N$-element set, containing a complete residue system modulo $m$, $x_1$, $x_2$, $\ldots$, $x_m$, with some other $n - \dfrac{1}{2}m$ $(d+1)$ numbers. If $d$ is odd, as shown in case 1, we may divide $x_1$, $x_2$, $\ldots$, $x_m$ into $\dfrac{1}{2}m_1(d+1)$ $d$-sets. Divide the remaining $n - \dfrac{1}{2}m(d+1)$ numbers arbitrarily into $n_1 - \dfrac{1}{2}m_1$ $(d+1)$ groups, each with $d$ numbers. Among each group of $d$ numbers, one may find a $d$-set, therefore, we have another $n_1 - \dfrac{1}{2}m_1(d+1)$ $d$-sets, and totally $n_1$ $d$-sets. If $d$ is even, as discussed in case 1, we may divide $x_1$, $x_2$, $\ldots$, $x_m$ into $\dfrac{1}{2}m_1$ $d + \left[\dfrac{m_1}{2}\right]$ $d$-sets. If $m_1$ is odd, we are left with a number $x_i$ with $d \mid x_i - \dfrac{d}{2}$. Divide the other $n - \dfrac{1}{2}m(d+1)$ numbers arbitrarily into $2n_1 - m_1(d+1)$ groups, each with $\dfrac{d}{2}$ numbers. From each group, we can find a $\dfrac{d}{2}$-set; from any two $\dfrac{d}{2}$-sets, we can find a $d$-set. If $m_1$ is even, then we can find another $n_1 - \dfrac{1}{2}m_1(d+1)$ $d$-sets, and totally $n_1$ $d$-sets. If $m_1$ is odd, $\{x_i\}$ is a

$\frac{d}{2}$-set, we have $2n_1 - m_1(d+1) + 1 \frac{d}{2}$-sets, and also $n_1 -$

$\frac{1}{2}m_1(d+1) + \frac{1}{2}d$-sets. Again, we can find $n_1$ $d$-sets. Finally, we can choose some of these $n_1$ $d$-sets, such that the sum of all their elements is divisible by $n_1 d (= n)$.  □

# $2013$ (Nanjing, Jiangsu)

## First Day

### 8:00 - 12:30, December 21, 2013

**1** In an acute triangle $ABC$, $AB >$ $AC$, the bisector of angle $BAC$ and side $BC$ intersect at point $D$, two points $E$ and $F$ are in sides $AB$ and $AC$, respectively, such that $B$, $C$, $F$, $E$ are concyclic. Prove that the circumcenter of triangle $DEF$ coincides with the innercenter of triangle $ABC$ if and only if $BE + CF = BC$.

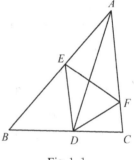

Fig. 1. 1

**Solution.** Let $I$ be the innercenter of $\triangle ABC$.

(*Sufficiency*) Suppose $BC = BE + CF$. Let $K$ be the point on $BC$ such that $BK = BE$, thus $CK = CF$. Since $BI$ bisects $\angle ABC$, $CI$ bisects $\angle ACB$, $\triangle BIK$ and $\triangle BIE$ are reflection with respect to $BI$, $\triangle CIK$ and $\triangle CIF$ are reflection with respect to $CI$, we have $\angle BEI = \angle BKI = \pi - \angle CKI = \pi -$

$\angle CFI = \angle AFI$. Therefore, $A$, $E$, $I$, $F$ are concyclic. Since $B$, $E$, $F$, $C$ are concyclic, we have $\angle AIE = \angle AFE = \angle ABC$, and hence $B$, $E$, $I$, $D$ are concyclic.

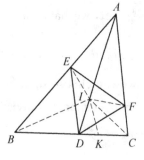

Fig. 1. 2

Since the bisector of $\angle EAF$ and the circumcircle of $\triangle AEF$ meet at $I$, $IE = IF$. Since the bisector of $\angle EBD$ and the circumcircle of $\triangle BED$ also meet at $I$, $IE = ID$. So, $ID = IE = IF$, that is, $I$ is also the circumcenter of $\triangle DEF$.

Q. E. D.

(*Necessity*) Suppose $I$ is the circumcenter of $\triangle DEF$. Since $B$, $E$, $F$, $C$ are concyclic, $AE \cdot AB = AF \cdot AC$, $AB > AC$, we have $AE < AF$. Therefore, the bisector of $\angle EAF$ and the perpendicular bisector of $EF$ meet at $I$, which lies on the circumcircle of $\triangle AEF$.

Since $BI$ bisects $\angle ABC$, let $K$ be the symmetric point of $E$ with respect to $BI$, then we have $\angle BKI = \angle BEI = \angle AFI > \angle ACI = \angle BCI$. Therefore, $K$ lies on $BC$, $\angle IKC = \angle IFC$, $\angle ICK = \angle ICF$, and $\triangle IKC \cong \triangle IFC$.

Hence $BC = BK + CK = BE + CF$. □

**2** For any integer $n$ with $n > 1$, let

$$D(n) = \{a - b \mid a, b \text{ are positive integers with } n$$
$$= ab \text{ and } a > b\}.$$

Prove that for any integer $k$ with $k > 1$, there exist $k$ pairwise distinct integers $n_1$, $n_2$, ..., $n_k$ with $n_i > 1(1 \leqslant i \leqslant k)$, such that $D(n_1) \cap D(n_2) \cap \cdots \cap D(n_k)$ has at least two elements.

**Proof.** Let $a_1$, $a_2$, $\ldots$, $a_{k+1}$ be $k+1$ distinct positive odds, where each of them is smaller than the product of other $k$ numbers. Write $N = a_1 a_2 \cdots a_{k+1}$. For each $i = 1, 2, \ldots, k+1$,

let $x_i = \dfrac{1}{2}\left(\dfrac{N}{a_i} + a_i\right)$, $y_i = \dfrac{1}{2}\left(\dfrac{N}{a_i} - a_i\right)$, then $x_i^2 - y_i^2 = N$.

Since $a_i a_j < N$ and $\dfrac{N}{a_i} > a_i$, $(x_i, y_i)(1 \leqslant i \leqslant k+1)$ are $k+1$ positive integer solutions of equation $x^2 - y^2 = N$. Without loss of generality, suppose $x_{k+1} = \min\{x_1, x_2, \ldots, x_{k+1}\}$. For each $i \in \{1, 2, \ldots, k\}$, since $x_i^2 - y_i^2 = x_{k+1}^2 - y_{k+1}^2$, we have

$$(x_i + x_{k+1})(x_i - x_{k+1}) = x_i^2 - x_{k+1}^2 = y_i^2 - y_{k+1}^2$$
$$= (y_i + y_{k+1})(y_i - y_{k+1}).$$

Let $n_i = (x_i + x_{k+1})(x_i - x_{k+1}) = (y_i + y_{k+1})(y_i - y_{k+1})$, then

$$2x_{k+1} = (x_i + x_{k+1}) - (x_i - x_{k+1}) \in D(n_i),$$
$$2y_{k+1} = (y_i + y_{k+1}) - (y_i - y_{k+1}) \in D(n_i).$$

So $x_{k+1} > y_{k+1}$, $2x_{k+1}$ and $2y_{k+1}$ are two different members of $D(n_1) \cap D(n_2) \cap \cdots \cap D(n_k)$. ☐

**3** Let $\mathbf{N}^*$ be the set of all positive integers. Prove that there exists a unique function $f: \mathbf{N}^* \to \mathbf{N}^*$ satisfying $f(1) = f(2) = 1$ and

$$f(n) = f(f(n-1)) + f(n - f(n-1)), \ n = 3, 4, \ldots.$$

For such $f$, find the value of $f(2^m)$ for integer $m \geqslant 2$.

**Solution.** Since $f(1) = 1$, we have $\dfrac{1}{2} \leqslant f(1) \leqslant 1$.

We show by induction that for any integer $n > 1$, $f(n)$ is uniquely determined by the value of $f(1)$, $f(2)$, $\ldots$, $f(n-1)$, and $\dfrac{n}{2} \leqslant f(n) \leqslant n$.

For $n = 2$, $f(2) = 1$, the claim is true.

Assume that for any $k$, $1 \leqslant k < n(n \geqslant 3)$, $f(k)$ is uniquely determined, and $\frac{k}{2} \leqslant f(k) \leqslant k$, then $1 \leqslant \frac{n-1}{2} \leqslant f(n-1) \leqslant n - 1$, and $1 \leqslant n - f(n-1) \leqslant n - 1$, hence by induction hypothesis, the value of $f(f(n-1))$ and $f(n - f(n-1))$ is determined, and the value of $f(n)$, by definition, is

$$f(n) = f(f(n-1)) + f(n - f(n-1)), \qquad \text{①}$$

which is uniquely determined. Furthermore, we have

$$\frac{1}{2} f(n-1) \leqslant f(f(n-1)) \leqslant f(n-1),$$

$$\frac{1}{2}(n - f(n-1)) \leqslant f(n - f(n-1)) \leqslant n - f(n-1).$$

Equality ① implies $\frac{n}{2} \leqslant f(n) \leqslant f(n-1) + (n - f(n-1)) = n$. The claim is also true for $n$. By induction, we proved that there exists a unique function $f: \mathbf{N}^* \to \mathbf{N}^*$ satisfying the required properties, and $\frac{n}{2} \leqslant f(n) \leqslant n$.

Next, we show by induction that for any positive integer $n$, we have

$$f(n+1) - f(n) \in \{0, 1\}. \qquad \text{②}$$

When $n = 1$, ② is true.

Assume that ② is true for $n \leqslant k$. By ①, we have

$$f(k+2) - f(k+1)$$
$$= (f(f(k+1)) + f(k+2 - f(k+1))) - (f(f(k)) + f(k+1 - f(k)))$$
$$= (f(f(k+1)) - f(f(k))) + (f(k+2 - f(k+1)) - f(k+1 - f(k))). \qquad \text{③}$$

By induction hypothesis, $f(k+1) - f(k) \in \{0, 1\}$.

If $f(k+1) = f(k) + 1$, since $1 \leqslant f(k) \leqslant k$, it follows, from ③ and induction hypothesis, that

$$f(k+2) - f(k+1) = f(f(k)+1) - f(f(k)) \in \{0, 1\}.$$

If $f(k+1) = f(k)$, since $1 \leqslant k+1-f(k) \leqslant k$, it follows from ③ and the induction hypothesis that

$$f(k+2) - f(k+1) = f(k+2-f(k)) -$$
$$f(k+1-f(k)) \in \{0, 1\}.$$

Thus, ② is true for $n = k+1$. By induction, ② is true for any positive integer $n$.

Finally, we show by induction that for any positive integer $m$, we have $f(2^m) = 2^{m-1}$.

For $m = 1$, the result is clear.

Assume that the result is true for $m = k$, i. e., $f(2^k) = 2^{k-1}$, consider the case for $m = k+1$.

Assume on the contrary that $f(2^{k+1}) \neq 2^k$, since $f(2^{k+1}) \geqslant 2^k$, and $f(2^{k+1})$ is an integer, we have $f(2^{k+1}) \geqslant 2^k + 1$. Since $f(1) = 1$, by ②, let $n$ be the smallest integer such that $f(n) = 2^k + 1$, we have $n \leqslant 2^{k+1}$, by the minimality of $n$, $f(n-1) = 2^k$. Notice that $n - 2^k \leqslant 2^k$, we have

$$2^k + 1 = f(n) = f(f(n-1)) + f(n - f(n-1))$$
$$= f(2^k) + f(n - 2^k) \leqslant 2f(2^k) = 2^k,$$

which is a contradiction. It follows that $f(2^{k+1}) = 2^k$, the result is also true for $m = k+1$. By induction, $f(2^m) = 2^{m-1}$ for any positive integer $m$. □

## Second Day

8:00 – 12:30 December 22, 2013

**④** For any integer $n$ with $n > 1$, let $n = p_1^{a_1} \cdots p_t^{a_t}$ be its standard factorization, write

$$\omega(n) = t, \quad \Omega(n) = \alpha_1 + \cdots + \alpha_t.$$

Prove or disprove the following statement: Given any positive integer $k$ and any positive real numbers $\alpha$ and $\beta$, there exists a positive integer $n$ with $n > 1$ such that

$$\frac{\omega(n+k)}{\omega(n)} > \alpha \text{ and } \frac{\Omega(n+k)}{\Omega(n)} < \beta.$$

**Solution.** The answer is YES.

From the definition of $\omega$ and $\Omega$, we have

$$\omega(ab) \leqslant \omega(a) + \omega(b), \qquad \qquad ①$$

$$\Omega(ab) = \Omega(a) + \Omega(b), \qquad \qquad ②$$

for any positive integers $a$, $b$. Given a fixed positive integer $k$ and positive real numbers $\alpha$, $\beta$, we take a positive integer $m > (\omega(k) + 1)\alpha$. As there are infinitely many prime numbers, we can take a sufficiently large prime $p$ such that $\dfrac{\Omega(k) + 1}{p^m} + \log_p 2$ $< \beta$, and take $m$ pairwise distinct prime numbers $q_1, q_2, \ldots,$ $q_m$ that are all greater than $p$. We will show that $n = 2^{q_1 q_2 \cdots q_m} k$ has the desired property.

First, we prove $\dfrac{\omega(n+k)}{\omega(n)} > \alpha$. Let $n_1 = \dfrac{n+k}{k} = 2^{q_1 q_2 \cdots q_m} + 1$. As $q_1, q_2, \ldots, q_m$ are all odd prime numbers, $2^{q_i} + 1 \mid n_1$ when $1 \leqslant i \leqslant m$, then $d_i = \dfrac{2^{q_i} + 1}{3}$ is an integer greater than 1.

Note that

$$(2^r - 1, 2^s - 1) = 2^{(r, s)} - 1 \text{ for all positive integers } r, s, \quad ③$$

and $(q_i, q_j) = 1 (i \neq j)$, we have

$$(d_i, d_j) = \frac{1}{3}(2^{q_i} + 1, 2^{q_j} + 1) \leqslant \frac{1}{3}(2^{2q_i} - 1, 2^{2q_j} - 1)$$

$$= \frac{2^{(2q_i, 2q_j)} - 1}{3} = \frac{2^2 - 1}{3} = 1.$$

$d_1, d_2, \ldots, d_m$ are the pairwise coprime factors of $n_1$, and each of them is greater than 1. Hence, $\omega(n_1) \geqslant m$. From ① and the choice of $m$, we have

$$\frac{\omega(n+k)}{\omega(n)} \geqslant \frac{\omega(n_1)}{\omega(n)} \geqslant \frac{\omega(n_1)}{\omega(k)+1} \geqslant \frac{m}{\omega(k)+1} > \alpha.$$

Next, we prove $\dfrac{\Omega(n+k)}{\Omega(n)} < \beta$. As $q_1 q_2 \cdots q_m$ is a odd number and cannot be divided by 3, we have $n_1 = 2^{q_1 q_2 \cdots q_m} + 1 \equiv \pm 3 \pmod 9$, that is, $3 \parallel n_1$. Suppose $q$ is a prime factor of $\dfrac{n_1}{3}$ and $q \leqslant p$, then

$$2^{2q_1 q_2 \cdots q_m} - 1 = (2^{q_1 q_2 \cdots q_m} - 1) \cdot n_1 \equiv 0 \pmod q.$$

From the Fermat's Little Theorem, $2^{q-1} \equiv 1 \pmod q$. From ③, $q \mid 2^{(2q_1 q_2 \cdots q_m, q-1)} - 1$. From $(q-1, 2q_1 q_2 \cdots q_m) = (q-1, 2) \leqslant 2$, $q - 1 < p < q_i (i = 1, 2, \ldots, m)$, hence $q \mid 2^2 - 1$, $q = 3$. This contradicts that $\dfrac{n_1}{3}$ is not a multiple of 3. Therefore, each prime factor of $\dfrac{n_1}{3}$ is larger than $p$. So $\dfrac{n_1}{3} > p^{\Omega(n_1/3)}$. From ② and the choice of primes $p$ and $q_1, q_2, \ldots, q_m$, we have

$$\Omega(n+k) = \Omega(k) + \Omega(3) + \Omega\left(\frac{n_1}{3}\right) < \Omega(k) + 1 + \log_p\left(\frac{n_1}{3}\right)$$

$$< \Omega(k) + 1 + \log_p(n_1 - 1)$$

$$= \Omega(k) + 1 + q_1 q_2 \cdots q_m \log_p 2,$$

$$\frac{\Omega(n+k)}{\Omega(n)} < \frac{\Omega(k)+1+q_1q_2\cdots q_m \log_p 2}{q_1q_2\cdots q_m} < \frac{\Omega(k)+1}{p^m} + \log_p 2 < \beta.$$

$\square$

**5** Given $X = \{1, 2, \ldots, 100\}$, consider function $f: X \to X$ satisfying both the following conditions:

(1) $f(x) \neq x$ for all $x \in X$;

(2) $A \cap f(A) \neq \varnothing$ for all $A \subseteq X$ with $|A| = 40$.

Find the smallest positive integer $k$, such that for any such function $f$ there exists a set $B \subseteq X$ satisfying $|B| = k$ and $B \cup f(B) = X$.

**Remark.** For a subset $T$ of $X$, we define $f(T) = \{x \mid$ there exists $t \in T$ such that $x = f(t)\}$.

**Solution.** First, we define a function $f: X \to X$ with

$$f(3i-2) = 3i-1, \ f(3i-1) = 3i, \ f(3i) = 3i-2,$$
$$i = 1, 2, \ldots, 30,$$
$$f(j) = 100, \ 91 \leqslant j \leqslant 99, \ f(100) = 99.$$

Obviously, $f$ satisfies condition (1). For any $A \subseteq X$ with $|A| = 40$, if

(i) there exists an integer $i$ with $1 \leqslant i \leqslant 30$ such that $|A \cap \{3i-2, 3i-1, 3i\}| \geqslant 2$, then $A \cap f(A) \neq \varnothing$; or

(ii) $91, 92, \ldots, 100 \in A$, then $A \cap f(A) \neq \varnothing$ also holds.

In both cases, $f$ satisfies condition (2). If a subset $B$ of $X$ satisfies $f(B) \cup B = X$, then we have $|B \cap \{3i-2, 3i-1, 3i\}| \geqslant 2$ for all $1 \leqslant i \leqslant 30$, $\{91, 92, \ldots, 98\} \subset B$, and $B \cap \{99, 100\} \neq \varnothing$. Hence, $|B| \geqslant 69$.

Next, we will show that, for any function $f$ satisfying the described conditions, there exists a subset $B \subseteq X$ with $|B| \leqslant 69$ such that $f(B) \cup B = X$.

Among all the subsets $U \subseteq X$ with $U \cap f(U) = \varnothing$, choose one such that $|U|$ is maximal. If there are many $U \subseteq X$ with $|U|$ being maximal, choose one such that $|f(U)|$ is maximal. The existence of $U$ is guaranteed by condition (1). Let $V = f(U)$, $W = X \backslash (U \cup V)$. Note that $U$, $V$, $W$ are pairwise disjoint and $X = U \cup V \cup W$. From condition (2), $|U| \leqslant 39$, $|V| \leqslant 39$, $|W| \geqslant 22$. We make the following assertions:

(i) $f(w) \in U$, for all $w \in W$. Otherwise, let $U' = U \cup \{w\}$, since $f(U) = V$, $f(w) \notin U$, $f(w) \neq w$, we have $U' \cap f(U') = \varnothing$. It is a contradiction to $|U|$ being maximal.

(ii) $f(w_1) \neq f(w_2)$ for all $w_1$, $w_2 \in W$, $w_1 \neq w_2$. Otherwise, let $u = f(w_1) = f(w_2)$ then by condition (1), $u \in U$. Let $U' = (U \backslash \{u\}) \cup \{w_1, w_2\}$, since $f(U') \subseteq V \cup \{u\}$, $U' \cap (V \cup \{u\}) = \varnothing$, we have $U' \cap f(U') = \varnothing$. It is a contradiction to $|U|$ being maximal.

Let $W = \{w_1, w_2, \ldots, w_m\}$, $u_i = f(w_i)$, $1 \leqslant i \leqslant m$ then by (i) and (ii), $u_1$, $u_2$, $\ldots$, $u_m$ are distinct elements of $U$.

(iii) $f(u_i) \neq f(u_j)$ for all $1 \leqslant i < j \leqslant m$. Otherwise, let $v = f(u_i) = f(u_j) \in V$, $U' = (U \backslash \{u_i\}) \cup \{w_i\}$, then $f(U') = V \cup \{u_i\}$, $U' \cap f(U') = \varnothing$. However, $|f(U')| > |f(U)|$. It is a contradiction to $|f(U)|$ being maximal.

Therefore, $f(u_1)$, $f(u_2)$, $\ldots$, $f(u_m)$ are distinct elements of $V$. In particular, $|V| \geqslant |W|$. As $|U| \leqslant 39$, we have $|V| + |W| \geqslant 61$ and $|V| \geqslant 31$. Let $B = U \cup W$, then $|B| \leqslant 69$ and $f(B) \cup B \supseteq V \cup B = X$. Overall, the desired smallest integer $k$ is 69. $\qquad\square$

**6** For non-empty sets $S$, $T$ of numbers, we define

$$S + T = \{s + t \mid s \in S, t \in T\}, \quad 2S = \{2s \mid s \in S\}.$$

Let $n$ be a positive integer, and $A$, $B$ be non-empty subsets of $\{1, 2, \ldots, n\}$. Prove that there exists a subset $D$ of $A + B$ such that

$$D + D \subseteq 2(A + B), \text{ and } |D| \geqslant \frac{|A| \cdot |B|}{2n},$$

where $|X|$ denotes the number of elements of a finite set $X$.

**Solution.** Let $S_y = \{(a, b) \mid a - b = y, a \in A, b \in B\}$. Since $\sum_{y=1-n}^{n-1} |S_y| = |A| \cdot |B|$, there exists an integer $y_0$ such that $1 - n \leqslant y_0 \leqslant n - 1$ and $|S_{y_0}| \geqslant \dfrac{|A| \cdot |B|}{2n - 1} > \dfrac{|A| \cdot |B|}{2n}$.

Let $D = \{2b + y_0 \mid (a, b) \in S_{y_0}\}$, then

$$|D| = |S_{y_0}| > \frac{|A| \cdot |B|}{2n}.$$

From the definition of $S_{y_0}$, for each $d \in D$, there exists $(a, b) \in S_{y_0}$ such that $d = 2b + y_0 = a + b \in A + B$. So $D \subseteq A + B$. For any $d_1$, $d_2 \in D$, let $d_1 = 2b_1 + y_0 = 2a_1 - y_0$, $d_2 = 2b_2 + y_0 (b_1, b_2 \in B, a_1 \in A)$, then

$$d_1 + d_2 = 2a_1 - y_0 + 2b_2 + y_0 = 2(a_1 + b_2) \subseteq 2(A + B).$$

Therefore, $D$ satisfies the condition. $\qquad\square$

# China National
# Team Selection Test

*2011* (Fuzhou, Fujian)

## First Day

8:00 – 12:30, March 27, 2011

**1** Given an integer $n \geqslant 3$, find the maximum real number $M$, such that for any positive numbers $x_1$, $x_2$, $\ldots$, $x_n$, there exists a permutation $y_1$, $y_2$, $\ldots$, $y_n$ of $x_1$, $x_2$, $\ldots$, $x_n$ that satisfies

$$\sum_{i=1}^{n} \frac{y_i^2}{y_{i+1}^2 - y_{i+1} y_{i+2} + y_{i+2}^2} \geqslant M,$$

where $y_{n+1} = y_1$, $y_{n+2} = y_2$. (posed by Qu Zhenhua)

**Solution.** Let

$$F(x_1, \ldots, x_n) = \sum_{i=1}^{n} \frac{x_i^2}{x_{i+1}^2 - x_{i+1} x_{i+2} + x_{i+2}^2}.$$

First, take $x_1 = x_2 = \cdots = x_{n-1} = 1$, $x_n = \varepsilon$, then all permutations are the same in the sense of circulation. In this case, we have

$$F(x_1, \ldots, x_n) = n - 3 + \frac{2}{1 - \varepsilon + \varepsilon^2} + \varepsilon^2.$$

Let $\varepsilon \to 0^+$, $F \to n - 1$, so $M \leqslant n - 1$.

Next, we show that for any positive numbers $x_1, \ldots, x_n$, there exists a permutation $y_1, \ldots, y_n$ satisfying $F(y_1, \ldots, y_n) \geqslant n - 1$. In fact, take the permutation $y_1, \ldots, y_n$ with $y_1 \geqslant y_2 \geqslant \cdots \geqslant y_n$ and by the inequality $a^2 - ab + b^2 \leqslant \max(a^2, b^2)$, we see that

$$F(y_1, \ldots, y_n) \geqslant \frac{y_1^2}{y_2^2} + \frac{y_2^2}{y_3^2} + \cdots + \frac{y_{n-1}^2}{y_1^2} \geqslant n - 1,$$

where the last inequality is obtained by AM-GM inequality. Summing up, $M = n - 1$. $\qquad \square$

**2** Let $n > 1$ be an integer, $k$ be the number of distinct prime factors of $n$. Prove that there exists an integer $a$, $1 < a < \dfrac{n}{k} + 1$, such that $n \mid a^2 - a$. (posed by **Yu Hongbing**)

**Solution.** Let $n = p_1^{a_1} \cdots p_k^{a_k}$ be the standard factorization of $n$.

Since $p_1^{a_1}, \ldots, p_k^{a_k}$ are pairwise coprime, by the Chinese Remainder Theorem, for each $i$, $1 \leqslant i \leqslant k$, congruence equations

$$\begin{cases} x \equiv 1 \pmod{p_i^{a_i}} \\ x \equiv 0 \pmod{p_j^{a_j}}, \ j \neq i \end{cases}$$

have solution $x_i$.

For any solution of $x_0^2 = x_0 \pmod{n}$, we see that $x_0(x_0 - 1) \equiv 0 \pmod{n}$. Then for each $i = 1, 2, \ldots, k$, either $x_0 \equiv 0 \pmod{p_i^{a_i}}$ or $x_0 \equiv 1 \pmod{p_i^{a_i}}$. Further, let $S(A)$ be the sum of elements of subset $A$ of $\{x_1, x_2, \ldots, x_k\}$ (particularly, $S(\emptyset) = 0$). Obviously, we have

$$S(A)(S(A) - 1) \equiv 0 \pmod{n}.$$

(This is because of the selection of $x_i$, such that $S(A) \bmod(p_i^{a_i})$ is either 0 or 1.) Moreover if $A \neq A'$, then $S(A) \not\equiv S(A') \pmod{n}$. Therefore, the sum of all subsets of $\{x_1, x_2, \ldots, x_n\}$ is exactly all solutions of $x(x - 1) \equiv 0 \pmod{n}$.

Let $S_0 = n$, $S_r$ be the least non-negative remainder of $x_1 + x_2 + \cdots + x_r$ module $n$, $r = 1, 2, \ldots, k$. Thus $S_k = 1$. For all $1 \leqslant r \leqslant k - 1$, $S_r \neq 0$. Since $k + 1$ numbers $S_0, S_1, \ldots, S_k$ are in $[1, n]$, by Dirichlet's Drawer Principle, there exist $0 \leqslant l < m \leqslant k$, such that $S_l$, $S_m$ in the same interval $\left( \dfrac{jn}{k}, \dfrac{(j+1)n}{k} \right]$, $(0 \leqslant j \leqslant k - 1)$, where $l = 0$ and $m = k$ do not hold simultaneously.

Thus, $| S_l - S_m | < \dfrac{n}{k}$. Denote $y_1 = S_1$, $y_r = S_r - S_{r-1}$ $(r = 2, 3, \ldots, k)$. So any sum of $y_r \equiv x_r \pmod{n}$ $(r = 1, 2, \ldots, k)$ meets the requirement.

If $S_m - S_l > 1$, then $a = y_{l+1} + y_{l+2} + \cdots + y_m = S_m - S_l \in \left( 1, \dfrac{n}{k} \right)$ is the solution of the equation $x^2 - x \equiv 0 \pmod{n}$.

If $S_m - S_l = 1$, then $n \mid (y_1 + y_2 + \cdots + y_l) + (y_{m+1} + y_{m+2} + \cdots + y_k)$, that is, $n \mid (x_1 + x_2 + \cdots + x_l) + (x_{m+1} + x_{m+2} + \cdots + x_k)$. Notice that $m > l$, which contradicts to the definition of $x_i$.

If $S_m - S_l = 0$, then $n \mid y_{l+1} + y_{l+2} + \cdots + y_m$, that is, $n \mid x_{l+1} + x_{l+2} + \cdots + x_m$, which contradicts the definition of $x_i$.

If $S_m - S_l < 0$, then

$$a = (y_1 + y_2 + \cdots + y_l) + (y_{m+1} + \cdots + y_k)$$
$$= S_k - (S_m - S_l) = 1 - (S_m - S_l)$$

is the solution of equation $x^2 - x \equiv 0 \pmod{n}$, and $1 < a < 1 + \dfrac{n}{k}$.

Summing up, there exists $a$ satisfying the condition.    $\square$

**③** Let $3n^2$ be the vertex number of a simple graph $G$ (integer $n \geqslant 2$). If the degree of each vertex is not greater than $4n$, there exists at least one vertex with degree 1, and there exists a route with length not greater than 3 between any two vertices. Prove that the minimum number of edges of $G$ is $\dfrac{7}{2}n^2 - \dfrac{3}{2}n$.

**Remark.** A route between two distinct vertices $u$ and $v$ with length $k$ is a sequence of vertices $u = v_0, v_1, \ldots, v_k = v$, where $v_i$ and $v_{i+1}$, $i = 0, 1, \ldots, k-1$, are adjacent. (posed by Leng Gangsong)

**Solution.** For any two distinct vertices $u$ and $v$, we say that the distance between $u$ and $v$ is the shortest length of the route between $u$ and $v$. Consider a graph $G^*$ with vertex set $\{x_1, x_2, \ldots, x_{3n^2-n}, y_1, y_2, \ldots, y_n\}$, where $y_i$ and $y_j$ are adjacent $(1 \leqslant i < j \leqslant n)$, $x_i$ and $x_j$ are not adjacent $(1 \leqslant i < j \leqslant 3n^2 - n)$, $x_i$ and $y_j$ are adjacent if and only if $i \equiv j \pmod{n}$. Thus, the degree of each $x_i$ is 1, and the degree of $y_i$ does not exceed

$$n - 1 + \frac{3n^2 - n}{n} = 4n - 2.$$

It is easy to see that the distance between $x_i$ and $x_j$ is not greater than 3. So graph $G^*$ satisfies the condition of the problem. $G^*$ has $N = 3n^2 - n + C_n^2 = \frac{7}{2}n^2 - \frac{3}{2}n$ edges.

In the following, we show that any graph $G = G(V, E)$ satisfying the condition of the problem has at least $N$ edges. Let $X \subseteq V$ be the set of vertices with degree 1, $Y \subseteq (V \setminus X)$ be the set of remaining vertices adjacent to $X$, and $Z \subseteq V \setminus (X \cup Y)$ be the set of remaining vertices adjacent to $Y$. Let $W = V \setminus (X \cup Y \cup Z)$. We will point out the following facts.

**Property 1.** Any two vertices in $Y$ are adjacent. This is because of the fact that if $y_1$, $y_2 \in Y$ are two vertices, there exist $x_1$, $x_2 \in X$ that are adjacent to $y_1$ and $y_2$, respectively; hence $y_1$ and $y_2$ are adjacent since the distance between $x_1$ and $x_2$ is not greater than 3.

**Property 2.** The distance between vertex in $W$ and vertex in $Y$ is 2. This is because of the fact that if the distance between $w_0 \in W$ and $y_0 \in Y$ is greater than 2 (obviously, distance $>1$), suppose that $x_0 \in X$ is adjacent to $y_0$, then the distance between $w_0$ and $x_0$ is greater than 3, which is a contradiction. Furthermore, we know this Property 2 means each vertex in $W$ is adjacent to some vertex in $Z$.

Denote by $x$, $y$, $z$ and $w$ the numbers of element in sets $X$, $Y$, $Z$ and $W$, respectively. Now count the number of edges: there are $C_y^2$ edges between points in $Y$, $x$ edges from points of $X$ to $Y$, at least $z$ edges from points of $Z$ to $Y$, and at least $w$ edges from points of $W$ to $Z$. So, if $y \geqslant n$, then

$$|E| \geqslant C_y^2 + x + z + w = 3n^2 + C_y^2 - y \geqslant 3n^2 + C_n^2 - n = N,$$

and if $y \leqslant n - 1$, since each degree of vertex is at most $4n$, we have

$$x + z \leqslant y(4n - (y - 1)) = y(4n + 1 - y)$$
$$\leqslant (n - 1)(3n + 2) = 3n^2 - n - 2,$$
$$w \geqslant 3n^2 - y - y(4n + 1 - y) \geqslant 3.$$

Select a vertex $P$ in $W$ such that $P$ is adjacent as less as possible to vertices in $Z$. Suppose the least number is $a$, $a > 0$ (by Property 2). Denote the set of these $a$ vertices by $N_P \subseteq Z$. Counting the number of the edges again, there are $C_y^2$ edges between points in $Y$, $x$ edges from points of $X$ to $Y$, at least $y$ edges from points of $N_P$ to $Y$ (by Property 2, the distance from $P$ to vertex in $Y$ is 2), at least $z - a$ edges from points of $Z \backslash N_P$ to $Y$, and at least $aw$ edges from points of $W$ to $Z$. Thus,

$$|E| \geqslant C_y^2 + x + y + z - a + aw$$
$$= 3n^2 - 1 + C_y^2 + (a - 1)(w - 1).$$

If $a > 1$, then

$$|E| \geqslant 3n^2 - 1 + C_y^2 + (w - 1)$$
$$\geqslant 3n^2 - 2 + C_y^2 + 3n^2 - y - y(4n + 1 - y) > N.$$

If $a = 1$, since the degree of each vertex in $W$ is at least 2, when we count the edges from points of $W$ to $Z$, we should add at least $w/2$ edges, so

$$|E| \geqslant 3n^2 - 1 + C_y^2 + \frac{1}{2}w$$
$$\geqslant 3n^2 - 1 + C_y^2 + \frac{1}{2}(3n^2 - y - y(4n + 1 - y)) > N.$$

Summing up, the least number of edges is $N = \dfrac{7}{2}n^2 - \dfrac{3}{2}n$.

□

# Second Day

8:00 – 12:30, March 28, 2011

**4** Let $H$ be the orthocenter of a cubit-angled $\triangle ABC$, $P$ be a point on $\overset{\frown}{BC}$ of the circumcircle of $\triangle ABC$. $PH$ intersects $\overset{\frown}{AC}$ at $M$. There exists a point $K$ on $\overset{\frown}{AB}$ such that the line is $KM$ parallel to the Simson line of $P$ with respect to $\triangle ABC$, the

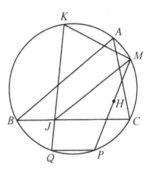

chord $QP \parallel BC$, the chord $KQ$ intersects $BC$ at point $J$. Prove that $\triangle KMJ$ is isosceles. (posed by Xiong Bin)

**Solution.** We show that $JK = JM$.

Draw line from $P$ and let it be perpendicular to $BC$ and intersect the circumcircle and $BC$ at point $S$ and $L$, respectively. Let $N$ be the project point of $P$ on $AB$. Since $B$, $P$, $L$ and $N$ are concyclic,

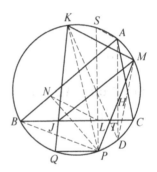

$$\angle SLN = \angle NBP = \angle ABP = \angle ASP.$$

Thus, $NL \parallel AS$. Since $NL \parallel KM$, then $KM \parallel SA$.

Let $T$ be the intersection point of $BC$ and $PH$. Since points $K$, $Q$, $P$ and $M$ are concyclic and $BC \parallel PQ$, so are $K$, $J$, $T$ and $M$. Suppose that the extension of $AH$ intersects the circumcircle at point $D$. Then we have

$$\angle JKM = \angle MTC, \quad \angle KMJ = \angle KTJ.$$

It suffices to show that $\angle MTC = \angle KTJ$.

It is easy to see that $D$ and $H$ are symmetric over line $BC$. Then

$$\angle SPM = \angle SPH = \angle THD = \angle HDT.$$

Furthermore $KS = AM$, so $\angle ADM = \angle KPS$. Consequently,

$$\angle TDM = \angle HDT + \angle ADM = \angle SPM + \angle KPS$$
$$= \angle KPM = \angle KDM,$$

which means that points $K$, $T$ and $D$ are collinear. Thus

$$\angle KTJ = \angle DTC = \angle MTC.$$

So $\angle JKM = \angle KMJ$. Consequently $JK = JM$.          □

**5** Let $a_1$, $a_2$, ... be a permutation of all positive integers. Prove that there exist infinite positive integers $i$'s, such that $(a_i, a_{i+1}) \leqslant \dfrac{3}{4}i$. (posed by Chen Yonggao)

**Solution.** We prove this problem by contradiction. If the conclusion of the problem is not true, then there exists $i_0$, and we have $(a_i, a_{i+1}) > \dfrac{3}{4}i$ for $i \geqslant i_0$.

Take a positive number $M > i_0$, so if $i \geqslant 4M$, then $(a_i, a_{i+1}) > \dfrac{3}{4}i \geqslant 3M$.

So, if $i \geqslant 4M$, $a_i \geqslant (a_i, a_{i+1}) > 3M$, then $\{1, 2, \ldots, 3M\} \subseteq \{a_1, a_2, \ldots, a_{4M-1}\}$.

Hence

$$| \{1, 2, \ldots, 3M\} \cap \{a_{2M}, a_{2M+1}, \ldots, a_{4M-1}\} |$$
$$\geqslant 3M - (2M - 1) = M + 1.$$

By Dirichlet's Drawer Principle, there exists $2M \leqslant j_0 < 4M -$

1 such that $a_{j_0}$, $a_{j_0+1} \leqslant 3M$. Thus,

$$(a_{j_0}, a_{j_0+1}) \leqslant \frac{1}{2}\max\{a_{j_0}, a_{j_0+1}\} \leqslant \frac{3M}{2} = \frac{3}{4} \cdot 2M \leqslant \frac{3}{4}j_0,$$

which is a contradiction.                                              $\square$

**6** We call a point sequence $(A_0, A_1, \ldots, A_n)$ *interesting*, if the abscissa and ordinate are equal for each $A_i$, and the slopes of segment $OA_0$, $OA_1$, $\ldots$, $OA_n$ strictly increase ($O$ is the origin), and the area of each $\triangle OA_iA_{i+1}$ ($0 \leqslant i \leqslant n - 1$) is $\frac{1}{2}$.

For a point sequence $(A_0, A_1, \ldots, A_n)$, insert a point $A$ adjacent to two points $A_i$, $A_{i+1}$ satisfying $\overrightarrow{OA} = \overrightarrow{OA_i} + \overrightarrow{OA_{i+1}}$, then we call thus obtained new point sequence $(A_0, \ldots, A_i, A, A_{i+1}, \ldots, A_n)$ an expansion of $(A_0, A_1, \ldots, A_n)$. Let $(A_0, A_1, \ldots, A_n)$ and $(B_0, B_1, \ldots, B_m)$ be any two interesting point sequences. Prove that if $A_0 = B_0$ and $A_n = B_m$, then we can expand both point sequences to some same point sequence $(C_0, C_1, \ldots, C_k)$. (posed by Qu Zhenhua)

**Solution.** We see that by the condition of the problem, an expansion of an interesting sequence is still interesting.

First, we construct the interesting sequence $(C_0, C_1, \ldots, C_k)$ containing all points of sequences $(A_0, A_1, \ldots, A_n)$ and $(B_0, B_1, \ldots, B_m)$, and $C_0 = A_0 = B_0$, $C_k = A_n = B_m$.

By the Pick Theorem, we know that the area of triangle equals $1/2$ if and only if there is no grid point on the triangle except triangle vertices. Hence there is no grid point on $\triangle OA_iA_{i+1}$ except its vertices. Therefore, if the slopes of $OA_i$ and $OB_j$ are equal, then $A_i = B_j$.

Denote the slopes of segments
from points of $\{A_i\}$ and $\{B_j\}$ to the
origin in strictly increasing order by
$D_0, D_1, \ldots, D_l$, where $D_0 = C_0 =$
$A_0 = B_0$ and $D_l = C_k = A_n = B_m$. If
a sequence $(D_i, D_{i+1})$ is not
interesting, then we can insert
several points $E_1, \ldots, E_s$ such that
the sequence $(D_i, E_1, \ldots, E_s,$

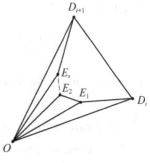

Fig. 6. 1

$D_{i+1})$ is interesting. In fact, consider the convex hull $P$ of the
grid points on $\triangle OD_iD_{i+1}$, except the origin. $P$ is a convex
polygon or segment $D_iD_{i+1}$ (a degenerated polygon). Then the
sequence of vertices of $P$ is interesting. Thus, we have
constructed the interesting sequence $(C_0, C_1, \ldots, C_k)$.

Finally, it suffices to show that the interesting sequence
$(A_0, A_1, \ldots, A_n)$ can be expanded to $(C_0, C_1, \ldots, C_k)$, and
the same is true for $(B_0, B_1, \ldots, B_m)$. We only need to prove
this for the case of $n = 1$, since we can apply the conclusion for
$n = 1$ to $(A_i, A_{i+1})$, $i = 0, 1, \ldots, n-1$ successively. Let $C_0 =$
$A_0$ and $C_k = A_1$. By induction on $k$, for $k = 1$, we need no
expansion. Suppose that the conclusion is true for all positive
integers less than $k$.

Then denote the grid point $A$
satisfying $\overrightarrow{OA} = \overrightarrow{OA_0} + \overrightarrow{OA_1}$, and we
see that $A$ must be a point of $C_1, \ldots,$
$C_{k-1}$. Since if not, there is no grid
point on interior of segment $OA$, and
there exists $i$, $0 \leqslant i < k$, such that $A$
locates in the angle made by rays $OC_i$
and $OC_{i+1}$. We may suppose that $i >$

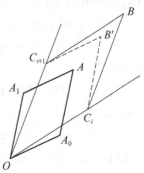

Fig. 6. 2

$0$, otherwise take the graph symmetric over the line $x = y$. Since the area of the parallelogram $\square OA_0AA_1$ is $1$, $C_i$ locates outside of $\square OA_0AA_1$, $C_{i+1}$ locates outside of $\square OA_0AA_1$ or $C_{i+1} = A_1$. In any way, we take $B$ such that $\overrightarrow{OB} = \overrightarrow{OC_i} + \overrightarrow{OC_{i+1}}$ and take $B'$ such that $C_{i+1}B' \parallel OA_0$, $C_iB' \parallel A_0A$, then $A$ locates on $\square OC_{i+1}B'C_i$. Thus, $A$ locates inside of $\square OC_iBC_{i+1}$, which contradicts the fact that the area of $\square OC_iBC_{i+1}$ is $1$. Therefore the inserted point $A$ to $(A_0, A_1)$ for expansion is some $C_i$. Then we use the induction hypotheses to $(A_0, A)$ and $(A, A_1)$, respectively. $\qquad\square$

# $2012$ (Nanchang, Jianxi)

## First Day

### 8:00 – 12:30, March 25, 2012

**1** Let $H$ be the orthocenter of an acute-angled $\triangle ABC$ with $\angle A > 60°$. Let points $M$ and $N$ be on sides $AB$ and $AC$, respectively, such that $\angle HMB = 60° = \angle HNC$. Let $O$ be the circumcenter of $\triangle HMN$. Let points $D$ and $A$ be on the same side of line $BC$, such that $\triangle DBC$ is regular (see

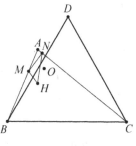

Fig. 1. 1

Fig. 1. 1). Prove that points $H$, $O$ and $D$ are collinear. (posed by Zhang Sihui)

**Solution.** Let $T$ be the orthocenter of $\triangle HMN$. Extended lines of $HM$ and $CA$ intersect at point $P$. Extended lines $HN$ and $BA$ intersect at point $Q$. It is easy to see that points $N$, $M$, $P$ and $Q$ are concyclic.

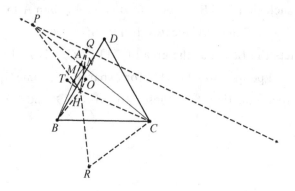

Fig. 1. 2

By $\angle THM = \angle OHN$, we see that $\angle PQH - \angle OHN = \angle NMH - \angle THM = 90°$, that is,

$$HO \perp PQ. \qquad \qquad ①$$

Let point $R$ be symmetric to point $C$ over $HP$. Then $HC = HR$.

By $\angle HPC + \angle HCP = (\angle BAC - 60°) + (90° - \angle BAC) = 30°$, we see that $\angle CHR = 60°$. Hence $\triangle HCR$ is regular.

By $\angle HPC = \angle HQB$ and $\angle HCP = \angle HBQ$, we see that $\triangle PHC \backsim \triangle QHB$. Then $\triangle PHR \backsim \triangle QHB$, so $\triangle QHP \backsim \triangle BHR$.

Denote by $\angle(UV, XY)$ the angle between $\overrightarrow{UV}$ and $\overrightarrow{XY}$ (positive anticlockwise).

Since $\angle PHR = 150°$, we have $\angle(PQ, RB) = \angle(HP, HR) = 150°$, $\triangle BCD$ and $\triangle RCH$ are regular. So $\triangle BRC \cong \triangle DHC$. Hence $\angle(RB, HD) = \angle(CR, CH) = -60°$.

Consequently,

$$\angle(PQ, HD) = \angle(PQ, RB) + \angle(RB, HD) = 150° - 60° = 90°,$$

that is

$$DH \perp PQ. \tag{②}$$

By ① and ②, points $H$, $O$ and $D$ are collinear. □

**2** Prove that, for any given integer $k \geqslant 2$, there exist $k$ distinct positive integers $a_1$, $a_2$, ..., $a_k$ such that for any integers $b_1$, $b_2$, ..., $b_k$ with $a_i \leqslant b_i \leqslant 2a_i$, $i = 1, 2, ...,$ $k$, and any non-negative integers $c_1$, $c_2$, ..., $c_k$, we have $k \prod_{i=1}^{k} b_i^{c_i} < \prod_{i=1}^{k} b_i$, provided $\prod_{i=1}^{k} b_i^{c_i} < \prod_{i=1}^{k} b_i$.
(posed by Chen Yonggao)

**Solution.** We will prove a stronger proposition: for any given real number $k$ and any positive integer $n$, there exist $n$ positive integers $a_1$, $a_2$, ..., $a_n$, satisfying $a_{i+1} > 2a_i$, $1 \leqslant i \leqslant n - 1$, and for any real numbers $b_1$, $b_2$, ..., $b_n$, $a_i \leqslant b_i \leqslant 2a_i$, $i = 1, 2, ..., n$, and any non-negative integers $c_1$, $c_2$, ..., $c_n$, we have $k \prod_{i=1}^{n} b_i^{c_i} < \prod_{i=1}^{n} b_i$, provided $\prod_{i=1}^{n} b_i^{c_i} < \prod_{i=1}^{n} b_i$.

Let $k > 1$. We prove the proposition by induction on $n$. If $n = 1$, $c_1 = 0$, then take $a_1 > k$, the conclusion is true. Suppose that the conclusion is true for $n \geqslant 1$. There are $x_1 < x_2 < \cdots < x_n$ satisfying $x_{i+1} > 2x_i$, $1 \leqslant i \leqslant n - 1$. Now consider the case of $n + 1$. Take a positive integer $x_{n+1} > 2x_n$, satisfying

$$\frac{x_{n+1}}{2x_n} > k \left( \frac{2x_n}{x_1} \right)^n. \tag{①}$$

Then let $a_i = tx_i$, $i = 1, 2, ..., n + 1$, $t$ is a sufficiently large integer, such that

$$a_1^{n+1} > 2^n a_2 \cdots a_{n+1},$$ ②

and

$$k \cdot 2^{n-1} a_{n+1}^{n-1} < a_1 \cdots a_n.$$ ③

In the following, we will show that these $a_i (1 \leqslant i \leqslant n + 1)$ meet the requirement. Let $b_i \in [a_i, 2a_i](1 \leqslant i \leqslant n + 1)$ be the real number (notice that we have $b_1 < b_2 < \cdots < b_{n+1}$), and $c_i (1 \leqslant i \leqslant n + 1)$ be non-negative integers, satisfying $\prod_{i=1}^{n+1} b_i^{c_i} < \prod_{i=1}^{n+1} b_i$.

If $\sum_{i=1}^{n+1} c_i \leqslant n$, then

$$k \prod_{i=1}^{n+1} b_i^{c_i} \leqslant k b_{n+1}^n \leqslant k \cdot 2^{n-1} a_{n+1}^{n-1} b_{n+1} < a_1 \cdots a_n b_{n+1}$$

$$\leqslant \prod_{i=1}^{n+1} b_i. \text{ (by using ③)}$$

If $\sum_{i=1}^{n+1} c_i \geqslant n + 2$, then

$$\prod_{i=1}^{n+1} b_i^{c_i} \geqslant b_1^{n+2} \geqslant a_1^{n+1} b_1 > 2^n a_2 \cdots a_{n+1} b_1 \geqslant \prod_{i=1}^{n+1} b_i. \text{ (by using ②)}$$

It is impossible.

If $\sum_{i=1}^{n+1} c_i = n + 1$, we consider three cases.

If $c_{n+1} \geqslant 2$, then

$$\frac{\prod_{i=1}^{n+1} b_i^{c_i}}{\prod_{i=1}^{n+1} b_i} \geqslant \frac{b_{n+1}}{b_n} \cdot \left(\frac{b_1}{b_n}\right)^{n-1} \geqslant \frac{a_{n+1}}{2a_n} \cdot \left(\frac{a_1}{2a_n}\right)^{n-1}$$

$$= \frac{x_{n+1}}{2x_n} \cdot \left(\frac{x_1}{2x_n}\right)^{n-1} > 1. \text{ (by using ①)}$$

It is impossible.

If $c_{n+1} = 1$, then

$$\frac{\prod_{i=1}^{n+1} b_i^{c_i}}{\prod_{i=1}^{n+1} b_i} = \frac{\prod_{i=1}^{n} b_i^{c_i}}{\prod_{i=1}^{n} b_i}.$$

Note that $\dfrac{b_i}{t} \in [x_i, 2x_i]$, $1 \leqslant i \leqslant n$. By the induction hypotheses, we see that

$$k \prod_{i=1}^{n+1} b_i^{c_i} < \prod_{i=1}^{n+1} b_i.$$

If $c_{n+1} = 0$, then

$$\frac{\prod_{i=1}^{n+1} b_i}{\prod_{i=1}^{n+1} b_i^{c_i}} \geqslant \frac{b_{n+1}}{b_n} \cdot \left(\frac{b_n}{b_1}\right)^n \geqslant \frac{a_{n+1}}{2a_n} \cdot \left(\frac{a_n}{2a_1}\right)^n \cdot$$

$$= \frac{x_{n+1}}{2x_n} \cdot \left(\frac{x_n}{2x_1}\right)^n > k. \text{ (by using ①)}$$

Thus, we have checked that $a_1, a_2, \ldots, a_{n+1}$ meet the requirement. □

**3** Let $P(x) = x^{2012} + a_{2011} x^{2011} + a_{2010} x^{2010} + \cdots + a_1 x + a_0$ be a polynomial of degree 2012 of real coefficients with 1 as its leading coefficient. Find the minimum of real number $c$ such that $|\operatorname{Im} z| \leqslant c |\operatorname{Re} z|$, where $\operatorname{Re} z$ and $\operatorname{Im} z$ are, respectively, the real and the imaginary parts of any root of a polynomial obtained by changing some of the coefficients of $P(x)$ to their opposite numbers. (posed by Zhu Huawei)

**Solution.** First, we point out that $c \geqslant \cot \dfrac{\pi}{4022}$. Consider the polynomial $P(x) = x^{2012} - x$. Changing the sign of coefficients of $P(x)$, we obtain four polynomials $P(x)$, $-P(x)$, $Q(x) = x^{2012} + x$ and $-Q(x)$. Note that $P(x)$ and $-P(x)$ have the same roots; one of the roots is $z_1 = \cos\dfrac{1006}{2011}\pi + i\sin\dfrac{1006}{2011}\pi$. $Q(x)$ and $-Q(x)$ have the same roots, and are the opposite number of the roots of $P(x)$. Thus $Q(x)$ has a root $z_2 = -z_1$. Then,

$$c \geqslant \min\left(\frac{|\operatorname{Im} z_1|}{|\operatorname{Re} z_1|}, \frac{|\operatorname{Im} z_2|}{|\operatorname{Re} z_2|}\right) = \cot\frac{\pi}{4022}.$$

Next, we show that the answer is $c = \cot\dfrac{\pi}{4022}$. For any

$$P(x) = x^{2012} + a_{2011}x^{2011} + a_{2010}x^{2010} + \cdots + a_1 x + a_0,$$

we obtain a polynomial

$$R(x) = b_{2012}x^{2012} + b_{2011}x^{2011} + b_{2010}x^{2010} + \cdots + b_1 x + b_0,$$

by changing sign of some coefficient of $P(x)$, where $b_{2012} = 1$, and for $j = 1, 2, \ldots, 2011$,

$$b_j = \begin{cases} |a_j|, & j \equiv 0, 1 \pmod 4, \\ -|a_j|, & j \equiv 2, 3 \pmod 4. \end{cases}$$

We show that, for each root $z$ of $R(x)$, we have $|\operatorname{Im} z| \leqslant c|\operatorname{Re} z|$.

We prove this result by contradiction. Suppose there is a root $z_0$ of $R(x)$, such that $|\operatorname{Im} z_0| > c|\operatorname{Re} z_0|$, then $z_0 \neq 0$ and either the angle of $z_0$ and $i$ is less than $\theta = \dfrac{\pi}{4022}$, or the angle of $z_0$ and $-i$ is less than $\theta$. Suppose that the angle of $z_0$ and $i$ is less than $\theta$; for the other case, we need only consider the conjugate of $z_0$. There are two cases:

If $z_0$ is on the first quadrant (or imaginary axis), suppose that $\angle(z_0, i) = \alpha < \theta$, where $\angle(z_0, i)$ is the least angle that rotates $z_0$ to $i$ anticlockwise. For $0 \leqslant j \leqslant 2012$, if $j \equiv 0, 2(\bmod 4)$, then $\angle(b_j z_0^j, 1) = j\alpha \leqslant 2012\alpha < 2012\theta$.

If $j \equiv 1, 3(\bmod 4)$, then $\angle(b_j z_0^j, i) = j\alpha < 2011\theta$ and $\angle(b_1 z_0, i) = \alpha$. Thus, the principal argument of $b_j z_0^j \in [2\pi - 2012\alpha, 2\pi) \cup \left[0, \frac{1}{2}\pi - \alpha\right]$. The vertex angle of this angle-domain is $2012\alpha + \frac{1}{2}\pi - \alpha = \frac{1}{2}\pi + 2011\alpha < \pi$. And $b_j z_0^j$, $0 \leqslant j \leqslant 2012$, are not all zero, so there sum cannot be zero.

If $z_0$ is at the second quadrant, suppose that $\angle(i, z_0) = \alpha < \theta$, if $j \equiv 0, 2(\bmod 4)$, then $\angle(1, b_j z_0^j) = j\alpha < 2012\theta$. If $j \equiv 1, 3(\bmod 4)$, then $\angle(i, b_j z_0^j) = j\alpha \leqslant 2011\alpha < \frac{\pi}{2}$. Thus, every principle argument of $b_j z_0^j \in \left[0, \frac{\pi}{2} + 2011\alpha\right]$. Since $\frac{\pi}{2} + 2011\alpha < \pi$, and $b_j z_0^j$, $0 \leqslant j \leqslant 2012$, are not all zero, so their sum cannot be zero.

Summing up, the least real number $c = \cot \dfrac{\pi}{4022}$. $\qquad\square$

## Second Day
### 8:00 – 12:30, March 26, 2012

**4** Given an integer $n \geqslant 4$, Let $A$, $B \subseteq \{1, 2, \ldots, n\}$. Suppose that $ab + 1$ is a perfect square number for any $a \in A$ and $b \in B$. Prove that

$$\min\{|A|, |B|\} \leqslant \log_2 n.$$

(posed by Xiong Bin)

**Solution.** First we prove a lemma.

**Lemma.** Given an integer $n \geq 4$, let $A$, $B \subseteq \{1, 2, \ldots, n\}$. Suppose that $ab + 1$ is a perfect square number for any $a \in A$ and $b \in B$. Let $a$, $a' \in A$, $b$, $b' \in B$, and $a < a'$, $b < b'$, then $a'b' > 5.5ab$.

**Proof of the lemma.** First notice that $(ab + 1)(a'b' + 1) > (ab' + 1)(a'b + 1)$. So

$$\sqrt{(ab + 1)(a'b' + 1)} > \sqrt{(ab' + 1)(a'b + 1)}.$$

Since two sides of above inequality are all integers, we have

$$(ab + 1)(a'b' + 1) \geq (\sqrt{(ab' + 1)(a'b + 1)} + 1)^2.$$

By expansion, we obtain

$$ab + a'b' \geq ab' + a'b + 2\sqrt{(ab' + 1)(a'b + 1)} + 1$$
$$> ab' + a'b + 2\sqrt{ab' \cdot a'b}.$$

By $a < a'$, $b < b'$, we have $ab' + a'b > 2ab$. Let $a'b' = \lambda ab$, and combining the above inequality, we have $(1 + \lambda)ab > (2 + 2\sqrt{\lambda})ab$, so $\lambda > 3 + 2\sqrt{2} > 5.5$.

Now turn to the origin problem. Let $A = \{a_1, a_2, \ldots, a_m\}$, $B = \{b_1, b_2, \ldots, b_n\}$, $a_1 < a_2 < \cdots < a_m$, $b_1 < b_2 < \cdots < b_n$. We may suppose that $2 \leq m \leq n$. Since $a_1b_1 + 1$ is a perfect square number, we have $a_1b_1 \geq 3$. By the lemma, $a_2b_2 > 5.5a_1b_1 > 4^2$, $a_{k+1}b_{k+1} > 4a_kb_k$, $k = 2, \ldots, m - 1$. Thus,

$$n^2 \geq a_mb_m \geq 4^{m-2}a_2b_2 > 4^m.$$

Therefore, $m \leq \log_2 n$. $\qquad\square$

**⑤** Find all integers $k \geq 3$, with the following properties: There exist integers $m$ and $n$ satisfying $(m, k) = (n, k) =$

1 and $k \mid (m-1)(n-1)$ with $1 < m < k$, $1 < n < k$ and $m+n > k$. (posed by Yu Hongbing)

**Solution.** If $k$ has a factor of square number greater than 1, let $t^2 \mid k$, $t > 1$, then taking $m = n = k - \dfrac{k}{t} + 1$, we see that such $k$ has the properties.

If $k$ has no factor of square number, if there are two primes $p_1$, $p_2$ such that $(p_1 - 2)(p_2 - 2) \geqslant 4$ and $p_1 p_2 \mid k$. Let $k = p_1 p_2 \cdots p_r$ and $p_1$, $p_2$, $\ldots$, $p_r$ be pairwise different, $r \geqslant 2$. Since there is at least one of $(p_1 - 1)p_2 p_3 \cdots p_r + 1$ and $(p_1 - 2)p_2 p_3 \cdots p_r + 1$ is coprime with $p_1$ (otherwise $p_1$ divides their difference $p_2 p_3 \cdots p_r$, which is a contradiction), taking this number as the number $m$, then, $1 < m < k$, $(m, k) = 1$. Similarly, we can take a number $n$ of $(p_2 - 1)p_1 p_3 \cdots p_r + 1$ or $(p_2 - 2)p_1 p_3 \cdots p_r + 1$ such that $1 < n < k$, $(n, k) = 1$. So $p_1 p_2 \cdots p_r \mid (m-1)(n-1)$, and

$$
\begin{aligned}
m + n &\geqslant (p_1 - 2)p_2 p_3 \cdots p_r + 1 + (p_2 - 2)p_1 p_3 \cdots p_r + 1 \\
&= k + ((p_1 - 2)(p_2 - 2) - 4)p_3 \cdots p_r + 2 > k.
\end{aligned}
$$

Such $m$, $n$ will satisfy the conditions.

If there are no two primes $p_1$, $p_2$ such that $(p_1 - 2)(p_2 - 2) \geqslant 4$, $p_1 p_2 \mid k$, then it is easy to verify such integer $k \geqslant 3$ can only be 15, 30 or as $p$, $2p$ (where $p$ is an odd prime). It is easy to see that, if $k = p$, $2p$, 30, then there are no $m$, $n$ satisfying the conditions; if $k = 15$, then $m = 11$, $n = 13$ satisfy the conditions.

Summing up, integer $k \geqslant 3$ satisfies the conditions if and only if $k$ is not an odd prime, nor double of an odd prime and nor 30. □

**6** Suppose there are beetles on a chessboard consisting of $2012 \times 2012$ unit squares. Each unit square can accommodate at most one beetle. At a moment, all beetles fly and land on the chessboard again. For a beetle, we call the vector from its flying unit to its landing unit the beetle's "displacement vector". We call the sum of all beetle's "displacement vectors" the "total displacement vectors".

Find the maximum length of "total displacement vector" considering the number of beetles and all possible positions of flying and landing. (posed by Qu Zhenhua)

**Solution.** Set up a coordinate with origin at the center of chessboard $O$ and the grid line as the coordinate line. Denote the set of the centers of squares by $S$, and the set where the beetles initially stand on by $M_1 \subseteq S$, and the set that the beetles land on by $M_2 \subseteq S$. Let $f : M_1 \to M_2$ be the one-to-one mapping defined by a beetle's position $v$ at the beginning to the position $u = f(v)$ of first landing. Thus, the total displacement vector is given by

$$V = \sum_{v \in M_1} (f(v) - v) = \sum_{u \in M_2} u - \sum_{v \in M_1} v. \qquad ①$$

Note that the right-hand side of ① is independent of $f$. We need only to find the maximum of $|V| = \left| \sum_{u \in M_2} u - \sum_{v \in M_1} v \right|$ for all $M_1$, $M_2 \subseteq S$, $|M_1| = |M_2|$. We may suppose that $M_1 \cap M_2 = \varnothing$, since element of $M_1 \cap M_2$ does not change $|V|$. Suppose that $|V|$ attains its maximum at $(M_1, M_2)$, obviously $V \neq 0$. Let line $l \perp V$ be at point $O$.

**Lemma 1.** Line $l$ does not pass any point of $S$. $M_1$ is the set of $S$ on one side of $l$, and $M_2$ is the set of $S$ on the other side of $l$.

**Proof of Lemma 1.** First, $M_1 \cup M_2 = S$. Otherwise, since $|S|$ is even, there are at least two points $a$ and $b$ which are not in $M_1 \cup M_2$. Suppose that the angle between $a - b$ and $V$ does not exceed $90°$, then $|V + (a - b)| > |V|$. So add $a$ into $M_2$, add $b$ into $M_1$, then $|V|$ will increase, which is a contradiction.

Second, $M_2 = -M_1$. Otherwise, there exist $a$, $b \in S$, such that $a$, $-a \in M_1$ and $b$, $-b \in M_2$. Suppose that the angle between $a - b$ and $V$ does not exceed $90°$. Put $a$ into $M_2$, and $b$ into $M_1$, $V$ changes to $V + 2(a - b)$, and $|V + 2(a - b)| > |V|$, which is a contradiction.

Third, $l$ does not pass any point of $S$. Otherwise, let $l$ pass $a$, $a \in M_1$, $-a \in M_2$, then change $a$ into $M_2$, $-a$ into $M_1$, $V$ changes to $V + 4a$. Notice that $a \perp V$, so $|V + 4a| > |V|$, which is a contradiction.

Fourth, we show that $M_2$ is the set of $S$ on one side of $l$ (the side that $V$ is pointing to), and $M_1$ is the set of $S$ on the other side of $l$. Otherwise, there is $a \in M_1$ at the side that $V$ is pointing to. And there is a $b \in M_2$ on the other side. Then the angle between $a - b$ and $V$ is less than $90°$. Change $a$ into $M_2$ and $b$ into $M_1$, then $V$ changes into $V + 2(a - b)$, the length of which is greater, which is a contradiction. Lemma 1 is now proved.

**Lemma 2.** Let $S_k = \left\{ (x, y) \in S \,\middle|\, |x| = k - \dfrac{1}{2} \text{ or } |y| = k - \dfrac{1}{2} \right\}$, $k = 1, 2, \ldots, 1006$, $l$ be a line passing $O$ and does not pass points of $S_k$. Denote all points of $S_k$ on one side of $l$ by $A_k$, all points of $S_k$ on the other side by $B_k$. Denote $V_k = \sum_{u \in A_k} u - \sum_{v \in B_k} v$, then the maximum of $|V_k|$ is obtained when $l$ is horizontal (or vertical), and $V_k$ is vertical (or horizontal).

**Proof of Lemma 2.** $S_k$ is located at the boundary of a square with $2k$ points on each side. Let the four vertices of the square be $A\left(k - \dfrac{1}{2}, k - \dfrac{1}{2}\right)$,

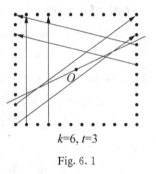

$k=6, t=3$

Fig. 6. 1

$B\left(-k + \dfrac{1}{2}, k - \dfrac{1}{2}\right)$, $C\left(-k + \dfrac{1}{2}, -k + \dfrac{1}{2}\right)$ and $D\left(k - \dfrac{1}{2}, -k + \dfrac{1}{2}\right)$. By

symmetricity, we may suppose that $l$ intersects $AD$ at point $P$ with non-negative slope. Let $P$ be located between the $t(1 \leqslant t \leqslant k)$th (from top to bottom) of $S_k$ on $AD$ and the $(t+1)$th of $S_k$ on $AD$ (see Fig. 6. 1, in case of $k = 6$, $t = 3$). Thus,

$$V_k = (2k - 2)(2k - 1)\vec{j} + (2k - t)(-(2k - 1)\vec{i} + t\vec{j}) +$$
$$t\,((2k - 1)\vec{i} + (2k - t)\vec{j})$$
$$= -2(2k - 1)(k - t)\vec{i} + 2(-(k - t)^2 + 3k^2 - 3k + 1)\vec{j},$$

where $\vec{i}$ and $\vec{j}$ are horizontal and vertical unit vectors, respectively. Denote $(k - t)^2 = u$, $0 \leqslant u \leqslant (k - 1)^2$, then

$$\frac{1}{4} \mid V_k \mid^2 = (2k - 1)^2 u + u^2 - 2(3k^2 - 3k + 1)u + (3k^2 - 3k + 1)^2$$
$$= u^2 - (2k^2 - 2k + 1)u + (3k^2 - 3k + 1)^2.$$

As a quadratic function of $u$, $u = k^2 - k + \dfrac{1}{2}$ is the symmetric axis. It is easy to know that $\mid V_k \mid^2$ takes its maximum at $u = 0$. So $t = k$, that is, $l$ is horizontal, so $V_k$ is vertical. Lemma 2 is proved.

Turn to the original problem. By symmetricity, we need to only consider that the slope of $l$ is non-negative and less than 1. Let $M_1$, $M_2$ be located on two sides of $l$. Denote $M_2 \cap S_k =$

$A_k$, $M_1 \cap S_k = B_k$, $V_k = \sum_{u \in A_k} u - \sum_{v \in B_k} v$, then

$$|V| = \left| \sum_{k=1}^{1006} V_k \right| \leqslant \sum_{k=1}^{1006} |V_k|. \qquad ②$$

So, $|V|_{\max} \leqslant \sum_{k=1}^{1006} |V_k|_{\max}$. If $l$ is horizontal, $M_2$ is all the points of $S$ on the upper half-plane, $M_1$ is all points of $S$ on the lower half-plane; each $|V_k|$ takes its maximum, and all $V_k$ point upward. And the equality of ② holds. So $|V|$ indeed takes the maximum $|V|_{\max} = 2 \times 1006^3$. □

# 2013 (Jiangyin, Jiangsu)

## First Day

### 8:00 – 12:30, March 24, 2013

**1** Given any $n(>1)$ coprime positive integers $a_1$, $a_2$, ..., $a_n$, denote $A = a_1 + a_2 + \cdots + a_n$. Let $d_i = (A, a_i)$ (the greatest common divisor), $i = 1, 2, \ldots, n$.

Let $D_i$ be the greatest common divisor of $\{a_1, a_2, \ldots, a_n\} \backslash \{a_i\}$, $i = 1, 2, \ldots, n$. Find the minimum of $\prod_{i=1}^{n} \dfrac{A - a_i}{d_i D_i}$. (posed by Zhang Sihui)

**Solution.** Consider

$$D_1 = (a_2, a_3, \ldots, a_n) \text{ and } d_2 = (a_2, A)$$
$$= (a_2, a_1 + a_2 + \cdots + a_n).$$

Let $(D_1, d_2) = d$. Then $d \mid a_2, d \mid a_3, \ldots, d \mid a_n, d \mid a_1 + a_2 + \cdots + a_n$. Thus, $d \mid a_1$. Consequently,

$$d \mid (a_1, a_2, \ldots, a_n).$$

Since $a_1, a_2, \ldots, a_n$ are coprime, we have $d = 1$. Note that $D_1 \mid a_2$, $d_2 \mid a_2$ and $(D_1, d_2) = 1$. We have $D_1 d_2 \mid a_2$. So $D_1 d_2 \leqslant a_2$. Similarly, we have $D_2 d_3 \leqslant a_3, \ldots, D_n d_1 \leqslant a_1$. Hence,

$$\prod_{i=1}^{n} d_i D_i = (D_1 d_2) \cdot (D_2 d_3) \cdot \cdots \cdot (D_n d_1)$$
$$\leqslant a_2 a_3 \cdots a_n a_1$$
$$= \prod_{i=1}^{n} a_i. \qquad \text{①}$$

Considering

$$\prod_{i=1}^{n} (A - a_i) = \prod_{i=1}^{n} \left( \sum_{j \neq i} a_j \right)$$
$$\geqslant \prod_{i=1}^{n} \left( (n-1) \left( \prod_{j \neq i} a_j \right)^{\frac{1}{n-1}} \right)$$
$$= (n-1)^n \cdot \prod_{i=1}^{n} a_i \qquad \text{②}$$

and by ① and ②, we see that

$$\prod_{i=1}^{n} \frac{A - a_i}{d_i D_i} \geqslant (n-1)^n.$$

On the other hand, if $a_1 = a_2 = \cdots = a_n = 1$,

$$\prod_{i=1}^{n} \frac{A - a_i}{d_i D_i} = (n-1)^n. \quad \bullet$$

Summing up, the minimum of $\prod_{i=1}^{n} \dfrac{A - a_i}{d_i D_i}$ is $(n-1)^n$.

□

**2**  Suppose that $O$ and $I$ are the centres of the circumcircle and incircle of $\triangle ABC$ with radius $R$ and $r$, respectively, $P$

is the midpoint of arc $\overset{\frown}{BAC}$. Let $QP$ be the diameter of $O$. Let $PI$ intersect $BC$ at point $D$, and let the circumcircle of $\triangle AID$ intersect the extended line of $PA$ at point $F$. Let point $E$ be on $PD$ such that $DE = DQ$. Prove that, if $\angle AEF = \angle APE$, then $\sin^2 \angle BAC = \dfrac{2r}{R}$. (posed by Xiong Bin)

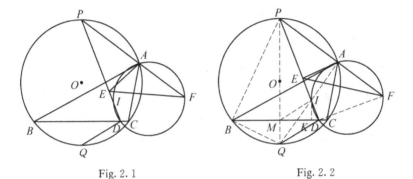

Fig. 2. 1 Fig. 2. 2

**Solution.** Since $\angle AEF = \angle APE$, then $\triangle AEF \backsim \triangle EPF$. So $AF \cdot PF = EF^2$. Since points $A$, $I$, $D$ and $F$ are concyclic, $PA \cdot PF = PI \cdot PD$. Thus,

$$PF^2 = AF \cdot PF + PA \cdot PF \qquad \qquad ①$$
$$= EF^2 + PI \cdot PD.$$

Since $PQ$ is the dimeter of circle $O$ and point $I$ is on $AQ$, we see that $AI \perp AP$. Consequently,

$$\angle IDF = \angle IAP = 90°.$$

Thus, we have

$$PF^2 - EF^2 = PD^2 - ED^2.$$

Combining ①, we have

$$PI \cdot PD = PD^2 - ED^2.$$

Thus

$$QD^2 = ED^2 = PD^2 - PI \cdot PD = ID \cdot PD.$$

Consequently, we have

$$\triangle QID \backsim \triangle PQD. \qquad ②$$

Since $PQ$ is the diameter of circle $O$, we see that $BP \perp BQ$. Suppose that $PQ$ is the perpendicular bisector of $BC$ at point $M$. Note that $I$ is the incentre of $\triangle ABC$. We have $QI^2 = QB^2 = QM \cdot QP$. Thus,

$$\triangle QMI \backsim \triangle QIP. \qquad ③$$

By ② and ③, we see that $\angle IQD = \angle QPD = \angle QPI = \angle QIM$. Hence, $MI \parallel QD$. Let $IK \perp BC$ be at $K$. Then $IK \parallel PM$; thus,

$$\frac{PM}{IK} = \frac{PD}{ID} = \frac{PQ}{MQ}.$$

By the Circle-Power Theorem and the Sine Theorem, we know that

$$PQ \cdot IK = PM \cdot MQ = BM \cdot MC$$
$$= \left(\frac{1}{2}BC\right)^2 = (R\sin\angle BAC)^2,$$

thus, $\sin^2 \angle BAC = \dfrac{PQ \cdot IK}{R^2} = \dfrac{2R \cdot r}{R^2} = \dfrac{2r}{R}.$  $\qquad \square$

**3** Suppose there are 101 persons sitting around a round table in an arbitrary order. The $k$th person possesses $k$ pieces of carts, $k = 1, \ldots, 101$. We call it a *transition* if one transits one of his carts to one of his adjacent persons. Find the minimum positive number $k$, such that whatever the order of the seating, there is way of no more than $k$

transitions so that each person possesses 51 carts. (posed by Qu Zhenhua)

**Solution.** The answer is $k = 42\,925$.

Let the circumference of the table be 101, and the distance between two adjacent persons be 1. In the following, we consider the least transition times.

Denote the person who initially possesses $i$ cards by $[i - 51]$. So, if $p > 0$, then we can think of person $[p]$ as the source who should send out $p$ cards; and if $p < 0$, person $[p]$ is the sink who should receive $-p$ cards.

Suppose that at the end of transitions, each person has 51 cards. If person $B$ possesses a card $u$ initially belonging to person $A$, we can think that $A$ carries card $u$ to $B$ by passing through the minor arc $\overset{\frown}{AB}$, the length of the route is the arc length $|\overset{\frown}{AB}|$ means the times of transitions.

Let seating order be as shown in the figure. Person $[i]$ transits all his $i$ cards to person $[-i](i = 1, 2, \ldots, 50)$ with route length $i$. So there are all together $1^2 + 2^2 + \cdots + 50^2 = 42\,925$ transitions. In the following, we will show that no less transitions can meet the requirement if persons are seated in this way.

We use the notion of "potential". Let potential at the highest position $[50]$ be 50; the neighbor positions $[49]$ and $[48]$ each has potential 49, ..., the lowest position $[-49]$ and $[-50]$ each has potential 0. Then, the total potential at the

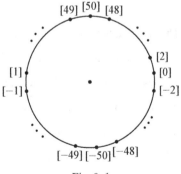

Fig. 3. 1

beginning is

$$S = 101 \times 50 + (100 + 99) \times 49 + \cdots + (2 + 1) \times 0.$$

And at the end of transitions, the total potential is

$$T = 51 \times 50 + (51 + 51) \times 49 + \cdots + (51 + 51) \times 0.$$

The difference is $S - T = 42,925$.

Since after each transition, the total potential changes at most 1, so at least 42,925 transitions are needed.

In the following, we show that whatever be the order of seating, there is always a way of no more than 42, 925 transitions such that each person possesses 51 cards. To show this, we give two lemmas.

**Lemma 1.** Let $c$, $a_0$, $a_1$, ..., $a_{n-1}$ be integers with their sum zero, and $c \geqslant 0$, $a_0 \leqslant a_1 \leqslant \cdots \leqslant a_{n-1}$.

If $n + 1$ persons denoted by $[c]$, $[a_0]$, $[a_1]$, ..., $[a_{n-1}]$, possess $N + c$, $N + a_0$, $N + a_1$, ... and $N + a_{n-1}$ cards, respectively, where $N$ is a positive integer, such that $N + a_0 > 0$. Let the persons stand on 0, 1, ..., $n$ of the number axis, such that $[c]$ stands at $n$. Then there is a way of no more than $cn + \sum_{i=0}^{n-1} i a_i$ transitions, such that each person possesses $N$ cards.

**Proof of Lemma 1.** Suppose that $a_{n-1} \geqslant \cdots \geqslant a_s > 0 \geqslant a_{s-1} \geqslant \cdots \geqslant a_0$, then induction on $M = a_{n-1} + \cdots + a_s$. If $M = 0$, then $[c]$ passes $c$ cards to $[a_i]$, such that $[a_i]$ obtains $-a_i$ cards ($0 \leqslant i \leqslant s - 1$). Suppose that person $[a_i]$ stands at $x_i$ ($0 \leqslant i \leqslant n - 1$), then $x_0$, $x_1$, ..., $x_{n-1}$ is a permutation of 0, 1, ..., $n - 1$. Thus, a card that passes from $[c]$ to $[a_i]$ needs $n - x_i$ transitions. So the total transitions needed are

$$\sum_{i=0}^{s-1} (n - x_i)(-a_i) = cn + \sum_{i=0}^{s-1} x_i a_i \leqslant cn + \sum_{i=0}^{s-1} i a_i$$

$$= cn + \sum_{i=0}^{n-1} i a_i.$$

Now suppose that the conclusion is true for integers smaller than $M$. Consider the case of integer $M$. Suppose that

$$a_{n-1} = \cdots = a_{n-s} > a_{n-s-1}, a_l > a_{l-1} = \cdots = a_0.$$

Then there is a person in $[a_{n-1}], \ldots, [a_{n-s}]$ and a person in $[a_0], \ldots, [a_{l-1}]$ such that their distance is no more than $n - s - l + 1$. We may suppose that the distance between $[a_{n-s}]$ and $[a_{l-1}]$ is no more than $n - s - l + 1$, then let person $[a_{n-s}]$ pass a card $u$ to $[a_{l-1}]$ by no more than $n - s - l + 1$ transitions. Next, by induction on $c$ and

$$a_{n-1} \geqslant \cdots \geqslant a_{n-s+1} > a_{n-s} - 1 \geqslant a_{n-s-1}$$
$$\geqslant \cdots \geqslant a_l \geqslant a_{l-1} + 1 > a_{l-2} \geqslant \cdots \geqslant a_0,$$

we see that by taking no more than

$$L = cn + (n-1)a_{n-1} + \cdots + (n - s + 1)a_{n-s+1} +$$
$$(n - s)(a_{n-s} - 1) + (n - s - 1)a_{n-s-1} + \cdots +$$
$$la_l + (l - 1)(a_{l-1} + 1) + (l - 2)a_{l-2} + \cdots + 0 \cdot a_0$$

transitions, each person possesses $N$ cards. Addition of the transitions of card $u$, we know that there are no more than

$$lL + (n - s - l + 1) = cn + \sum_{i=0}^{n-1} i a_i$$

transitions. The proof of Lemma 1 is completed.

**Lemma 2.** For any permutation of $[-50], [-49], \ldots, [49], [50]$ on a circle, there always exists a person $[c]$, denote the line passing through $[c]$ and the origin by $l$, such that the sum of each side of $l$ (include $c$) in the brackets has the same sign of $c$.

**Proof of Lemma 2.** Let the permutation on the circle be $[a_1]$, $[a_2]$, ..., $[a_{101}]$ clockwise. Then there is a directed diameter $l$ of the circle such that $[a_1]$, $[a_2]$, ..., $[a_{50}]$ are on one side of $l$ and $[a_{51}]$, $[a_{52}]$, ..., $[a_{101}]$ are on the other side.

If $\sum_{i=1}^{50} a_i = 0$, then take $c = a_{51}$; else, if $\sum_{i=1}^{50} a_i \neq 0$, then the sums of each side of $l$ have different sign. If we rotate $l$ $180°$ clockwise, we see that the sum of each side changes sign. So there exists a $[c]$ which meeting the requirement of Lemma 2.

Take $[c]$ in Lemma 2. Suppose that $c \geqslant 0$ (else change each $[a_i]$ by $[-a_i]$ and change all the directions of the arc. Let $c = c_1 + c_2$, $c_1$, $c_2 \geqslant 0$, such that the sum of $c_1$ and the numbers on one side of (not include $c$) $l$ is zero. Denote 50 numbers on this side by $a_0 \leqslant a_1 \leqslant \cdots \leqslant a_{49}$, and denote 50 numbers on the other side by $b_0 \leqslant b_1 \leqslant \cdots \leqslant b_{49}$. Then the sum of $c_2$ and $b_0$, $b_1$, ..., $b_{49}$ is also zero. Using Lemma 1 to $c_1$, $a_0$, $a_1$, ..., $a_{49}$ and $c_2$, $b_0$, $b_1$, ..., $b_{49}$, respectively, we obtain that the transition times are no more than

$$L = 50c_1 + \sum_{j=0}^{49} j a_j + 50c_2 + \sum_{j=0}^{49} j b_j.$$

Since $c$, $a_0$, ..., $a_{49}$, $b_0$, ..., $b_{49}$ is a permutation of $-50$, $-49$, ..., $50$, by the order inequality

$$L = 50c + \sum_{j=0}^{49} j(a_j + b_j)$$

$$\leqslant 50^2 + \sum_{j=0}^{49} j(2j - 50 + 2j - 49)$$

$$= 42\,925.$$

Summing up, the least positive integer $k = 42\,925$.   $\square$

# Second Day

8:00 - 12:30, March 25, 2013

**④** Let $p$ be a prime, $a$ and $k$ be positive integers, satisfying $p^a < k < 2p^a$. Prove that there exists positive integer $n$, $n < p^{2a}$ such that $C_n^k \equiv n \equiv k \pmod{p^a}$. (posed by Yu Hongbing)

**Solution.** We are to prove an extended problem.

Let $p$ be a prime, $a$ and $k$ be positive integers, satisfying $p^a < k < 2p^a$. Prove that for any non-negative integer $b$, there exists positive integer $n$, $n < p^{a+b}$ such that $n \equiv k \pmod{p^a}$ and $C_n^k \equiv k \pmod{p^b}$.

If $b = 0$, $p^b = 1$, take $n = k - p^a$. We prove by induction. Suppose that the conclusion is true for integer $b \geqslant 0$. That is, there exists a positive integer $n < p^{a+b}$ $n \equiv k \pmod{p^a}$, and $C_n^k \equiv k \pmod{p^b}$.

Let $1 \leqslant t \leqslant p - 1$. Consider

$$C_{n+tp^{a+b}}^k = \prod_{i=0}^{k-1} \frac{n + tp^{a+b} - i}{k - i}.$$

For integer $m$, let $P(m) = p^{v_p(m)}$, $r(m) = \dfrac{m}{P(m)}$, where $v_p(m)$ is the number of $p$ in the standard factorization of $m$.

Since $k - i < 2p^a \leqslant p^{a+1}$, we see that $v_p(k - i) \leqslant a$ and $n - i \equiv k - i \pmod{p^a}$. Hence,

$$P(k - i) \mid n + tp^{a+b} - i.$$

Consequently,

$$C_{n+tp^{a+b}}^k = \prod_{i=0}^{k-1} \frac{\dfrac{n - i}{P(k - i)} + tp^{a+b-v_p(k-i)}}{r(k - i)}.$$

If $k - i \neq p^a$, then $v_p(k - i) \leqslant a - 1$ and $a + b - v_p(k - i)$ $\geqslant b + 1$. If $k - i = p^a$, then $v_p(k - i) = a$. Hence,

$$
\begin{aligned}
C_{n+tp^{a+b}}^k &\equiv \prod_{i=0}^{k-1} \frac{n-i}{P(k-i)r(k-i)} + \left( \prod_{\substack{0 \leqslant i \leqslant k-1 \\ i \neq k-p^a}} \frac{n-i}{P(k-i)r(k-i)} \right) \cdot tp^b \\
&\equiv C_n^k + \left( \prod_{\substack{0 \leqslant i \leqslant k-1 \\ i \neq k-p^a}} \frac{r(n-i)}{r(k-i)} \right) \cdot tp^b \pmod{p^{b+1}}.
\end{aligned}
$$

This is because that if $k - i \neq p^a$, then $p^a \mid (n - i) - (k - i)$. So

$$
v_p(n - i) = v_p(k - i).
$$

Since $\prod_{\substack{0 \leqslant i \leqslant k-1 \\ i \neq k-p^a}}^{k-1} \frac{r(n-i)}{r(k-i)}$ is coprime to $p$, we see that $C_{n+tp^{a+b}}^k$ $(0 \leqslant t \leqslant p - 1)$ goes through the following remainders of modular $p^{b+1}$:

$$
C_n^k + jp^b, \ j = 0, 1, \ldots, p - 1.
$$

Since $C_n^k \equiv k \pmod{p^b}$, there exists $j \, (0 \leqslant j \leqslant p - 1)$, such that $C_n^k + jp^b \equiv k \pmod{p^{b+1}}$. That is, there exists $t \, (0 \leqslant t \leqslant p - 1)$, such that $C_{n+tp^{a+b}}^k \equiv k \pmod{p^{b+1}}$. Let $N = n + tp^{a+b}$. Then $N < p^{a+b+1}$, $N \equiv n \equiv k \pmod{p^a}$ and $C_N^k \equiv k \pmod{p^{b+1}}$.

The extended problem is proved by induction.    $\square$

**5**   Let $n \geqslant 2$ and $a_1, a_2, \ldots, a_n, b_1, b_2, \ldots, b_n$ be non-negative integers. Prove that

$$
\left( \frac{n}{n-1} \right)^{n-1} \left( \frac{1}{n} \sum_{i=1}^n a_i^2 \right) + \left( \frac{1}{n} \sum_{i=1}^n b_i \right)^2 \geqslant \prod_{i=1}^n (a_i^2 + b_i^2)^{\frac{1}{n}}.
$$

$$\textcircled{1}$$

(posed by Leng Gangsong)

**Solution.** Denote $\lambda = \left(\dfrac{n}{n-1}\right)^{n-1}$, $n \geqslant 2$. Obviously, $\lambda > 1$.

For given $i \in \{1, \ldots, n\}$, fix $p = a_k^2 + b_k^2$ for $k = 1$, $2, \ldots, n$ and fix $a_j$ and $b_j$ ($j \neq i$). Then the left-hand side of

① $= \dfrac{\lambda}{n}(p - b_i^2 + \sum_{j \neq i} a_j^2) + \dfrac{1}{n^2}(b_i + \sum_{j \neq i} b_j)^2$ is a quadratic

function of $b_i$, $b_i \in [0, \sqrt{p}]$, with leading coefficient $-\dfrac{\lambda}{n} + \dfrac{1}{n^2} <$

0. Thus, its minimum is taken at the endpoints, that is, $b_i = 0$ or $a_i = 0$.

So, we can suppose that $a_i b_i = 0$, $i = 1, 2, \ldots, n$.

*Case* 1. Each $a_i = 0$, then by the mean value inequality, we have

$$\left(\frac{1}{n}\sum_{i=1}^{n} b_i\right)^2 \geqslant \prod_{i=1}^{n} b_i^{\frac{2}{n}}.$$

*Case* 2. Each $b_i = 0$, then by the mean value inequality, we have

$$\lambda\left(\frac{1}{n}\sum_{i=1}^{n} a_i^2\right) \geqslant \frac{1}{n}\sum_{i=1}^{n} a_i^2 \geqslant \prod_{i=1}^{n} a_i^{\frac{2}{n}}.$$

*Case* 3. We may suppose that $b_1 = \cdots = b_k = 0$, $a_{k+1} = \cdots = a_n = 0$, $1 \leqslant k < n$.

Let $a_1 a_2 \cdots a_k = a^k$, $b_{k+1} \cdots b_n = b^{n-k}$, $a$, $b \geqslant 0$. Then by the mean value inequality, we have

$$a_1^2 + a_2^2 + \cdots + a_k^2 \geqslant ka^2, \quad b_{k+1} + \cdots + b_n \geqslant (n-k)b.$$

It suffices to prove that

$$\frac{\lambda k}{n}a^2 + \frac{(n-k)^2}{n^2}b^2 \geqslant a^{\frac{2k}{n}} \cdot b^{\frac{2(n-k)}{n}}. \qquad ②$$

By the mean value inequality, we see that

The left-hand side of

$$② = \underbrace{\frac{\lambda}{n}a^2 + \cdots + \frac{\lambda}{n}a^2}_{k\,\text{terms}} + \underbrace{\frac{n-k}{n^2}b^2 + \cdots + \frac{n-k}{n^2}b^2}_{n-k\,\text{terms}}$$

$$\geqslant \lambda^{\frac{k}{n}} a^{\frac{2k}{n}} \cdot \left(\frac{n-k}{n}\right)^{\frac{n-k}{n}} \cdot b^{\frac{2(n-k)}{n}}.$$

So, it suffices to show that $\lambda^{\frac{k}{n}}\left(\dfrac{n-k}{n}\right)^{\frac{n-k}{n}} \geqslant 1$, that is, to show $\left(\dfrac{n}{n-k}\right)^{n-k} \leqslant \lambda^k$.

In fact,

$$\underbrace{\frac{n}{n-k} \cdot \frac{n}{n-k} \cdot \cdots \cdot \frac{n}{n-k}}_{n-k\,\text{terms}} \cdot \underbrace{1 \cdot 1 \cdot \cdots \cdot 1}_{nk-n\,\text{terms}}$$

$$\leqslant \left(\frac{n+(nk-n)}{nk-k}\right)^{nk-k}$$

$$= \left(\frac{n}{n-1}\right)^{(n-1)k} = \lambda^k. \qquad \square$$

**6** In a plane with cartesian coordinates, let $P$ and $Q$ be two regions of convex polygon (including boundary and interior) whose vertices are all integer points (i. e. , their coordinates are all integers) and $T = P \cap Q$. Prove that if $T$ is not empty and does not contain integer point, then $T$ is a non-degenerate convex quadrilateral. (posed by Qu Zhenhua)

**Solution.** Since the non-empty intersection $T$ of two convex closed polygons is a closed convex polygon or degenerated polygon, there are three possible cases.

(1) $T$ is a point. Then $T$ must be the vertex of $P$ or $Q$, contradicting the fact that $T$ contains no integer point.

(2) $T$ is a segment. Then $T$ must be the intersection of an edge of $P$ and an edge of $Q$, which contains the vertex of $P$ or $Q$, which is a contradiction.

(3) $T$ is a closed convex polygon.

So, it remains to be shown that $T$ is a quadrilateral.

First, we note that if $T$ has two adjacent edges on the edges of $P$ (or $Q$), then the common vertex of this two edges must be the vertex of $P$ (or $Q$), which is a contradiction. Thus, the boundary of $T$ is formed alternately by a part of an edge of $P$ and then a part of an edge of $Q$, and each vertex of $T$ is the intersect point of edges of $P$ and $Q$. Thus, the number of edges of $T$ is even.

We see that if an edge $e$ of $P$ intersects an edge $f$ of $Q$, then $e$ must intersect another edge of $Q$, otherwise $T$ will contain an integer point.

In the following, we show by contradiction that the number of edges of $T$ can be 6 or more.

If $T$ has edges no less than 6, then suppose $P$ contains $k$ integer points except the vertices of $P$.

*Case* 1. If $k = 0$, then $P$ is an element integer triangle, or a parallelogram with area 1. So, $P$ can be located between two parallel lines $l_1$, $l_2$. And there is no integer point in the open domain $\Omega$ between $l_1$ and $l_2$. At least three edges of $P$ are the edges of $T$, because $T$ has at least six edges.

(a) In case of $P$ being $\triangle ABC$ (See Fig. 6. 1), $DE$, $FG$ and $HI$ are edges of $Q$. $D$, $G$ and $E$ may coincide with $F$, $H$ and $I$, respectively. Lines $FG$, $HI$, $l_1$ and $l_2$ form a convex quadrilateral. Since line $DE$ does not intersect segment $BC$, we see that the

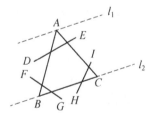

Fig. 6. 1

intersect point of line $DE$ and $FG$ or of line $DE$ and $HI$ is in $\Omega$. Thus, $Q$ has integer vertex in $\Omega$, a contradiction.

(b) In case of $P$ being a parallelogram $\square ABCD$. Let $AD$, $AB$ and $BC$ be three edges of $P$ (see Fig. 6. 2). The intersection point of line $EF$ and $HG$ locates in $\Omega$. Let $AB$, $BC$ and $CD$ be three edges of $P$ (see Fig. 6. 3), and there is no edge of $Q$ on $AD$. Similar to the case of (a), we can see the intersection point of line $EF$ and $HG$ or of line $EF$ and $IJ$ locates in $\Omega$, which is a contradiction.

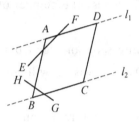

Fig. 6. 2

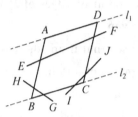

Fig. 6. 3

*Case* 2. $k \geqslant 1$. Consider integer point $X$ on $P$ other than the vertices. Since $X \notin T$, there exists an edge $MN$ of $T$ such that $T$ and $X$ are separated by line $MN$ (denote by $l$ ) (see Fig. 6. 4). Thus, $MN$ is a part

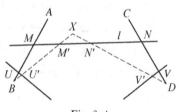

Fig. 6. 4

of boundary of $Q$. $M$ is on the edge $AB$ of $P$, $N$ is on the edge $CD$ of $P$. $A$, $C$ and $X$ are on the same side of $l$ ( $A$ and $C$ may coincide, but $B$ and $D$ do not by the hypothesis that $T$ has at least six edges). Thus, there is another vertex $U$ of $T$ on $AB$, and there is another vertex $V$ of $T$ on $CD$.

Denote the convex hull of points $X$ and vertices of $P$ below $l$ by $P'$. Then $P'$ and $P$ coincide below $BD$, and the part of $P'$ up $BD$ is $\triangle BXD$. Comparing $T' = P' \cap Q$ with $T$, we see that $T' \subset T$, and the boundary of $T'$ is the boundary of $T$ with $MN$,

$MU$ and $NV$ replaced by $M'N'$, $M'U'$ and $N'V'$, respectively. That is, $T'$ and $T$ have the same number of edges, and the number of integer points of $P'$ other than vertices is less than that of $P$. By a finite procedure like this, we can obtain a convex polygon with no interior integer point. By Case 1, it is impossible. $\square$

## 2014 (Nanjing, Jiangsu)

### First Day
8:00 – 12:30, March 23, 2014

**1** Let $O$ be the circumcenter of $\triangle ABC$ and $H_A$ be the projection of $A$ onto $BC$. The extension of $AO$ intersects the circumcircle of $\triangle BOC$ at $A'$. The projections of $A'$ onto $AB$ and $AC$ are $D$ and $E$, respectively. Let $O_A$ be the circumcenter of $\triangle DH_AE$. Define $H_B$, $O_B$, $H_C$ and $O_C$ similarly.

Prove that $H_AO_A$, $H_BO_B$ and $H_CO_C$ are concurrent. (posed by Zhang Sihui)

**Solution.** Let $T$ be the symmetry point of $A$ over $BC$, $F$ be the projection of $A'$ onto $BC$, and $M$ be the projection of $T$ onto $AC$.

Since $AC = CT$, we have $\angle TCM = 2\angle TAM$. Since $\angle TAM = \dfrac{\pi}{2} - $

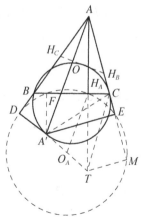

Fig. 1. 1

$\angle ACB = \angle OAB$, we have

$$\angle TCM = 2\angle OAB = \angle A'OB = \angle A'CF,$$

and

$$\angle TCH_A = \angle A'CF + \angle A'CT = \angle TCM + \angle A'CT = \angle A'CE.$$

Because of the fact that $\angle CH_AT$, $\angle CMT$, $\angle CEA'$, $\angle CFA'$ are right angles, we have,

$$\begin{aligned}\frac{CH_A}{CM} &= \frac{CH_A}{CT} \cdot \frac{CT}{CM} = \frac{\cos\angle TCH_A}{\cos\angle TCM} = \frac{\cos\angle A'CE}{\cos\angle A'CF} \\ &= \frac{CE}{CA'} \cdot \frac{CA'}{CF} = \frac{CE}{CF},\end{aligned}$$

i.e., $CH_A \cdot CF = CM \cdot CE$, so $H_A$, $F$, $M$ and $E$ are on the same circle $\omega_1$.

Similarly, let $N$ be the projection of $T$ onto $AB$, then $H_A$, $F$, $N$ and $D$ are on the same circle $\omega_2$. Since $A'FH_AT$ and $A'EMT$ are both right trapezoid, the perpendicular bisector of the segments $H_AF$ and $EM$ meet at the midpoint $K$ of the segment $A'T$, i.e., $K$ is the center of circle $\omega_1$ and $KF$ is the radius of circle $\omega_1$. Similarly, $K$ and $KF$ are also the center and the radius of circle $\omega_2$, respectively. Thus, $\omega_1$ and $\omega_2$ are the same, $D$, $N$, $F$, $H_A$, $E$ and $M$ are on the same circle. So $O_A$ is the midpoint $K$ of $A'T$, $O_AH_A \parallel AA'$.

Since $\angle H_CAO + \angle AH_CH_B = \frac{\pi}{2} - \angle ACB + \angle ACB = \frac{\pi}{2}$,

we have $AA' \perp H_BH_C$, thus $O_AH_A \perp H_BH_C$, therefore, $O_AH_A$, $O_BH_B$ and $O_CH_C$ all pass through the orthocenter of $\triangle H_AH_BH_C$. □

**2** Let $A_1A_2\cdots A_{101}$ be a regular $101$-gon, and color every vertex red or blue. Let $N$ be the number of obtuse triangles satisfying the following conditions: The three vertices of the triangle must be vertices of the $101$ – gon, both the vertices with acute angles have the same color, and the vertex with obtuse angle has different color.

(1) Find the largest possible value of $N$.

(2) Find the number of ways to color the vertices such that maximum $N$ is achieved. (Two colorings are different if for some $A_i$ the colors are different on the two coloring schemes.) (posed by Qu Zhenhua)

**Solution.** Defining $x_i = 0$ or $1$ depends on whether $A_i$ is red or blue. For obtuse triangle $A_{i-a}A_iA_{i+b}$ (vertex $A_i$ is the vertex of the obtuse angle, i.e., $a + b \leqslant 50$), these three vertices satisfy the conditions of the problem if and only if

$$(x_i - x_{i-a})(x_i - x_{i+b}) = 1, \qquad \qquad \text{①}$$

otherwise equal $0$, here the subscript modules $101$. Thus,

$$N = \sum_{i=1}^{101} \sum_{(a, b)} (x_i - x_{i-a})(x_i - x_{i+b}).$$

Here $\sum_{(a, b)}$ stands for the summation of all positive integer pairs $(a, b)$ satisfying $a + b \leqslant 50$. There are $49 + 48 + \cdots + 1 = 1225$ such positive integer pairs. Expanding ①, we have

$$\begin{aligned}
N &= \sum_{i=1}^{101} \sum_{(a, b)} (x_i - x_{i-a})(x_i - x_{i+b}) \\
&= \sum_{i=1}^{101} \sum_{(a, b)} (x_i^2 - x_i x_{i-a} - x_i x_{i+b} + x_{i-a} x_{i+b}) \\
&= 1225 \sum_{i=1}^{101} x_i^2 + \sum_{i=1}^{101} \sum_{k=1}^{50} (k - 1 - 2(50 - k)) x_i x_{i+k} \\
&= 1225n + \sum_{i=1}^{101} \sum_{k=1}^{50} (3k - 101) x_i x_{i+k}. \qquad \qquad \text{②}
\end{aligned}$$

Here $n$ is the number of the blue vertices. For any two vertexes

$A_i$ and $A_j$, $1 \leqslant i \leqslant j \leqslant 101$, let

$$d(A_i, A_j) = d(A_j, A_i) = \min\{j - i, 101 - j + i\}.$$

Let $B \subseteq \{A_1, A_2, \ldots, A_{101}\}$ be the set of all blue vertices. Then ② can be written as

$$N = 1225n - 101C_n^2 + 3 \sum_{\{P, Q\} \subseteq B} d(P, Q), \qquad ③$$

where $\{P, Q\}$ pass through all the two-element subsets of $B$. Without loss of generality, we assume that $n$ is even. Otherwise, change the color of all vertices, the value of $N$ does not change. Write $n = 2t$, $0 \leqslant t \leqslant 50$, from one point, renumber all the blue vertices by $P_1, P_2, \ldots, P_{2t}$ clockwise. Then

$$\sum_{\{P, Q\} \subseteq B} d(P, Q) = \sum_{i=1}^{t} d(P_i, P_{i+t}) + \frac{1}{2} \sum_{i=1}^{t} \sum_{j=1}^{t-1} (d(P_i, P_{i+j}) +$$

$$d(P_{i+j}, P_{i+t}) + d(P_{i+t}, P_{i-j}) + d(P_{i-j}, P_i))$$

$$\leqslant 50t + \frac{101}{2} t(t - 1).$$

$$④$$

Here the subscript of $P_i$ modules $2t$, and we use the inequality $d(P_i, P_{i+t}) \leqslant 50$ and

$$d(P_i, P_{i+j}) + d(P_{i+j}, P_{i+t}) + d(P_{i+t}, P_{i-j}) + d(P_{i-j}, P_i)$$
$$\leqslant 101.$$

$$⑤$$

Combining ③ and ④, we have

$$N \leqslant 1225n - 101C_n^2 + 3\left(50t + \frac{101}{2}t(t - 1)\right)$$

$$= -\frac{101}{2}t^2 + \frac{5099}{2}t. \qquad ⑥$$

The right-hand side of ⑥ attains its maximum value when

$t = 25$, thus $N \leqslant 32\,175$.

Next, consider the necessary and sufficient condition of $N = 32\,175$. First, $t = 25$ means that there are 50 blue vertices. Secondly, for $1 \leqslant i \leqslant t$, we have $d(P_i, P_{i+t}) = 50$. When $d(P_i, P_{i+t}) = 50$, equality of ⑤ holds. Therefore, the number of ways to choose 50 blue vertices such that $N$ achieves maximum is equal to the number of ways to choose 25 diagonals from the longest 101 diagonals such that any two diagonals do not have common vertices.

Edging $A_i$ and $A_{i+50}$, $i = 1, 2, \ldots, 101$, we get a graph $G$. Note that 50 and 101 are relatively prime, so $1 + 50n\,(0 \leqslant n \leqslant 100)$ form a complete residue system module 101, i. e. $G$ is a circle with 101 edges. Therefore, the number of ways (written as $S$) to color 50 vertices blue and $N$ achieves maximum is equal to the number of ways to choose 25 edges of $G$ such that any two of them do not have common vertices. Now, fix one edge $e$ of $G$. Since the number of ways to choose edges having $e$ is $C_{75}^{24}$, not having $e$ is $C_{76}^{25}$, so $S = C_{75}^{24} + C_{76}^{25}$. Similarly, the number of ways to have 50 red vertices is also $S$, thus the number of ways to color the vertices such that the largest value of $N$ is achieved is $2S$.

In summary, the largest possible value of $N$ is $32\,175$, the number of ways to color is $2S = 2(C_{75}^{24} + C_{76}^{25})$. ☐

**3** Show that there are no 2 – tuples $(x, y)$ of positive integers satisfying the equation

$$(x + 1)(x + 2) \cdots (x + 2014) = (y + 1)(y + 2) \cdots (y + 4028).$$

(posed by Li Weigu)

**Proof.** For $n = 2^k \cdot m$ ($k$ is a non-negative integer, $m$ is odd), let $v(n) = 2^k$.

We prove this by contradiction. Assume $(x, y)$ is one

positive integer solution of the equation. Let

$$v(x + i) = \max_{1 \leqslant j \leqslant 2014} \{v(x + j)\}.$$

If $1 \leqslant j \leqslant 2014$, $j \neq i$, then

$$v(x + j) = v(x + i + (j - i)) = v(j - i),$$

so

$$v\left(\prod_{1 \leqslant j \leqslant 2014, j \neq i} (x + j)\right) = v((2014 - j)! \cdot (j - 1)!) \leqslant v(2013!).$$

Since $\prod_{j=1}^{2014} (x + j) = \prod_{j=1}^{4028} (y + j)$ is a multiple of $4028!$, thus

$$x + i \geqslant v(x + i) \geqslant v\left(\frac{4028!}{2013!}\right) > 2^{1007}.$$

Therefore, $x > 2^{1006}$. So $(y + 4028)^{4028} > \prod_{j=1}^{4028} (y + j) =$

$\prod_{j=1}^{2014} (x + j) > 2^{1006 \cdot 2014}$. We have $y + 4028 > 2^{503}$, $y > 2^{502}$.

**Lemma.** Let $0 \leqslant x_i < \frac{1}{2} (1 \leqslant i \leqslant n)$. If $x = \frac{1}{n} \sum_{i=1}^{n} x_i$, $y =$

$2 \max_{1 \leqslant i \leqslant n} \{x_i^2\}$, then

$$1 - x \geqslant \left(\prod_{i=1}^{n} (1 - x_i)\right)^{\frac{1}{n}} \geqslant 1 - x - y.$$

**Proof of lemma.** By the AM – GM inequality, the inequality on the left is easy to prove. The inequality on the left holds, since

$$\left(\prod_{i=1}^{n} (1 - x_i)\right)^{\frac{1}{n}} \geqslant \frac{n}{\sum_{i=1}^{n} \frac{1}{1 - x_i}} \geqslant \frac{n}{\sum_{i=1}^{n} (1 + x_i + 2x_i^2)}$$

$$= \frac{n}{n + nx + 2 \sum_{i=1}^{n} x_i^2}$$

$$\geqslant \frac{1}{1 + x + y}$$

$$\geqslant 1 - x - y.$$

The lemma is proved.

Back to the problem, let $w = x + 2015 > 2^{1007}$, $z = y + \dfrac{4029}{2}$

$> 2^{502}$; then the equation is equivalent to

$$w \cdot \left( \left(1 - \frac{1}{w}\right)\left(1 - \frac{2}{w}\right)\cdots\left(1 - \frac{2014}{w}\right) \right)^{\frac{1}{2014}}$$

$$= z^2 \cdot \left( \left(1 - \frac{1}{4z^2}\right)\left(1 - \frac{9}{4z^2}\right)\cdots\left(1 - \frac{4027^2}{4z^2}\right) \right)^{\frac{1}{2014}}.$$

By the lemma,

$$w\left(1 - \frac{2015}{2w}\right) > w \cdot \left( \left(1 - \frac{1}{w}\right)\left(1 - \frac{2}{w}\right)\cdots\left(1 - \frac{2014}{w}\right) \right)^{\frac{1}{2014}}$$

$$> w\left(1 - \frac{2015}{2w} - \frac{2 \cdot 2014^2}{w^2}\right)$$

$$> w\left(1 - \frac{2015}{2w}\right) - \frac{1}{8}.$$

Therefore, the decimal of $w \cdot \left( \left(1 - \dfrac{1}{w}\right)\left(1 - \dfrac{2}{w}\right)\cdots \right.$

$\left. \left(1 - \dfrac{2014}{w}\right) \right)^{\frac{1}{2014}}$ belongs to $\left(\dfrac{3}{8}, \dfrac{1}{2}\right)$.

On the other hand, by the lemma

$$z^2\left(1 - \frac{1^2 + 3^2 + \cdots + 4027^2}{4z^2 \cdot 2014}\right)$$

$$> z^2 \cdot \left( \left(1 - \frac{1}{4z^2}\right)\left(1 - \frac{9}{4z^2}\right)\cdots\left(1 - \frac{4027^2}{4z^2}\right) \right)^{\frac{1}{2014}}$$

$$> z^2\left(1 - \frac{1^2 + 3^2 + \cdots + 4027^2}{4z^2 \cdot 2014} - \frac{2 \cdot 4027^4}{(4z^2)^2}\right),$$

thus

$$z^2 - \frac{4 \cdot 2014^2 - 1}{12} > z^2 \cdot \left( \left(1 - \frac{1}{4z^2}\right)\left(1 - \frac{9}{4z^2}\right)\cdots\left(1 - \frac{4027^2}{4z^2}\right) \right)^{\frac{1}{2014}}$$

$$> z^2 - \frac{4 \cdot 2014^2 - 1}{12} - \frac{4027^4}{8z^2}$$

$$> z^2 - \frac{4 \cdot 2014^2 - 1}{12} - \frac{1}{8}.$$

Since $z^2 - \dfrac{4 \cdot 2014^2 - 1}{12}$ is an integer, so the decimal part of

$$z^2 \cdot \left(\left(1 - \dfrac{1}{4z^2}\right)\left(1 - \dfrac{9}{4z^2}\right)\cdots\left(1 - \dfrac{4027^2}{4z^2}\right)\right)^{\frac{1}{2014}} \text{ belongs to } \left(\dfrac{7}{8}, 1\right),$$

which is a contradiction.

Therefore, there are no 2-tuples $(x, y)$ of positive integers satisfying

$$\prod_{j=1}^{2014} (x + i) = \prod_{j=1}^{4028} (y + i). \qquad \square$$

## Second Day

### 8:00 – 12:30, March 24, 2014

**4** . Given an odd integer $k > 3$. Prove that there exist infinitely many positive integers $n$, such that there are two positive integers $d_1$, $d_2$ that all divide $\dfrac{n^2 + 1}{2}$, and $d_1 + d_2 = n + k$. (posed by Yu Hongbing)

**Solution.** Consider the Diophantine equation

$$((k - 2)^2 + 1)xy = (x + y - k)^2 + 1, \qquad ①$$

we prove that ① has infinitely many positive odd solutions $(x, y)$.

Obviously, $(1, 1)$ is a positive odd solution, let $(x_1, y_1) = (1, 1)$. Assume that $(x_i, y_i)$ is a positive odd solution of ①, and $x_i \leqslant y_i$, let $x_{i+1} = y_i$, $y_{i+1} = (k - 1)(k - 3)y_i + 2k - x_i$. Since ① can be written as

$$x^2 - ((k - 1)(k - 3)y + 2k)x + (y - k)^2 + 1 = 0,$$

by Vieta's theorem, $(x_{i+1}, y_{i+1})$ is also a integer solution of ①. Since $x_i$, $y_i$ and $k$ are all positive odd integers, and $k \geqslant 5$, so $x_{i+1}$ is a positive odd integer, and

$$y_{i+1} = (k-1)(k-3)y_i + 2k - x_i \equiv -x_i \equiv 1(\bmod 2),$$

$$y_{i+1} \geqslant 8y_i + 2k - x_i > y_i > 0.$$

Thus, $(x_{i+1}, y_{i+1})$ is a positive odd solution of ①, and $x_i + y_i < x_{i+1} + y_{i+1}$. By $(x_1, y_1)$ and the construction above, we get a series of positive odd solutions of ①: $(x_i, y_i)$, $i = 1, 2, \ldots$, such that $x_1 + y_1 < x_2 + y_2 < \cdots$.

For any integer $i$ greater than $k$, $x_i + y_i > k$. Let $n = x_i + y_i - k$, $d_1 = x_i$, $d_2 = y_i$. Then $n$ is a positive odd integer, and $d_1 + d_2 = n + k$. Since $(k-2)^2 + 1$ is even, we can show that $d_1$, $d_2$ are both divisors of $\dfrac{n^2 + 1}{2}$, and $d_1 + d_2 = n + k$. Thus such $n$

satisfies all the conditions, therefore there exist infinitely many positive odd integers $n$ satisfying the conditions. $\qquad\Box$

**5** Let $n$ be a given integer which is greater than 1. Find the greatest constant $\lambda(n)$ such that for any non-zero complex $z_1, z_2, \ldots, z_n$, we have

$$\sum_{k=1}^{n} |z_k|^2 \geqslant \lambda(n) \min_{1 \leqslant k \leqslant n} \{ |z_{k+1} - z_k|^2 \},$$

where $z_{n+1} = z_1$. (posed by Leng Gangsong)

**Solution.** Let

$$\lambda_0(n) = \begin{cases} \dfrac{n}{4}, & 2 \mid n, \\[2mm] \dfrac{n}{4\cos^2 \dfrac{\pi}{2n}}, & \text{otherwise}, \end{cases}$$

we prove $\lambda_0(n)$ is the greatest constant.

If there exists $k (1 \leqslant k \leqslant n)$ such that $|z_{k+1} - z_k| = 0$, the inequality holds obviously. So without loss of generality, we can assume that

$$\min_{1 \leqslant k \leqslant n} \{ | z_{k+1} - z_k |^2 \} = 1. \qquad ①$$

Under this assumption, it suffices to show that the minimum value of $\sum_{k=1}^{n} | z_k |^2$ is $\lambda_0(n)$.

When $n$ is even. Since

$$\sum_{k=1}^{n} | z_k |^2 = \frac{1}{2} \sum_{k=1}^{n} (| z_k |^2 + | z_{k+1} |^2) \geqslant \frac{1}{4} \sum_{k=1}^{n} | z_{k+1} - z_k |^2$$

$$\geqslant \frac{n}{4} \min_{1 \leqslant k \leqslant n} \{ | z_{k+1} - z_k |^2 \} = \frac{n}{4},$$

and equality holds when $(z_1, z_2, \ldots, z_n) = \left( \frac{1}{2}, -\frac{1}{2}, \ldots, \frac{1}{2}, -\frac{1}{2} \right)$, thus the minimum value of $\sum_{k=1}^{n} | z_k |^2$ is $\frac{n}{4} = \lambda_0(n)$.

Next, consider the condition when $n$ is odd. Let

$$\theta_k = \arg \frac{z_{k+1}}{z_k} \in [0, 2\pi), k = 1, 2, \ldots, n.$$

For all $k = 1, 2, \ldots, n$, if $\theta_k \leqslant \frac{\pi}{2}$ or $\theta_k \geqslant \frac{3\pi}{2}$, then by ①,

$$| z_k |^2 + | z_{k+1} |^2 = | z_k - z_{k+1} |^2 + 2 | z_k | | z_{k+1} | \cos \theta_k$$

$$\geqslant | z_k - z_{k+1} |^2 \geqslant 1. \qquad ②$$

If $\theta_k \in \left( \frac{\pi}{2}, \frac{3\pi}{2} \right)$, then by $\cos \theta_k < 0$ and ②,

$$1 \leqslant | z_k - z_{k+1} |^2$$

$$= | z_k |^2 + | z_{k+1} |^2 - 2 | z_k | | z_{k+1} | \cos \theta_k$$

$$\leqslant (| z_k |^2 + | z_{k+1} |^2)(1 + (-2\cos \theta_k))$$

$$= (| z_k |^2 + | z_{k+1} |^2) \cdot 2\sin^2 \frac{\theta_k}{2}.$$

Therefore,

$$|z_k|^2 + |z_{k+1}|^2 \geqslant \frac{1}{2\sin^2 \frac{\theta_k}{2}}. \qquad ③$$

Now consider the following two conditions.

(1) If for all $k (1 \leqslant k \leqslant n)$, $\theta_k \in \left(\frac{\pi}{2}, \frac{3\pi}{2}\right)$, by ③

$$\sum_{k=1}^{n} |z_k|^2 = \frac{1}{2}\sum_{k=1}^{n}(|z_k|^2 + |z_{k+1}|^2) \geqslant \frac{1}{4}\sum_{k=1}^{n} \frac{1}{\sin^2 \frac{\theta_k}{2}}. \qquad ④$$

Since $\prod_{k=1}^{n} \frac{z_{k+1}}{z_k} = \frac{z_{n+1}}{z_1} = 1$, we have

$$\sum_{k=1}^{n} \theta_k = \arg\left(\prod_{k=1}^{n} \frac{z_{k+1}}{z_k}\right) + 2m\pi = 2m\pi, \qquad ⑤$$

where $m$ is a positive integer, and $m < n$. Note that $n$ is odd, so

$$0 < \sin \frac{m\pi}{n} \leqslant \sin \frac{(n-1)\pi}{2n} = \cos \frac{\pi}{2n}. \qquad ⑥$$

Let $f(x) = \frac{1}{\sin^2 x}$, $x \in \left[\frac{\pi}{4}, \frac{3\pi}{4}\right]$. It is easy to show that $f(x)$ is a convex function. By ④ and Jensen's Inequality, and combining ⑤ and ⑥, we have

$$\sum_{k=1}^{n} |z_k|^2 \geqslant \frac{1}{4}\sum_{k=1}^{n} \frac{1}{\sin^2 \frac{\theta_k}{2}} \geqslant \frac{n}{4} \cdot \frac{1}{\sin^2\left(\frac{1}{n}\sum_{k=1}^{n} \frac{\theta_k}{2}\right)}$$

$$= \frac{n}{4} \cdot \frac{1}{\sin^2 \frac{m\pi}{n}} \geqslant \frac{n}{4} \cdot \frac{1}{\cos^2 \frac{\pi}{2n}} = \lambda_0(n).$$

(2) If there exists $j (1 \leqslant j \leqslant n)$, such that $\theta_j \notin \left(\frac{\pi}{2}, \frac{3\pi}{2}\right)$, let

$$I = \left\{j \,\Big|\, \theta_j \notin \left(\frac{\pi}{2}, \frac{3\pi}{2}\right), j = 1, 2, \ldots, n\right\}.$$

By ②, for $j \in I$, we have $|z_j|^2 + |z_{j+1}|^2 \geqslant 1$; and by ③, for $j \notin I$, we have

$$|z_j|^2 + |z_{j+1}|^2 \geqslant \frac{1}{2\sin^2 \frac{\theta_j}{2}} \geqslant \frac{1}{2}.$$

Therefore,

$$\sum_{k=1}^{n} |z_k|^2 = \frac{1}{2}\left(\sum_{j \in I}(|z_j|^2 + |z_{j+1}|^2) + \sum_{j \notin I}(|z_j|^2 + |z_{j+1}|^2)\right)$$

$$\geqslant \frac{1}{2}|I| + \frac{1}{4}(n - |I|)$$

$$= \frac{1}{4}(n + |I|) \geqslant \frac{n+1}{4}. \qquad ⑦$$

Note that

$$\frac{n+1}{4} \geqslant \frac{n}{4} \cdot \frac{1}{\cos^2 \frac{\pi}{2n}} \Leftrightarrow \cos^2 \frac{\pi}{2n} \geqslant \frac{n}{n+1}$$

$$\Leftrightarrow \sin^2 \frac{\pi}{2n} = 1 - \cos^2 \frac{\pi}{2n} \leqslant 1 - \frac{n}{n+1} = \frac{1}{n+1}.$$

The equality holds when $n = 3$; when $n \geqslant 5$,

$$\sin^2 \frac{\pi}{2n} < \left(\frac{\pi}{2n}\right)^2 < \frac{\pi^2}{2n} \cdot \frac{1}{n+1} < \frac{1}{n+1},$$

the inequality also holds. So for odd integer $n \geqslant 3$, $\frac{n+1}{4} \geqslant \frac{n}{4} \cdot \frac{1}{\cos^2 \frac{\pi}{2n}}$. Combining ⑦, we have

$$\sum_{k=1}^{n} |z_k|^2 \geqslant \frac{n}{4} \cdot \frac{1}{\cos^2 \frac{\pi}{2n}} = \lambda_0(n).$$

On the other hand, when $z_k = \frac{1}{2\cos \frac{\pi}{2n}} \cdot e^{\frac{i(n-1)k\pi}{n}}$, $k = 1$,

$2, \ldots, n$, we have $|z_k - z_{k+1}| = 1, k = 1, 2, \ldots, n$, and $\sum_{k=1}^{n} |z_k|^2$ achieves its minimum value $\lambda_0(n)$.

Summing up, the greatest $\lambda(n)$ is

$$\lambda_0(n) = \begin{cases} \dfrac{n}{4}, & 2 \mid n, \\[2mm] \dfrac{n}{4\cos^2 \dfrac{\pi}{2n}}, & \text{otherwise.} \end{cases}$$

$\qquad\qquad\qquad\qquad\qquad\qquad\qquad\qquad\qquad\qquad$ $\square$

**6** For positive integer $k > 1$, let $f(k)$ be the number of ways of factoring $k$ into product of positive integers greater than 1. (The order of factors are not countered, for example $f(12) = 4$, as 12 can be factored in these four ways: $12, 2 \times 6, 3 \times 4, 2 \times 2 \times 3$.)

Prove that, if $n$ is a positive integer greater than 1, $p$ is a prime factor of $n$, then $f(n) \leqslant \dfrac{n}{p}$. (posed by Yao Yijun)

**Solution.** We use $P(n)$ to stand for the biggest prime divisor of $n$, and define $P(1) = f(1) = 1$. We first prove two lemmas.

**Lemma 1.** For positive integer $n$ and prime $p \mid n$, we have $f(n)$ $\leqslant \sum_{d \mid \frac{n}{p}} f(d)$.

**Proof of Lemma 1.** For convenience, a factoring satisfying the condition is called a factoring for short.

For any factoring of $n$, write $n = n_1 n_2 \cdots n_k$, since $p \mid n$, so there exists $i \in \{1, \ldots, k\}$, such that $p \mid n_i$ (if there is more than one such $i$, choose any one of them), without loss of generality, assume that $i = 1$. Map this factoring to a factoring of $d = \dfrac{n}{n_1}$, $d = n_2 n_3 \cdots n_k$.

For two different factorings of $n$, $n = n_1 n_2 \cdots n_k$ and $n =$

$n'_1 n'_2 \cdots n'_l$ (where $p$ divides $n_1$ and $n'_1$). If $n_1 = n'_1$, then $d = n_2 \cdots n_k$ and $d = n'_2 \cdots n'_k$ are two different factorings of $d$ ($d$ is a divisor of $\frac{n}{p}$); if $n_1 \neq n'_1$, then $d = \frac{n}{n_1} \neq \frac{n}{n'_1} = d'$, so these two factorings map to a factoring of $d$ and $d'$, respectively ($d$ and $d'$ are divisors of $\frac{n}{p}$). Thus, $f(n) \leqslant \sum_{d \mid \frac{n}{p}} f(d)$. Lemma 1 is proved.

**Lemma 2.** For positive integer $n$, let $g(n) = \sum_{d \mid n} \dfrac{d}{P(d)}$, then $g(n) \leqslant n$.

**Proof of Lemma 2.** Induction on the number of different prime divisors of $n$. If $n = 1$, then $g(1) = 1$. If $n = p^a$, $p$ is a prime, then

$$g(n) = 1 + 1 + p + \cdots + p^{a-1} = 1 + \frac{p^a - 1}{p - 1} \leqslant 1 + p^a - 1 = n.$$

Assume that when the number of different prime divisors of $n$ is $k$, we have $g(n) \leqslant n$. Consider the situation when $n$ has $k + 1$ different prime divisors. Let the prime factorization of $n$ be $n = p_1^{a_1} \cdots p_k^{a_k} p_{k+1}^{a_{k+1}}$, where $p_1 < \cdots < p_k < p_{k+1}$, and write $n = m p_{k+1}^{a_{k+1}}$. So

$$g(n) = g(m) + \sum_{d \mid m} \sum_{i=1}^{a_{k+1}} \frac{d p_{k+1}^i}{p_{k+1}} = g(m) + \sigma(m) \frac{p_{k+1}^{a_{k+1}} - 1}{p_{k+1} - 1},$$

where $\sigma(m)$ stands for the sum of positive divisors of $m$. By assumption, $g(m) \leqslant m$, since

$$\sigma(m) \frac{p_{k+1}^{a_{k+1}} - 1}{p_{k+1} - 1} = \left( \prod_{i=1}^{k} \frac{p_i^{a_i+1} - 1}{p_i - 1} \right) \frac{p_{k+1}^{a_{k+1}} - 1}{p_{k+1} - 1}$$

$$\leqslant \left( \prod_{i=1}^{k} \frac{p_i^{a_i+1} - 1}{p_{i+1} - 1} \right) (p_{k+1}^{a_{k+1}} - 1)$$

$$\leqslant \left( \prod_{i=1}^{k} \frac{p_i^{a_i+1} - 1}{p_i} \right) (p_{k+1}^{a_{k+1}} - 1)$$

$$\leqslant \left( \prod_{i=1}^{k} p_i^{a_i} \right) (p_{k+1}^{a_{k+1}} - 1)$$

$$= n - m,$$

so $g(n) \leqslant n$. Lemma 2 is proved.

Back to the problem, it suffices to prove, for positive

integer $n$, $f(n) \leqslant \dfrac{n}{P(n)}$ holds.

By induction on $n$, if $n = 1$, the equality holds. Assume that

for $n = 1, 2, \ldots, k$, we have $f(n) \leqslant \dfrac{n}{P(n)}$. Then, if $n = k + 1$, by Lemmas 1 and 2 and the assumption,

$$f(k+1) \leqslant \sum_{d \mid \frac{k+1}{P(k+1)}} f(d) \leqslant \sum_{d \mid \frac{k+1}{P(k+1)}} \frac{d}{P(d)}$$

$$= g\left( \frac{k+1}{P(k+1)} \right) \leqslant \frac{k+1}{P(k+1)}.$$

□

# China Girls' Mathematical Olympiad

$$2010 \quad \text{Shijiazhuang}$$

## First Day 8:00～12:00

August 10, 2010

**1** Let $n$ be an integer greater than two, and let $A_1$, $A_2$, ..., $A_{2n}$ be pairwise distinct nonempty subsets of $\{1, 2, ..., n\}$. Determine the maximum value of $\sum_{i=1}^{2n} \dfrac{|A_i \cap A_{i+1}|}{|A_i| \cdot |A_{i+1}|}$.

(Here, we set $A_{2n+1} = A_1$. For a set $X$, let $|X|$ denote the number of elements in $X$.)

**Solution** The answer is $n$.

We consider each summand $s_i = \dfrac{|A_i \cap A_{i+1}|}{|A_i| \cdot |A_{i+1}|}$.

If $A_i \cap A_{i+1}$ is empty set, then $s_i = 0$.

If $A_i \cap A_{i+1}$ is nonempty, because $A_i \neq A_{i+1}$, at least one of $A_i$ and $A_{i+1}$ has more than one element, that is, $\max\{|A_i|, |A_{i+1}|\} \geqslant 2$. Because $A_i \cap A_{i+1}$ is a subset of each of $A_i$ and $A_{i+1}$, $|A_i \cap A_{i+1}| \leqslant \min\{|A_i|, |A_{i+1}|\}$ and

$$
\begin{aligned}
s_i &= \frac{|A_i \cap A_{i+1}|}{|A_i| \cdot |A_{i+1}|} \\
&\leqslant \frac{\min\{|A_i|, |A_{i+1}|\}}{\max\{|A_i|, |A_{i+1}|\} \cdot \min\{|A_i|, |A_{i+1}|\}} \leqslant \frac{1}{2}.
\end{aligned}
$$

It follows that

$$
\sum_{i=1}^{2n} \frac{|A_i \cap A_{i+1}|}{|A_i| \cdot |A_{i+1}|} \leqslant \sum_{i=1}^{2n} \frac{1}{2} = n.
$$

This upper bound can be achieved with sets

$$
A_1 = \{1\}, A_2 = \{1, 2\}, A_3 = \{2\}, A_4 = \{2, 3\}, \ldots,
$$
$$
A_{2n-2} = \{n-1, n\}, A_{2n-1} = \{n\}, A_{2n} = \{n, 1\}.
$$

□

**2** In triangle $ABC$, $AB = AC$. Point $D$ is the midpoint of side $BC$. Point $E$ lies outside the triangle $ABC$ such that $CE \perp AB$ and $BE = BD$. Let $M$ be the midpoint of segment $BE$. Point $F$ lies on the minor arc $\overset{\frown}{AD}$ of the circumcircle of triangle $ABD$ such that $MF \perp BE$. Prove that $ED \perp FD$.

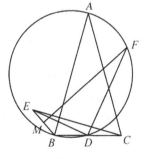

Fig. 2. 1

**Solution 1.** Construct point $F_1$ such that $EF_1 = BF_1$ and ray $DF_1$ is perpendicular to line $ED$. It suffices to show that $F = F_1$ or $ABDF_1$ is cyclic; that is,

$$\angle BAD = \angle BF_1D. \qquad \qquad ①$$

Set $\angle BAD = \angle CAD = x$. Because $EC \perp AB$ and $AD \perp BC$,

$$\angle ECB = 90° - \angle ABD = \angle BAD = x.$$

Note that $MD$ is a midline of triangle $BCE$. In particular, $MD \parallel EC$ and

$$\angle MDB = \angle ECD = x. \qquad \qquad ②$$

In isosceles triangle $EF_1M$, we may set $\angle EF_1M = \angle BF_1M = y$. Because $EM \perp MF_1$ and $MD \perp DF_1$,

$$\angle EMF_1 = \angle EDF_1 = 90°,$$

implying that $EMDF_1$ is cyclic. Consequently, we have

$$\angle EDM = \angle EF_1M = y. \qquad ③$$

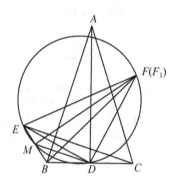

Fig. 2. 2

Combining ② and ③, we obtain

$$\angle BDE = \angle EDM + \angle MDB = x + y.$$

Because $BE = BD$, we conclude that triangle $BED$ is isosceles with $\angle MED = \angle BED = \angle BDE = x + y$. Because $EMDF_1$ is cyclic, we have $\angle MF_1D = \angle MED = x + y$. It is then clear that

$$\angle BF_1D = \angle MF_1D - \angle MF_1B = x = \angle BAD,$$

which is ①.

**Solution 2.** (Based on work by Sherry Gong and Inna Zakharevich) We maintain the notations used in Solution 1. Let $\omega$ and $O$ denote the circumcircle and the circumcenter of triangle $ABD$, and let $T$ be second intersection (other than $B$) between line $BE$ and $\omega$. Extend segment $DE$ through $E$ to meet $\omega$ at $S$. Point $F_2$ lies on $\omega$ such that $DF_2 \perp DE$. We will show that $F_2 = F$ or $F_2 B = F_2 E$. Let $M_2$ denote the foot of the perpendicular from $F_2$ to line $BE$. It suffices to show that $M_2$ is the midpoint of segment $BE$, that is, $EM_2 = M_2 B$.

Because $\angle SDF_2 = \angle EDF_2 = 90°$, $O$ is the midpoint of $SF_2$. Because $BTAD$ is cyclic, $\angle BTD = \angle BAD = \angle BCE = x$. Note also that $BD = BE$ and $\angle EBC = \angle DBT$. We conclude that triangle $BDT$ and $BEC$ are congruent to each other, implying that $BE = ET$. Hence, $OE \perp$

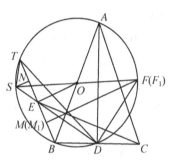

Fig. 2.3

$BT$. Let $N$ be the foot of the perpendicular from $S$ to line $BE$. Note that segments $NE$ and $EM_2$ are the respective perpendicular projections of segments $SO$ and $OF_2$ onto line $BE$. Because $SO = OF_2$, $NE = EM_2$. Because $TE = EB$, it suffices to show that $TN = NE$, which is evident since

$$\angle STE = \angle SEB = \angle SDB = \angle EDB = \angle BED = \angle SET$$

and so triangle $SET$ is isosceles with $SE = ST$. $\qquad \square$

---

**3** Prove that for every given positive integer $n$, there exist a prime $p$ and an integer $m$ such that

(a) $p \equiv 5 \pmod 6$;

(b) $p \nmid n$;

(c) $n \equiv m^3 \pmod{p}$.

**Solution 1.** There are infinitely many primes $p$ that are congruent to 5 modulo 6. (This is a special case of the Dirichlet's theorem on primes in an arithmetic progression that can be easily proven directly. There is at least one such prime (namely, 5). Suppose there were only finitely many primes congruent to 5 modulo 6, and let their product be $P$. Then 6 $P - 1$, which is larger than $P$ and congruent to 5 modulo 6, must have another prime divisor congruent to 5 modulo 6, which is a contradiction. Thus, the original assumption was wrong, and there are infinitely many odd primes that are congruent to 5 modulo 6.) In particular, one such prime $p$ is larger than $n$. Thus, $p$ satisfies both conditions (a) and (b). We can write $p = 6k + 5$ for some positive integer $k$. We set $m = n^{4k+3}$. Then the Fermat's Little Theorem, we have

$$m^3 \equiv n^{12k+9} \equiv n^{6k+4} \cdot n^{6k+4} \cdot n \equiv n^{p-1} \cdot n^{p-1} \cdot n \equiv n \pmod{p}.$$

**Solution 2.** Set $m = n - 1$. Then $m^3 - n = n^3 - 3n^2 + 2n - 1$, which is relatively prime to $n$. Any prime divisor $p$ of $m^3 - n$ is relatively prime to $n$, that is, $p$ satisfies the conditions (b) and (c). It remains to find a $p$ that is congruent to 5 modulo 6. Note that

$$m^3 - m = m(m^2 - 1) = (m - 1)m(m + 1),$$

which is divisible by 6. Hence, $m^3 - n \equiv m - n \equiv -1 \equiv 5 \pmod{6}$. Thus, there is a prime divisor $p$ of $m^3 - n$ that is congruent to 5 modulo 6, and this is the prime we sought.          $\square$

**Comment:** The original version of the problem is as follows.

Prove that for every given positive integer $n$, there exist a prime $p$ and an integer $m$ such that

    (a) $p \equiv 3 \pmod 4$;

    (b) $p \nmid n$;

    (c) $n \equiv m^2 \pmod p$.

It was inspired by a solution to the problem 3 of IMO 2010.

**4** Let $x_1, x_2, \ldots, x_n$ (where $n \geqslant 2$) be real numbers with
$$x_1^2 + x_2^2 + \cdots + x_2^2 = 1.$$

    Prove that

$$\sum_{k=1}^{n} \left(1 - \frac{k}{\sum_{i=1}^{n} i x_i^2}\right)^2 \cdot \frac{x_k^2}{k} \leqslant \left(\frac{n-1}{n+1}\right)^2 \sum_{k=1}^{n} \frac{x_k^2}{k}.$$

Determine when the equality holds.

**Comment:** Expanding the left-hand side of the desired inequality gives

$$\sum_{k=1}^{n} \left(1 - \frac{k}{\sum_{i=1}^{n} i x_i^2}\right)^2 \cdot \frac{x_k^2}{k}$$

$$= \sum_{k=1}^{n} \frac{x_k^2}{k} - \sum_{k=1}^{n} \frac{2 x_k^2}{\sum_{i=1}^{n} i x_i^2} + \sum_{k=1}^{n} \frac{k x_k^2}{\left(\sum_{i=1}^{n} i x_i^2\right)^2}$$

$$= \sum_{k=1}^{n} \frac{x_k^2}{k} - \frac{1}{\sum_{i=1}^{n} i x_i^2} \sum_{k=1}^{n} \frac{1}{2 x_k^2} + \frac{1}{\left(\sum_{i=1}^{n} i x_i^2\right)^2} \sum_{k=1}^{n} k x_k^2$$

$$= \sum_{k=1}^{n} \frac{x_k^2}{k} - \frac{2}{\sum_{i=1}^{n} i x_i^2} + \frac{1}{\sum_{i=1}^{n} i x_i^2}$$

$$= \sum_{k=1}^{n} \frac{x_k^2}{k} - \frac{1}{\sum_{i=1}^{n} i x_i^2}.$$

We want to show that

$$\sum_{k=1}^{n} \frac{x_k^2}{k} - \frac{1}{\sum_{i=1}^{n} i x_i^2} \leqslant \left(\frac{n-1}{n+2}\right)^2 \sum_{k=1}^{n} \frac{x_k^2}{k} = \sum_{k=1}^{n} \frac{x_k^2}{k} - \frac{4n}{(n+1)^2} \sum_{k=1}^{n} \frac{x_k^2}{k}$$

or

$$\frac{4n}{(n+1)^2} \sum_{k=1}^{n} \frac{x_k^2}{k} \leqslant \frac{1}{\sum_{i=1}^{n} i x_i^2},$$

that is,

$$\left( \sum_{k=1}^{n} \frac{x_k^2}{k} \right) \left( \sum_{k=1}^{n} k x_k^2 \right) \leqslant \frac{(n+1)^2}{4n}. \qquad \qquad ①$$

We present two proofs of the above inequality.

**Solution 1.** We rewrite ① as

$$4n \left( \sum_{k=1}^{n} \frac{x_k^2}{k} \right) \left( \sum_{k=1}^{n} k x_k^2 \right) \leqslant (n+1)^2.$$

By the AM – GM inequality, we have

$$4n \left( \sum_{k=1}^{n} \frac{x_k^2}{k} \right) \left( \sum_{k=1}^{n} k x_k^2 \right) = 4 \left( \sum_{k=1}^{n} \frac{n x_k^2}{k} \right) \left( \sum_{k=1}^{n} k x_k^2 \right)$$

$$\leqslant \left( \sum_{k=1}^{n} \frac{n x_k^2}{k} + \sum_{k=1}^{n} k x_k^2 \right)^2$$

$$= \left( \sum_{k=1}^{n} \left( \frac{n}{k} + k \right) x_k^2 \right)^2.$$

It suffices to show that

$$\frac{n}{k} + k \leqslant n + 1$$

or

$$0 \leqslant nk + k - k^2 - n = (n - k)(k - 1),$$

which is evident.

Now, we consider the equality case. Note that the last inequality is strict for $1 < k < n$. Hence, we must have $x_2 = \cdots = x_{n-1} = 0$. For the AM – GM inequality to hold, we must have

$$\sum_{k=1}^{n} \frac{nx_k^2}{k} = \sum_{k=1}^{n} kx_k^2 \text{ or } nx_1^2 + x_n^2 = x_1^2 + nx_n^2$$

with $x_1^2 + x_n^2 = 1$. We must have $x_1^2 = x_n^2 = \frac{1}{2}$ and $x_2 = \cdots = x_{n-1} = 0$.

**Solution 2.** (By Lynnelle Ye) We write ④ as

$$(n+1)^2 - 4n \left( \sum_{k=1}^{n} \frac{x_k^2}{k} \right) \left( \sum_{k=1}^{n} kx_k^2 \right) \geqslant 0.$$

Note that the left-hand side of the above inequality is the discriminant of the quadratic (in $t$):

$$f(t) = n(x_1^2 + 2x_2^2 + \cdots + nx_n^2)t^2 - (n+1)t + \left( x_1^2 + \frac{x_2^2}{2} + \cdots + \frac{x_n^2}{n} \right).$$

It suffices to show that $f(t)$ has a real root. Because the leading coefficient of $f(x)$ is $n(x_1^2 + 2x_2^2 + \cdots + nx_n^2)$, which is positive, it remains to be shown that $f(t) \leqslant 0$ for some $t$. Because $x_1^2 + \cdots + x_n^2 = 1$, we have

$$f(t) = n(x_1^2 + 2x_2^2 + \cdots + nx_n^2)t^2 - (n+1)(x_1^2 + x_2^2 + \cdots + x_n^2)t + \left( x_1^2 + \frac{x_2^2}{2} + \cdots + \frac{x_n^2}{n} \right)$$

$$= \sum_{k=1}^{n} \left( nkx_k^2 t^2 - (n+1)x_k^2 t + \frac{x_k^2}{k} \right)$$

$$= \sum_{k=1}^{n} x_k^2 (nkt - 1) \left( t - \frac{1}{k} \right).$$

It is easy to see that $f\left( \frac{1}{n} \right) \leqslant 0$ because for $t = \frac{1}{n}$, each summand

$$x_k^2 (nkt - 1) \left( t - \frac{1}{k} \right) = \frac{x_k^2(k-1)(k-n)}{nk} \leqslant 0,$$

completing our proof. For the equality case of the given inequality, we must have equality case for the above inequality for every $k$; that is, $x_k = 0$ for $2 \leqslant k \leqslant n - 1$. It is then not difficult to obtain that

$$x_1^2 = x_2^2 = \frac{1}{2}.$$            □

## Second Day
### 8:00 – 12:00, August 11, 2010

**5** Let $f(x)$ and $g(x)$ be strictly increasing linear functions from **R** to **R** such that $f(x)$ is an integer if and only if $g(x)$ is an integer. Prove that for any real number $x$, $f(x) - g(x)$ is an integer.

**Solution.** We can write $f(x) = ax + b$ and $g(x) = cx + d$ for some real numbers $a$, $b$, $c$, $d$ with $a$, $c > 0$.

By symmetry, we may assume that $a \geqslant c$.

We claim that $a = c$. Assume on the contrary that $a > c$. Because $a > c > 0$, the ranges of $f$ and $g$ are both **R**. There is a $x_0$ such that $f(x_0) = ax_0 + b$ is an integer. Hence $g(x_0) = cx_0 + d$ is also an integer. But then,

$$f\left(x_0 + \frac{1}{a}\right) = ax_0 + b + 1$$

and

$$g\left(x_0 + \frac{1}{a}\right) = cx_0 + b + \frac{c}{a}.$$

But this is impossible because we cannot have two integers $g(x_0)$ and $g(x_1)$ that have positive difference $\dfrac{c}{a}$ which is less than 1.

Therefore, we can write $f(x) = ax + b$ and $g(x) = ax + d$ for some real numbers $a$, $b$, $d$ with $a > 0$.

Then $b - d = f(x_0) - g(x_0)$ must be an integer, that is,

$$f(x) - g(x) = b - d$$

is an integer. □

> **6** In acute triangle $ABC$, $AB > AC$. Let $M$ be the midpoint of side $BC$. The exterior angle bisector of $\angle BAC$ meets ray $BC$ at $P$. Points $K$ and $F$ lie on

Fig. 6.1

the line $PA$ such that $MF \perp BC$ and $MK \perp PA$. Prove that $BC^2 = 4PF \cdot AK$.

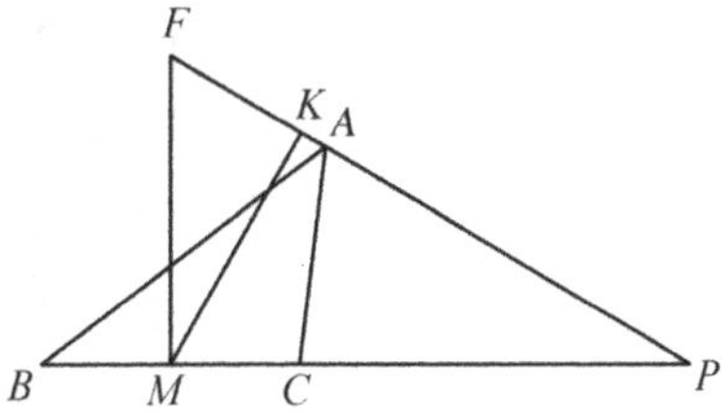

**Solution.** Let $\omega$ and $O$ denote the circumcircle and the circumcenter of triangle $ABC$, respectively, and let $N$ be the midpoint of arc $\overset{\frown}{BC}$ (not containing $A$). Note that line $MN$ is the perpendicular bisector of segment $BC$. In particular, both $F$ and $O$ lie on line $MN$. Note also that ray $AN$ is the interior bisector of $\angle BAC$ implying that $NA \perp FP$ or $\angle NAF = 90°$. It follows that $NF$ is a diameter of $\omega$. It is clear that $MK \parallel AN$, from which it follows that

$$\frac{AK}{MN} = \frac{FK}{FM}.$$

Hence,

$$PK \cdot AK = \frac{PK \cdot FK \cdot MN}{FM}.$$

Note that $MK$ is the altitude to the hypotenuse in right triangle $FMP$, implying that triangle $FMK$ and $FPM$ are similar to each other and $PF \cdot FK = FM^2$. Combining the last two equalities yields

$$PF \cdot AK = \frac{PF \cdot FK \cdot MN}{FM} = FM \cdot MN.$$

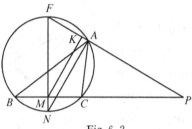

Fig. 6. 2

By the power-of-a-point theorem (or cross-chord theorem), we have $FM \cdot MN = BM \cdot MC = \dfrac{BC^2}{4}$. Combining the last two equalities gives the desired result. ☐

7 Let $n$ be an integer greater than or equal to 3. For a permutation $p = (x_1, x_2, \ldots, x_n)$ of $(1, 2, \ldots, n)$, we say that $x_j$ lies in between $x_i$ and $x_k$ if $i < j < k$. (For example, in the permutation $(1, 3, 2, 4)$, 3 lies in between 1 and 4, and 4 does not lie in between 1 and 2.) Set $S = \{p_1, p_2, \ldots, p_m\}$ consists of (distinct) permutations $p_i$ of $(1, 2, \ldots, n)$. Suppose that among every three distinct numbers in $\{1, 2, \ldots, n\}$, one of these numbers does not lie in between the other two numbers in every permutations $p_i \in S$. Determine the maximum value of $m$.

**Solution.** The answer is $2^{n-1}$.

We first show that $m \leqslant 2^{n-1}$. We induct on $n$. The base case $n = 3$ is trivial. (Indeed, say 3 does not lie in between 1 and 2, then we can have $S = \{(1, 2, 3), (3, 1, 2), (2, 1, 3), (3, 2, 1)\}$.) Assume that the statement is true for $n = k$ (where $k \geqslant 3$). Now consider $n = k + 1$ and a set $S_{k+1}$ satisfies the conditions of the problem. Note that if the element $k + 1$ is deleted from each permutation $p_i$ in $S_{k+1}$, the resulting permutations $q_i$ form a set $S_k$ that satisfies the conditions of the problem (for $n = k$). It suffices to show the following claim: there are at most two distinct permutations $p$ and $q$ in $S_{k+1}$ that can map to the same permutation $r$ in $S_k$ (by deleting the element $k + 1$ in the permutations $p$ and $q$).

Indeed, assume that for

$$p_1 = (x_1, x_2, \ldots, x_{k+1}), \quad p_2 = (y_1, y_2, \ldots, y_{k+1}),$$
$$p_3 = (z_1, z_2, \ldots, z_{k+1})$$

in $S_{k+1}$, $q_1 = q_2 = q_3 = q$. By symmetry, we may assume that $q = (1, 2, \ldots, k)$. Assume that $x_a = y_b = z_c = k + 1$. Again by symmetry, we may assume that $1 \leqslant a < b < c \leqslant k + 1$. (Note that because $q_1 = q_2 = q_3 = q$, $a$, $b$, $c$ are distinct.) We consider three numbers $a$, $b$, $k + 1$. We have $p_1 = (\ldots, k + 1, a, \ldots, b, \ldots)$ (in particular, $a$ lies in between $k + 1$ and $b$), $p_2 = (\ldots, a, \ldots, k + 1, b, \ldots)$ (in particular, $k + 1$ lies in between $a$ and $b$), and $p_3 = (\ldots, a, \ldots, b, \ldots, k + 1, \ldots)$ (in particular, $b$ lies in between $a$ and $k + 1$). Hence each one of the numbers $a$, $b$, $k + 1$ lie in between the other two numbers in some permutations in $S_k$, violating the conditions of $S_k$. Thus our assumption was wrong and at most two elements in $S_{k+1}$ can be mapped to a element in $S_k$, establishing our claim.

It remains to be shown that $m = 2^{n-1}$ is achievable. We

construct permutation $p$ inductively: (1) place 1; (2) after numbers 1, 2, ..., $l$ are placed, we place $l + 1$ either to the left or the right of the all the numbers placed so far. Because there are two possible places for each of the numbers 2, 3, ..., $n$, we can construct $2^{n-1}$ such permutations. For any three numbers $1 \leqslant a < b < c \leqslant n$, $c$ does not lie in between $a$ and $b$. Hence, this set of $2^{n-1}$ permutations satisfies the conditions of the problem, completing our proof.                                          $\square$

**Comment:** The original version of the problem is as follows.

There are $n$ books arranged in a row on a shelf. A librarian comes periodically and rearranges the books in a new order. It turns out that, among any three books, there is one that is never placed anywhere between the other two. Prove that the total number of different orders that occur is at most $2^{n-1}$.

To enhance the level of difficulty of the test paper, the problem decides to ask contestants to find this maximum value.

**8** Determine the least odd number $a > 5$ satisfying the following conditions: There are positive integers $m_1$, $m_2$, $n_1$, $n_2$ such that $a = m_1^2 + n_1^2$, $a^2 = m_2^2 + n_2^2$, and $m_1 - n_1 = m_2 - n_2$.

**Solution.** The answer is 261.

Note that

$$261 = 15^2 + 6^2, \quad 261^2 = 189^2 + 180^2, \quad 15 - 6 = 189 - 180.$$

We know that there is no number in between 5 and 261 that satisfies the condition of the problem. Assume on the contrary that $a$ is such a number. We may set $d = m_1 - n_1 > 0$. Because $a$ is odd, $m_1$ and $n_1$ have different parity, and so $d$ is odd. Because $m_1 < 261$, $d \leqslant 15$; that is, the possible values of $d$ are

1, 3, 5, 7, 9, 11, 13, 15. We will eliminate every one of them.

We can write $m_2 = n_2 + d$ and $a^2 = (n_2 + d)^2 + n_2^2$ or

$$2a^2 - d^2 = (2n_2 + d)^2. \qquad \qquad ①$$

If $d = 1$, then ① becomes a Pell's equation $x^2 - 2y^2 = -1$ with $(x, y) = (2n_2 + 1, a)$. This Pell's equation has minimal solution $(x, y) = (1, 1)$ and $x + y\sqrt{2} = (1 + \sqrt{2})^{2k-1}$ for positive integers $k$. The $y$ values of the solutions of this Pell's equations are 5, 29, 169, 985, .... Thus, the only possible values for $a$ are 29 and 169. It is easy to check that neither 29 nor 169 can be written in the form of $(n_1 + 1)^2 + n_1^2$. Hence $d \neq 1$.

If $d$ is a multiple of 3, then $m_2 \equiv n_2 \pmod 3$ and $a^2 \equiv 2m_2^2 \pmod 3$. Because 2 is not a quadratic residue modulo 3, we conclude that $0 \equiv m_2 \equiv n_2 \equiv a \pmod 3$. Hence, $m_1^2 + n_1^2 \equiv 0 \pmod 3$, from which it follows that $m_1 \equiv n_1 \equiv 0 \pmod 3$. Thus, $a = m_1^2 + n_1^2$ is a multiple of 9 and $m_2^2 + n_2^2 = a^2$ is a multiple of 81. We can write $m_2 = 3m'$, $n_2 = 3n'$, and $a = 9a'$ for integers $m_2$, $n_3$, $a'$. We have $m'^2 + n'^2 = 9a_2'$. Again, because $-1$ is a not a quadratic residue modulo 3, we must have $m' \equiv n' \equiv 0 \pmod 3$. It follows that $d = 3(m' - n')$ is divisible by 9. Thus, $d = 9$.

Because $a < 261 = 15^2 + 6^2$, $n_1 < 6$. Therefore, $n_1 = 3$ and $a = 12^2 + 3^2 = 153$. But then ① becomes $9^2 \cdot 577 = (2n_2 + 9)^2$, which is impossible. Hence, $d \neq 3$, or 9, or 15.

If $d = 11$ or $d = 13$, because 2 is not a quadratic residue modulo $d$, from $a^2 \equiv 2m_2^2 \pmod d$, we conclude that $0 \equiv m_2 \equiv n_2 \equiv a \pmod d$. It follows that $2m_1^2 \equiv a \equiv 0 \pmod d$ and $0 \equiv m_1 \equiv n_1 \equiv a \pmod d$. In particular, $n_1 \geqslant d$, $m_1 \geqslant 2d$ and

$$a = m_1^2 + n_1^2 \geqslant 5d^2 > 261.$$

Hence, $d \neq 11$ or $13$.

If $d = 5$, we also note that 2 is not a quadratic reside modulo 5. By the same reasoning before, we have

$$m_1 \equiv n_1 \equiv m_2 \equiv n_2 \equiv 0(\mathrm{mod}\,5).$$

If $n_1 \geqslant 10$, then $m_1 \geqslant 15$ and $a = m_1^2 + n_1^2 \geqslant 10^2 + 15^2 > 261$. Thus, $n_1 = 5$, $m_1 = 10$, and $a = 125$. But then ① becomes

$$5^2 \cdot 1249 = (2n_2 + 5)^2,$$

which is impossible. Hence, $d \neq 5$.

If $d = 7$, then because $a < 261 = 15^2 + 6^2$, we have $n_1 \leqslant 8$. The possible values of $a$ are then 65, 85, 109, 137, 169, 205, 245. But then ① becomes $(2n_2 + 7)^2 = 2a^2 - 49$. It is easy to check that there is no solution in this case. Hence, $d \neq 7$.

Combining the above, we conclude that 261 is the answer of this question. $\qquad\square$

# *2011* (Shenzhen, Guangdong)

The 10th China Girls' Mathematical Olympiad was held during July 28 – August 3, 2011 at the No. Three Senior High School of Shenzhen in Shenzhen, Guangdong Province, China. Around 39 teams from Mainland China, plus 9 teams from Russia, the United States, Singapore, Japan, etc., totally 188 girl students attended the competition. The competition consists of two rounds — each lasts four hours and contains four problems. The team from Shanghai High School won the first place in team total score. 20 participants won gold medals, 40 won silver medals, and 80 bronze medals.

# First Day

## 8:00 – 12:00, August 1, 2011

**1** Find all positive integers $n$ such that equation $\dfrac{1}{x} + \dfrac{1}{y} = \dfrac{1}{n}$ has exactly 2011 positive integer solutions $(x, y)$ with $x \leqslant y$. (posed by Xiong Bin)

**Solution.** From the given equation, we have $xy - nx - ny = 0 \Rightarrow (x - n)(y - n) = n^2$. Then, besides $x = y = 2n$, for any $x - n$ equal to a proper divisor of $n^2$, we will get a positive integer solution $(x, y)$ satisfying the required condition. Therefore, $n^2$ should have exactly 2010 proper divisors that are less than $n$.

Suppose $n = p_1^{\alpha_1} \cdots p_k^{\alpha_k}$, where $p_1, \ldots, p_k$ are prime numbers different from each other. Then the number of proper divisors of $n^2$ less than $n$ is

$$\frac{(2\alpha_1 + 1) \cdots (2\alpha_k + 1) - 1}{2}.$$

So $(2\alpha_1 + 1) \cdots (2\alpha_k + 1) = 4021$. Since 4021 is prime, we get $k = 1$, $2\alpha_1 + 1 = 4021$, and $\alpha_1 = 2010$.

Therefore, $n = p^{2010}$, where $p$ is any prime number. ☐

**2** As shown in Fig. 2.1, the diagonals $AC$, $BD$ of quadrilateral $ABCD$ intersect at point $E$, the midperpendiculars of $AB$, $CD$ (with $M$, $N$ being their midpoints, respectively) intersect at point $F$, and line $EF$ intersects with $BC$, $AD$ at points $P$, $Q$,

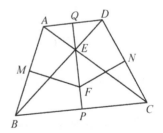

Fig. 2.1

respectively. Suppose $MF \cdot CD = NF \cdot AB$, $DQ \cdot BP = AQ \cdot CP$. Prove $PQ \perp BC$. (posed by Zheng Huan)

**Solution.** As shown in Fig. 2. 2, connect points $A - F$, $B - F$, $C - F$, and $D - F$. By the given condition, $\triangle AFB$ and $\triangle CFD$ are both isosceles triangles with $FM$ and $FN$ being the altitudes to each triangle's base.

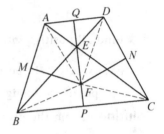

Fig. 2. 2

Since $MF \cdot CD = NF \cdot AB$, $\triangle AFB \backsim \triangle DFC$. Then $\angle AFB = \angle CFD$ and $\angle FAB = \angle FDC$. Moreover, $\angle BFD = \angle CFA$. From $FB = FA$, $FD = FC$, we have $\triangle BFD \cong \triangle AFC$, which means that $\angle FAC = \angle FBD$ and $\angle FCA = \angle FDB$. Therefore, points $A$, $B$, $F$, $E$ and points $C$, $D$, $E$, $F$ are each concyclic.

From the above result, we get

$$\angle FEB = \angle FAB = \angle FDC = \angle FEC,$$

which implies that line $EP$ is the angle bisector of $\angle BEC$. Then we have $\dfrac{EB}{EC} = \dfrac{BP}{CP}$. In the same way, we have $\dfrac{ED}{EA} = \dfrac{QD}{AQ}$.

If $DQ \cdot BP = AQ \cdot CP$, then $EB \cdot ED = EC \cdot EA$, which means $ABCD$ is a cyclic quadrilateral with $F$ as the center of its circumcircle. At this time, since

$$\angle EBC = \frac{1}{2} \angle DFC = \frac{1}{2} \angle AFB = \angle ECB,$$

we then get $PQ \perp BC$. ☐

**3** Suppose positive real numbers $a$, $b$, $c$, $d$ satisfy $abcd = 1$. Prove

$$\frac{1}{a} + \frac{1}{b} + \frac{1}{c} + \frac{1}{d} + \frac{9}{a + b + c + d} \geqslant \frac{25}{4}. \text{ (posed by Zhu}$$

Huawei)

**Solution 1.** First, we will prove that, whenever there are two numbers among $a$, $b$, $c$, $d$ that are equal, the inequality holds. We may assume that $a = b$ and let $s = a + b + c + d$. Then we have

$$\frac{1}{a} + \frac{1}{b} + \frac{1}{c} + \frac{1}{d} + \frac{9}{a + b + c + d}$$

$$= \frac{2}{a} + \frac{c + d}{cd} + \frac{9}{s} = \frac{2}{a} + a^2(s - 2a) + \frac{9}{s}$$

$$= \frac{2}{a} - 2a^3 + \left(a^2 s + \frac{9}{s}\right).$$

We define the expression above as $f(s)$. Then we see that $f(s)$ reaches the minimum for $s = \frac{3}{a}$.

When $a \geqslant \frac{\sqrt{2}}{2}$, however, we have

$$s = a + b + c + d \geqslant 2a + \frac{2}{a} \geqslant \frac{3}{a}.$$

At this time, $f(s)$ reaches the minimum for $s = 2a + \frac{2}{a}$. We then have

$$\frac{2}{a} - 2a^3 + \left(a^2 s + \frac{9}{s}\right) = \frac{2}{a} - 2a^3 + a^2\left(2a + \frac{2}{a}\right) + \frac{9}{s}$$

$$= \frac{2}{a} + 2a + \frac{9}{s} = s + \frac{9}{s}$$

$$= \frac{7}{16}s + \frac{9}{16}s + \frac{9}{s} \geqslant \frac{7}{16} \times 4 + 2\sqrt{\frac{9}{16}s \cdot \frac{9}{s}}$$

$$= \frac{7}{4} + \frac{9}{2} = \frac{25}{4}\left(\because s = 2a + \frac{2}{a} \geqslant 4\right).$$

When $0 < a < \frac{\sqrt{2}}{2}$, we have

$$\frac{2}{a} - 2a^3 + \left(a^2 s + \frac{9}{s}\right) \geq \frac{2}{a} - 2a^3 + 6a = \frac{2}{a} + 5a + (a - 2a^3)$$

$$> \frac{2}{a} + 5a \geq 2\sqrt{\frac{2}{a} \cdot 5a} = 2\sqrt{10} > \frac{25}{4}.$$

Second, we consider the case that $a$, $b$, $c$, $d$ are different from each other. We may assume that $a > b > c > d$. If $\frac{ad}{c} \cdot b \cdot c \cdot c = abcd = 1$, by using the result above, we have

$$\frac{1}{\frac{ad}{c}} + \frac{1}{b} + \frac{1}{c} + \frac{1}{c} + \frac{9}{\frac{ad}{c} + b + c + c} \geq \frac{25}{4}.$$

Therefore, we only need to prove that

$$\frac{1}{a} + \frac{1}{b} + \frac{1}{c} + \frac{1}{d} + \frac{9}{a + b + c + d}$$

$$\geq \frac{1}{\frac{ad}{c}} + \frac{1}{b} + \frac{1}{c} + \frac{1}{c} + \frac{9}{\frac{ad}{c} + b + c + c}. \qquad ①$$

We have

$$① \Leftrightarrow \frac{1}{a} + \frac{1}{d} + \frac{9}{a + b + c + d} \geq \frac{c}{ad} + \frac{1}{c} + \frac{9}{\frac{ad}{c} + b + 2c}$$

$$\Leftrightarrow \frac{ac + cd - c^2 - ad}{acd} \geq \frac{9}{(a + b + c + d)\left(\frac{ad}{c} + b + 2c\right)} \cdot$$

$$\left(a + d - \frac{ad}{c} - c\right)$$

$$\Leftrightarrow \frac{(a - c)(c - d)}{acd} \geq \frac{9}{(a + b + c + d)\left(\frac{ad}{c} + b + 2c\right)} \cdot$$

$$\frac{(a - c)(c - d)}{c}$$

$$\Leftrightarrow \frac{1}{ad} \geq \frac{9}{(a + b + c + d)\left(\frac{ad}{c} + b + 2c\right)}$$

$$\Leftrightarrow (a+b+c+d)\left(\frac{ad}{c}+b+2c\right) \geqslant 9ad$$

$$\Leftarrow \frac{ad}{c}+b+2c \geqslant \sqrt{9ad} \left(\because a+b+c+d > \frac{ad}{c}+b+2c\right)$$

$$\Leftarrow \frac{ad}{c}+3c \geqslant \sqrt{9ad}.$$

The proof is complete.                                    □

**Solution 2.** We prove it by using the adjustment method. We may assume that $a \leqslant b \leqslant c \leqslant d$, and define

$$f(a,b,c,d) = \frac{1}{a}+\frac{1}{b}+\frac{1}{c}+\frac{1}{d}+\frac{9}{a+b+c+d}.$$

Firstly, we will prove

$$f(a,b,c,d) \geqslant f(\sqrt{ac},b,\sqrt{ac},d). \qquad \textcircled{2}$$

As a matter of fact, expression ② is equivalent to

$$\frac{1}{a}+\frac{1}{c}+\frac{9}{a+b+c+d} \geqslant \frac{1}{\sqrt{ac}}+\frac{1}{\sqrt{ac}}+\frac{9}{2\sqrt{ac}+b+d}$$

$$\Leftrightarrow \frac{(\sqrt{a}-\sqrt{c})^2}{ac} \geqslant \frac{9(\sqrt{a}-\sqrt{c})^2}{(a+b+c+d)(2\sqrt{ac}+b+d)}$$

$$(\because (\sqrt{a}-\sqrt{c})^2 \geqslant 0) \qquad \textcircled{3}$$

$$\Leftarrow (a+b+c+d)(2\sqrt{ac}+b+d) \geqslant 9ac$$

$$\left(\because b+d \geqslant 2\sqrt{bd} = \frac{2}{\sqrt{ac}}\right)$$

$$\Leftarrow \left(a+c+\frac{2}{\sqrt{ac}}\right)\left(2\sqrt{ac}+\frac{2}{\sqrt{ac}}\right) \geqslant 9ac.$$

From $1 = abcd \geqslant a \cdot a \cdot c \cdot c \Rightarrow ac \leqslant 1 \Rightarrow \frac{2}{\sqrt{ac}} \geqslant 2\sqrt{ac}$ and

$a+c \geqslant 2\sqrt{ac}$, we have the following condition:

The left-hand side of ③ $\geqslant \left(2\sqrt{ac}+\frac{2}{\sqrt{ac}}\right)\left(2\sqrt{ac}+\frac{2}{\sqrt{ac}}\right)$

$\geqslant 4\sqrt{ac} \cdot 4\sqrt{ac} = 16ac > 9ac =$ the right-hand side of ③.

Therefore, ② holds, which means $f(a, b, c, d)$ (with $a \leqslant b \leqslant c \leqslant d$ ) reaches its minimum if and only if $a = c$ (i. e. $a = b = c$). So we may assume that $(a, b, c, d) = \left(\dfrac{1}{t}, \dfrac{1}{t}, \dfrac{1}{t}, t^3\right)$ ($t \geqslant 1$). Then we only need to prove that, for all $t \geqslant 1$,

$$f\left(\frac{1}{t}, \frac{1}{t}, \frac{1}{t}, t^3\right) \geqslant \frac{25}{4}. \qquad ④$$

We have

$$f\left(\frac{1}{t}, \frac{1}{t}, \frac{1}{t}, t^3\right) \geqslant \frac{25}{4}$$

$$\Leftrightarrow 3t + \frac{1}{t^3} + \frac{9}{t^3 + \dfrac{3}{t}} \geqslant \frac{25}{4} \qquad ⑤$$

$$\Leftrightarrow 12t^8 - 25t^7 + 76t^4 - 75t^3 + 12 \geqslant 0$$

$$\Leftrightarrow (t-1)^2(12t^6 - t^5 - 14t^4 - 27t^3 + 36t^2 + 24t + 12) \geqslant 0$$

$$\Leftarrow 12t^6 - t^5 - 14t^4 - 27t^3 + 36t^2 + 24t + 12 \geqslant 0$$

$$\Leftrightarrow (t-1)(12t^5 + 11t^4 - 3t^3 - 30t^2 + 6t + 30) + 42 \geqslant 0.$$

Since $t \geqslant 1$, $12t^5 + 6t \geqslant 2\sqrt{12t^5 \cdot 6t} = 12\sqrt{2}\, t^3 > 3t^3$, and

$$11t^4 + 30 \geqslant 2\sqrt{11t^4 \cdot 30} = 2\sqrt{330}\, t^2 > 30t^2,$$

then $(t-1)(12t^5 + 11t^4 - 3t^3 - 30t^2 + 6t + 30) + 42 > 0$. Therefore, ⑤ holds, which justifies ④.

The proof is complete.      □

**4**    $n(n \geqslant 3)$ table tennis players have a round-robin tournament — each player will play all the others exactly once, and there is no draw game. Suppose, after the tournament, all the players can be arranged in a circle such that, for any three players $A$, $B$, $C$, if $A$, $B$ are adjacent, then at least one of them defeated $C$. Please find all possible values of $n$. (posed by Fu Yunhao)

**Solution.** We will prove that $n$ can be any odd numbers not less than 3. Suppose $n = 2k + 1$, an odd number greater than 3, and $n$ players are represented by $A_1$, $A_2$, ..., $A_{2k+1}$. Let us arrange the competition result as follows: $A_i (1 \leqslant i \leqslant 2k + 1)$ defeated $A_{i+2}$, $A_{i+4}$, ..., $A_{i+2k}$ (we stipulate that $A_{2k+1+j} = A_j$, $j = 1$, 2, ..., $2k + 1$) but lost to the other players. Then these players can be arranged in a circle in order $A_1$, $A_2$, ..., $A_{2k+1}$, $A_1$. Now, given any three players $A$, $B$, $C$ with $A$, $B$ being adjacent in the circle, we may assume that $A = A_t$, $B = A_{t+1}$, $C = A_{t+r} (1 \leqslant t \leqslant 2k + 1, 2 \leqslant r \leqslant 2k)$. Then either $r$ or $r - 1$ is an even number not less than $2k$, which implies that at least one of the players $A$, $B$ defeated $C$.

On the other hand, suppose $n$ is an even number not less than 4, and the $n$ players can be arranged in a circle $A_1$, $A_2$, ..., $A_n$, $A_1$ that meets the required condition. We may assume that $A_1$ defeated $A_2$. According to the requirement, at least one of $A_2$, $A_3$ defeated $A_1$, and then $A_3$ defeated $A_1$ ; but at least one of $A_1$, $A_2$ defeated $A_3$, so $A_2$ defeated $A_3$, and so forth. We then get that, for any $1 \leqslant i \leqslant n$, $A_i$ defeated $A_{i+1}$ and lost to $A_{i-1}$ (stipulate that $A_{n+1} = A_1$, $A_0 = A_n$). We now divide the players after $A_i$ and before $A_{i-1}$ into $\dfrac{n-2}{2}$ pairs — each consists of two adjacent players. Then there is at least one player in each pair who defeated $A_i$, and that means, besides $A_{i-1}$, there are at least $\dfrac{n-2}{2}$ players who defeated $A_i$. Then $A_i$ lost at least $\dfrac{n}{2}$ games. So $n$ players lost totally at least $\dfrac{n^2}{2}$ games.

But the number of total games is $C_n^2 = \dfrac{n(n-1)}{2} < \dfrac{n^2}{2}$. This is a

contradiction. Therefore, the possible values of $n$ are all the odd numbers not less than 3.        □

## Second Day

### 8:00 – 12:00, August 2, 2011

**5** Given real number $\alpha$, please find the minimum real number $\lambda = \lambda(\alpha)$, such that for any complex numbers $z_1$, $z_2$ and real number $x \in [0, 1]$, if $|z_1| \leqslant \alpha |z_1 - z_2|$ then $|z_1 - xz_2| \leqslant \lambda |z_1 - z_2|$. (posed by Li Shenghong)

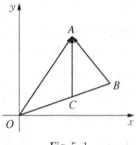

Fig. 5.1

**Solution.** As shown in the figure, in the complex plane, points $A$, $B$ and $C$ denote the complex numbers $z_1$, $z_2$ and $xz_2$, respectively. $C$ is obviously on the segment $OB$. The Vectors $\overrightarrow{BA}$ and $\overrightarrow{CA}$ are represented by complex numbers $z_1 - z_2$ and $z_1 - xz_2$, respectively. As $|z_1| \leqslant \alpha |z_1 - z_2|$, we have $|\overrightarrow{OA}| \leqslant \alpha |\overrightarrow{BA}|$. Then

$$\begin{aligned}
|z_1 - xz_2|_{\max} &= |\overrightarrow{AC}|_{\max} = \max\{|\overrightarrow{OA}|, |\overrightarrow{BA}|\} \\
&= \max\{|z_1|, |z_1 - z_2|\} \\
&= \max\{\alpha |z_1 - z_2|, |z_1 - z_2|\}.
\end{aligned}$$

Therefore, $\lambda(\alpha) = \max\{\alpha, 1\}$.        □

**6** Are there any positive integers $m$, $n$ such that $m^{20} + 11^n$ is a square number? Prove your conclusion. (posed by Yuan Hanhui)

**Solution.** Assuming there are positive integers $m$, $n$ such that $m^{20} + 11^n = k^2$ with $k \in \mathbf{Z}$, we then have

$$11^n = k^2 - m^{20} = (k - m^{10})(k + m^{10}),$$

which means that there are integers $\alpha, \beta \geqslant 0$ such that

$$\begin{cases} k - m^{10} = 11^\alpha, & \text{①} \\ k + m^{10} = 11^\beta. & \text{②} \end{cases}$$

It is obvious that $\alpha < \beta$. Subtracting ① from ②, we have

$$2m^{10} = 11^\alpha (11^{\beta-\alpha} - 1).$$

Let $m = 11^\gamma m_1$, where $\gamma, m_1 \in \mathbf{N}^*$, $11 \nmid m_1$. Then

$$11^{10\gamma} \cdot 2m_1^{10} = 11^\alpha (11^{\beta-\alpha} - 1),$$

which implies $10\gamma = \alpha$ and $2m_1^{10} = 11^{\beta-\alpha} - 1$.

By Fermat's Little Theorem, we have $m_1^{10} \equiv 1 \pmod{11}$, then $2m_1^{10} \equiv 2 \pmod{11}$. But $11^{\beta-\alpha} - 1 \equiv 10 \pmod{11}$, which is a contradiction. So the proved conclusion is that there are no positive integers $m$, $n$ such that $m^{20} + 11^n$ is a square number. $\quad\square$

**7** Suppose $n$ small balls have been placed into $n$ numbered boxes $B_1$, $B_2$, ..., $B_n$. Each time we can select a box $B_k$ and do the following operations:

(1) If $k = 1$ and there is at least one ball in $B_1$, move one ball from $B_1$ into $B_2$.

(2) If $k = n$ and there is at least one ball in $B_n$, move one ball from $B_n$ into $B_{n-1}$.

(3) If $2 \leqslant k \leqslant n - 1$ and there are at least two balls in $B_k$, move one ball from $B_k$ into $B_{k+1}$ and one ball into $B_{k-1}$, respectively.

Prove the following: no matter how the balls are

distributed among the boxes originally, it is always realizable to let each box contain exactly one ball by finite operations. (posed by Wang Xinmao)

**Solution.** For any two vectors $\mathbf{x} = (x_1, x_2, \ldots, x_n)$ and $\mathbf{y} = (y_1, y_2, \ldots, y_n)$, if there exists $1 \leqslant k \leqslant n$ such that

$$x_1 = y_1, \ldots, x_{k-1} = y_{k-1}, x_k > y_k,$$

we then denote it as $\mathbf{x} > \mathbf{y}$. Let $\mathbf{x} = (x_1, x_2, \ldots, x_n)$ represent the distribution of the balls among the boxes. Then $\mathbf{x}$ is a non-negative integer vector. The operation defined in the question, if executable, can be expressed as $\mathbf{x} + \alpha_k$, where $\alpha_1 = (-1, 1, 0, \ldots, 0)$, $\alpha_k = (\underbrace{0, \ldots, 0}_{k-2}, 1, -2, 1, 0, \ldots, 0)(2 \leqslant k \leqslant n-1)$, $\alpha_n = (0, \ldots, 0, 1, -1)$. Then for $k \geqslant 2$, we always have $\mathbf{x} + \alpha_k > \mathbf{x}$. So for any initial distribution of the balls, after a finite number of operations on every $B_k (k \geqslant 2)$ that contains at least two balls, we can arrive at a ball distribution $\mathbf{y} = (y_1, y_2, \ldots, y_n)$ satisfying $y_k \leqslant 1$ for all $k \geqslant 2$. If at this time $y_2 = \cdots = y_n = 1$, the problem is solved; otherwise, we have $y_1 \geqslant 2$. Assuming $i$ is the smallest number such that $y_i = 0$, we can then do a series of operations on $B_1, B_2, \ldots, B_{i-1}$:

$$(y_1, 1, \ldots, 1, 0, y_{i+1}, \ldots, y_n) \xrightarrow{B_1, B_2, \ldots, B_{i-1}}$$

$$(y_1, 1, \ldots, 1, 0, 1, y_{i+1}, \ldots, y_n) \xrightarrow{B_1, B_2, \ldots, B_{i-2}}$$

$$(y_1, 1, \ldots, 1, 0, 1, 1, y_{i+1}, \ldots, y_n) \to \cdots \to$$

$$(y_1, 0, 1, \ldots, 1, y_{i+1}, \ldots, y_n) \xrightarrow{B_1}$$

$$(y_1 - 1, 1, \ldots, 1, y_{i+1}, \ldots, y_n)$$

to get $(y_1 - 1, 1, \ldots, 1, y_{i+1}, \ldots, y_n)$. Repeating the operations above, we can finally arrive at the ball distribution vector that meets the requirement. $\qquad\square$

**8** As shown in Fig. 8.1, $\odot O$ is the escribed circle touching side $BC$ of $\triangle ABC$ at point $M$, and points $D$, $E$ are on the segments $AB$, $AC$, respectively, satisfying $DE \parallel BC$ ; $\odot O_1$ is the inscribed circle of $\triangle ADE$ tangent to side $DE$ at point $N$ ; $O_1B$, $DO$ intersect at point $F$, and $O_1C$, $EO$ intersect at point $G$. Please prove that $MN$ divides segment $FG$ equally. (posed by Bian Hongping)

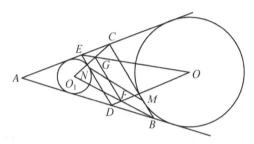

Fig. 8. 1

**Solution 1.** If $AB = AC$, then the graph is symmetric about the bisector of $\angle BAC$ and the conclusion is obvious. So we may assume that $AB > AC$. As shown in Fig. 8.2, let $L$ be the midpoint of $BC$, line $O_1L$ intersecting with $FG$ at $R$, and $O_1N$ be extended to intersect with $BC$ at $K$. Draw line $AT$ that is perpendicular to $BC$ at $T$ and intersects with line $DE$ at $S$. Connecting $AO$, obviously $O_1$ is on the segment $AO$. By Menelaus' theorem, we have

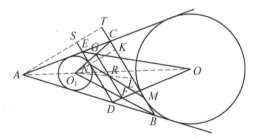

Fig. 8. 2

$$\frac{O_1F}{FB} \cdot \frac{BD}{DA} \cdot \frac{AO}{OO_1} = 1, \frac{O_1G}{GC} \cdot \frac{CE}{EA} \cdot \frac{AO}{OO_1} = 1. \qquad ①$$

Since $DE \parallel BC$, we have $\dfrac{BD}{DA} = \dfrac{CE}{EA}$. So $\dfrac{O_1F}{FB} = \dfrac{O_1G}{GC}$, which

means that $FG \parallel BC$. Then $\dfrac{FR}{GR} = \dfrac{BL}{CL} = 1$. Therefore, $R$ is the

midpoint of $FG$. We now only need to prove that $M$, $R$, $N$ are
collinear. For this purpose, by the inverse of Menelaus'
theorem, we only need to prove that

$$\frac{O_1R}{RL} \cdot \frac{LM}{MK} \cdot \frac{KN}{NO_1} = 1. \qquad ②$$

Since $FR \parallel BL$, we have $\dfrac{O_1R}{RL} = \dfrac{O_1F}{FB} = \dfrac{OO_1}{AO} \cdot \dfrac{AD}{DB}$ (the

second equality is justified by ①). So we only need to
prove that

$$\frac{OO_1}{AO} \cdot \frac{AD}{DB} \cdot \frac{LM}{MK} \cdot \frac{KN}{NO_1} = 1. \qquad ③$$

Since $O_1K \perp DE$, $OM \perp BC$, $AT \perp BC$, $DE \parallel BC$, then
lines $O_1K$, $OM$, $AT$ are parallel. By the theorem of dividing

the segments into proportional by parallel lines, we have $\dfrac{OO_1}{AO} =$

$\dfrac{MK}{MT}$. Substituting it into ③, we then only need to prove

$$\frac{AD}{DB} \cdot \frac{LM}{MT} \cdot \frac{KN}{NO_1} = 1. \qquad ④$$

Since $DE \parallel BC$, $KN \perp DE$, $ST \perp BC$, quadrilateral
$KNST$ is a rectangle and then $KN = ST$. Furthermore, from

$DS \parallel BT$ we have $\dfrac{AD}{DB} = \dfrac{AS}{ST}$. Substituting these results into ④,

we now only need to prove

$$\frac{LM}{MT} = \frac{NO_1}{AS}. \qquad \text{⑤}$$

Let $BC = a$, $AC = b$, $AB = c$. We have

$$BM = \frac{a+b-c}{2} \text{ (the property of an escribed circle)}, \quad BL = \frac{a}{2},$$

$$BT = c\cos\angle ABC = c \cdot \frac{a^2+c^2-b^2}{2ac} = \frac{a^2+c^2-b^2}{2a}.$$

Then

$$\frac{LM}{MT} = \frac{BL-BM}{BT-BM} = \frac{\dfrac{c-b}{2}}{\dfrac{c^2-b^2+a(c-b)}{2a}} = \frac{a}{a+b+c}.$$

On the other hand,

$$\frac{NO_1}{AS} = \frac{\dfrac{2S_{\triangle ADE}}{AD+DE+AE}}{\dfrac{2S_{\triangle ADE}}{DE}} = \frac{DE}{AD+DE+AE} = \frac{a}{a+b+c}.$$

Therefore, ⑤ holds. The proof is complete.

**Solution 2.** Let the radii of $\odot O$ and $\odot O_1$ be $r$ and $r_1$, respectively. Obviously, $O_1$, $O$ and $A$ are collinear.

$$\left. \begin{array}{l} DE \parallel BC \Rightarrow \dfrac{AB}{BD} = \dfrac{AC}{CE} \\[2mm] \dfrac{OF}{FD} = \dfrac{S_{\triangle BOO_1}}{S_{\triangle DBO_1}} = \dfrac{\dfrac{1}{2}\left(AB\sin\dfrac{A}{2}\right)\cdot OO_1}{\dfrac{1}{2}r_1 \cdot BD} \\[4mm] \dfrac{OG}{GE} = \dfrac{S_{\triangle OO_1C}}{S_{\triangle EOO_1}} = \dfrac{\dfrac{1}{2}\left(AC\sin\dfrac{A}{2}\right)\cdot OO_1}{\dfrac{1}{2}r_1 \cdot CE} \end{array} \right\} \Rightarrow \frac{OF}{FD} = \frac{OG}{GE}$$

$$\Rightarrow FG \parallel DE \parallel BC.$$

Connect $ON$ and extend it to intersect with $BC$ at $K$. If $\angle ABC = \angle ACB$, then by symmetry, the proposition holds. So we may assume that $\angle ABC < \angle ACB$ in the following.

We connect $OM$, $O_1M$, $OB$, $MD$, $DO_1$, respectively (see Fig. 8. 3). Since $O_1N \parallel OM$, we have

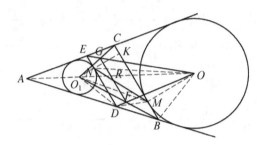

Fig. 8. 3

$$S_{\triangle ONM} = S_{\triangle MOO_1} = \frac{1}{2}r \cdot OO_1 \cdot \sin \frac{C-B}{2}, \qquad ①$$

$$\frac{OG}{GE} = \frac{OF}{DF} = \frac{S_{\triangle BOO_1}}{S_{\triangle BDO_1}} = \frac{\frac{1}{2}BO \cdot OO_1 \cdot \sin \frac{C}{2}}{\frac{1}{2} \cdot BD \cdot DO_1 \cdot \sin \frac{B}{2}}$$

$$= \frac{r \cdot OO_1 \cdot \sin \frac{C}{2}}{r_1 \cdot BD \cdot \cos \frac{B}{2}} \qquad ②$$

$$\left( r = BO \cdot \cos \frac{B}{2}, \ r_1 = DO_1 \cdot \sin \frac{B}{2} \right),$$

$$S_{\triangle DMN} - S_{\triangle MEN} = \frac{1}{2}NK \cdot DN - NE$$

$$= \frac{1}{2}BD \cdot \sin B \cdot \left( r_1 \cot \frac{B}{2} - r_1 \cot \frac{C}{2} \right). \qquad ③$$

From ② and ③, we have

$$\frac{OG}{GE} S_{\triangle DMN} - S_{\triangle MEN}$$

$$= \frac{1}{2} r_1 \cdot BD \cdot \sin B \cdot \left( \cot \frac{B}{2} - \cot \frac{C}{2} \right) \cdot \frac{r \cdot OO_1 \cdot \sin \frac{C}{2}}{r_1 \cdot BD \cdot \cos \frac{B}{2}}$$

$$= r \cdot OO_1 \cdot \sin \frac{B}{2} \cdot \sin \frac{C}{2} \cdot \left( \cot \frac{B}{2} - \cot \frac{C}{2} \right)$$

$$= r \cdot OO_1 \cdot \sin \frac{C - B}{2}.$$

Combining it with ①, we get

$$\frac{OG}{GE} (S_{\triangle DMN} - S_{\triangle MEN}) = 2 S_{\triangle MON}.$$

Since $\dfrac{OG}{GE} : 2 = \dfrac{OG}{OE} : \left( \dfrac{DF}{OD} + \dfrac{EG}{OE} \right)$, then

$$\frac{OF}{OD} \cdot S_{\triangle MND} - \frac{OG}{OE} \cdot S_{\triangle MEN} = \left( \frac{DF}{OD} + \frac{EG}{OE} \right) \cdot S_{\triangle MON},$$

$$\frac{OF \cdot S_{\triangle MND} - DF \cdot S_{\triangle MON}}{OD} = \frac{OG \cdot S_{\triangle MEN} + EG \cdot S_{\triangle MON}}{OE}. \quad ④$$

On the other hand,

$$S_{\triangle NMG} = \frac{S_{\triangle MEN} \cdot OG + EG \cdot S_{\triangle MON}}{OE}.$$

Therefore, $S_{\triangle NMG} = \dfrac{OF \cdot S_{\triangle MND} - DF \cdot S_{\triangle MON}}{OD}$.

In the same way,

$$S_{\triangle NMF} = \frac{OF \cdot S_{\triangle MND} - DF \cdot S_{\triangle MON}}{OD}.$$

By ④, we have $S_{\triangle NMG} = S_{\triangle NMF}$. Therefore, $MN$ divides segment $FG$ equally.

The proof is complete.     $\square$

## 2012 (Guangzhou, Guangdong)

### First Day
8:00 – 12:00, August, 10, 2012

**1** Let $a_1$, $a_2$, ..., $a_n$ be $n$ non-negative real numbers. Prove that

$$\frac{1}{1+a_1} + \frac{a_1}{(1+a_1)(1+a_2)} + \cdots + \frac{a_1 a_2 \cdots a_{n-1}}{(1+a_1)(1+a_2)\cdots(1+a_n)} \leqslant 1.$$

(posed by Ai Yinghua)

**Solution.** Let $a_0 = 1$. We prove the following identity:

$$\sum_{k=1}^{n}\prod_{j=1}^{k}\frac{a_{j-1}}{1+a_j} = 1 - \prod_{j=1}^{n}\frac{a_j}{1+a_j} \qquad \text{①}$$

by induction on $n$.

It is evident that ① is true for $n = 1$. Suppose that ① is true for $n-1$, $n \geqslant 2$, then for $n$,

$$\sum_{k=1}^{n}\prod_{j=1}^{k}\frac{a_{j-1}}{1+a_j} = \sum_{k=1}^{n-1}\prod_{j=1}^{k}\frac{a_{j-1}}{1+a_j} + \prod_{j=1}^{n}\frac{a_{j-1}}{1+a_j}$$

$$= 1 - \left(\prod_{j=1}^{n-1}\frac{a_j}{1+a_j} - \prod_{j=1}^{n}\frac{a_{j-1}}{1+a_j}\right)$$

$$= 1 - \prod_{j=1}^{n}\frac{a_j}{1+a_j}. \qquad \square$$

**2** As shown in the figure, circles $\Gamma_1$ and $\Gamma_2$ are tangent externally at point $T$. Points $A$ and $E$ are on circle $\Gamma_1$, lines $AB$ and $DE$ are tangent to circle $\Gamma_2$ at points $B$ and $D$, respectively. Lines $AE$ and $BD$ meet at point $P$. Prove that

(1) $\dfrac{AB}{AT} = \dfrac{ED}{ET}$ ;

(2) $\angle ATP + \angle ETP = 180°$.

(posed by Xiong Bin)

**Solution.** Let the extension of $AT$ meet circle $\Gamma_2$ at point $H$, and the extension of $ET$ meet circle $\Gamma_2$ at point $G$. Then it is easy to see that $AE \parallel GH$, thus $\triangle ATE \backsim \triangle HTG$, consequently $\dfrac{AT}{TH} = \dfrac{ET}{TG}$. And further we have $\dfrac{AH}{TH} = \dfrac{EG}{TG}$.

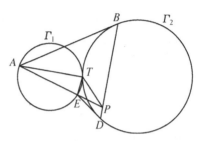

Fig. 2. 1

By the Circle Power Theorem, we have

$$\frac{AB^2}{TH^2} = \frac{AT \cdot AH}{TH \cdot TH} = \frac{ET \cdot EG}{TG \cdot TG} = \frac{ED^2}{TG^2},$$

Hence $\dfrac{TH}{TG} = \dfrac{AB}{ED}$ ;

therefore,

$$\frac{AT}{ET} = \frac{HT}{GT} = \frac{AB}{ED}.$$

By the Sine Law, we have

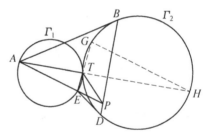

Fig. 2. 2

$$\frac{AP}{AB} = \frac{\sin \angle ABP}{\sin \angle APB} = \frac{\sin \angle EDP}{\sin \angle EPD} = \frac{EP}{ED}.$$

Hence,

$$\frac{AP}{EP} = \frac{AB}{ED} = \frac{AT}{ET},$$

that is, $PT$ is the bisector of the exterior angle of $\angle ATE$. Therefore,

$$\angle ATP + \angle ETP = 180°.$$  □

**3** Find all the pairs of integers $(a, b)$ satisfying the following condition: there exists an integer $d \geq 2$ such that $a^n + b^n + 1$ is divisible by $d$ for any positive integer $n$.

(posed by Chen Yonggao)

**Solution.** If $a + b$ is odd, then $a^n + b^n + 1$ is even, thus $d = 2$.

If $a + b$ is even, then $a^n + b^n + 1$ is odd, so $d$ is odd.

Since $d \mid a + b + 1$, $(a + b + 1)^2 = a^2 + b^2 + 1 + 2(a + b + 1) + 2(ab - 1)$, and $d \mid a^2 + b^2 + 1$, then, $d \mid 2(ab - 1)$, so we have $d \mid ab - 1$.

We see that $a^3 + b^3 + 1 \equiv (a + b)(a^2 + b^2 - ab) + 1 \equiv (-1)(-1-1) + 1 \equiv 3 \pmod{d}$, and $d \mid a^3 + b^3 + 1$, hence $d \mid 3$, so $d = 3$.

Since

$$(a - b)^2 = a^2 + b^2 - 2ab \equiv -1 - 2 \equiv 0 \pmod 3,$$

we have $a \equiv b \pmod 3$. Thus, $0 \equiv a + b + 1 \equiv 2a + 1 \pmod 3$, we see that $a \equiv 1 \pmod 3$.

Therefore, $a \equiv b \equiv 1 \pmod 3$. Consequently, for any positive integer $n$, we have

$$a^n + b^n + 1 \equiv 1 + 1 + 1 \equiv 0 \pmod 3.$$

Summing up, the required integer pairs are of the forms: $(2k, 2l + 1), (2k + 1, 2l), (3k + 1, 3l + 1)$, where $k$ and $l$ are integers.  □

**4** There is a stone (of go game) at each vertex of a given regular 13-gon, and the color of each stone is black or white. Prove that we may exchange the position of two stones such that the coloring of these stones are symmetric with respect to some symmetric axis of the 13-gon. (posed by Fu Yunhao)

**Solution.** Take any vertex $A$ and the symmetric axis $l$ passing it. There are six pairs of vertices symmetric to $l$. If the color of stones at each pair of vertices is the same, then the coloring is symmetric to $l$.

If there is only one pair of symmetric vertices that has different color stones, then exchange the stone at one vertex of the pair in different color to $A$ with the stone on $A$.

If there are exactly two pairs of vertices in different colors, then exchange the white stone in vertex of a pair with the black stone in vertex of the other pair.

Now suppose that for any vertex $A$ and the symmetric axis passing $A$, there are at least three pairs of symmetric vertices in different colors. We shall show this is impossible.

If there are $x$ black stones and $y$ white stones, then $x + y = 13$; without loss of generality, let $x$ be odd and $y$ be even.

If the stone on $A$ is in black, then the remaining stones are even in black and white, respectively. Hence, there are even pairs of vertices in different colors, that is, there are at least four pairs of vertices in different colors.

Similarly, if the stone on $A$ is in white, then there are at least three pairs of symmetric vertices in different colors.

Since each pair of vertices is symmetric to one axis, we see that the number of pairs in different colors is at least $4x + 3y$.

On the other hand, the number of pairs in different colors is exactly $xy$, thus

$$xy \geqslant 4x + 3y = x + 39,$$

that is, $x(y - 1) \geqslant 39$, but it contradicts to

$$x(y-1) \leqslant \left(\frac{x+(y-1)}{2}\right)^2 = 36.$$ □

## Second Day
### 8:00 – 12:00, August, 11, 2012

**5** As shown in Fig. 5.1, the inscribed circle $\odot I$ of $\triangle ABC$ is tangent to the sides $AB$ and $AC$ at points $D$ and $E$, respectively. And $O$ is the circumcentre of $\triangle BCI$. Prove that $\angle ODB = \angle OEC$. (posed by Zhu Huawei)

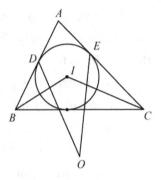

Fig. 5. 1

**Solution.** Since $O$ is the circumcentre of $\triangle BCI$, we see that

$$\angle BOI = 2\angle BCI = \angle BCA.$$

In the same manner, $\angle COI = \angle CBA$. Hence,

$$\angle BOC = \angle BOI + \angle COI = \angle BCA + \angle CBA = \pi - \angle BAC.$$

Thus, four points $A$, $B$, $O$ and $C$ are concyclic. By $OB = OC$, we know that $\angle BAO = \angle CAO$. (It can also be seen from the well-known conclusion that point $O$ is the midpoint of arc $\overset{\frown}{BC}$ (not contain point $A$) on circumcircle of $\triangle ABC$.) Combined with the fact that $AD = AE$, $AO = AO$, we have $\triangle OAD \cong \triangle OAE$. Hence, $\angle ODA = \angle OEA$, therefore, $\angle ODB = \angle OEC$. □

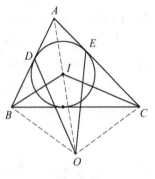

Fig. 5. 2

**6** There are $n$ cities ($n > 3$) and two airline companies in a country. Between any two cities, there is exactly one 2-way flight connecting them which is operated by one of the two companies. A female mathematician plans a travel route, so that it starts and ends at the same city, passes through at least two other cities, and each city on the route is visited once. She finds out that wherever she starts and whatever route she chooses, she must take flights of both companies. Find the maximum value of $n$.

(posed by Liang Yingde)

**Solution.** Consider each city as a vertex and each airline as an edge in a color corresponding to the airline company it belongs. Then the airline route chart of this country can be regarded as a two-color complete graph with $n$ vertices. By the problem states, any circle contains edges of both colors. That is, the subgraph of each color has no circle. It is well-known that for the simple graph without circle, the number of edges is less than the number of vertices. Thus, the number of edges of the same color is no more than $n - 1$, that is, the total number of edges is no more than $2(n - 1)$. On the other hand, the number of a complete graph with $n$ vertices is $n(n-1)/2$. Hence, $n(n-1) \leqslant 2(n - 1)$, that is, $n \leqslant 4$.

If $n = 4$, denote four cities by $A$, $B$, $C$ and $D$. Let three routes $AB$, $BC$ and $CD$ be operated by the first company and the other three routes $AC$, $AD$ and $BD$ by the second company. We see that the route chart of each company has no circle. So, the maximum number of $n$ is 4. ☐

**7** Let $a_1 \leqslant a_2 \leqslant \cdots$ be a sequence of positive integers such that $r/a_r = k + 1$ for some positive integers $k$ and $r$. Prove that there exists a positive integer $s$ such that $s/a_s = k$.

(posed by Jacek Fabrykowski, USA)

**Solution.** Let $g(t) = t - ka_t$. Then $g(r) = r - ka_r = a_r > 0$.
Note that $g(1) = 1 - ka_1 \leqslant 0$. So the set $\{t \mid t = 1, 2, \ldots, r,$
$g(t) \leqslant 0\}$ is not empty. Let $s$ be the maximal element of the
set; then $s < r$. Hence, $g(s + 1) > 0$. On the other hand,

$$g(s + 1) = s + 1 - ka_{s+1} \leqslant s + 1 - ka_s = g(s) + 1 \leqslant 1. \quad ①$$

Thus, $0 < g(s + 1) \leqslant 1$. Consequently, $g(s + 1) = 1$. And
by ①, $1 = g(s + 1) \leqslant g(s) + 1 \leqslant 1$. We have $g(s) = 0$, that is,
$s/a_s = k$. $\qquad\square$

**8** Find the number of integers $k$ in the set $\{0, 1, 2, \ldots, 2012\}$ such that the combination number $\dbinom{2012}{k} = \dfrac{2012!}{k!(2012 - k)!}$ is a multiple of 2012. (posed by Wang Bin)

**Solution.** Factorizing $2012 = 4 \times 503$, we see that $p = 503$ is a
prime. If $k$ is not a multiple of $p$, then

$$\binom{2012}{k} = \binom{4p}{k} = \frac{(4p)!}{k! \times (4p - k)!} = \frac{4p}{k} \times \binom{4p - 1}{k - 1}.$$

Hence $\dbinom{2012}{k} = \dfrac{2012!}{k!(2012 - k)!}$ is a multiple of $p$.

If $k$ is a multiple of $p$, there are only five cases: $k = 0, p,$
$2p, 3p, 4p$. And we see that in all cases, the combination
numbers

$$\binom{4p}{0} = \binom{4p}{4p} = 1,$$

$$\binom{4p}{p} = \binom{4p}{3p} = \frac{4(3p + 1)(3p + 2) \cdots (3p + (p - 1))}{(p - 1)!} \equiv 4 \pmod{p},$$

and

$$6[(2p+1)(2p+2)\cdots(2p+(p-1))]$$

$$\binom{4p}{2p} = \frac{[(2p+(p+1))(2p+(p+2))\cdots(2p+(2p-1))]}{(p-1)![(p+1)(p+2)\cdots(2p-1)]}$$

$$\equiv 6 \pmod{p}$$

are not multiple of $p$.

Now we denote the binary numeral of non-negative integer $n$ by

$$n = (a_r a_{r-1} \cdots a_0)_2 = \sum_{j=0}^{r} a_j 2^j, \text{ and } s(n) = \sum_{j=0}^{r} a_j,$$

where $a_j = 0$ or $1$ for $j = 0, 1, \ldots, r$.

Then the power $m$ of factor $2^m$ in factorization of $n$ ! can be expressed by

$$\left\lfloor \frac{n}{2} \right\rfloor + \left\lfloor \frac{n}{4} \right\rfloor + \cdots + \left\lfloor \frac{n}{2^m} \right\rfloor + \cdots$$

$$= (a_r a_{r-1} \cdots a_2 a_1)_2 + (a_r a_{r-1} \cdots a_3 a_2)_2 + \cdots + a_r$$

$$= a_r \times (2^r - 1) + a_{r-1} \times (2^{r-1} - 1) + \cdots + a_1 \times$$

$$(2^1 - 1) + a_0 \times (2^0 - 1)$$

$$= n - s(n).$$

If the combination number $\binom{2012}{k} = \dfrac{2012!}{k! \times (2012-k)!} =$

$\dfrac{(k+m)!}{k! \times m!}$ is odd ($m = 2012 - k$), then it means that the powers

of 2 in factors of the numerator and the denominator of the fraction above are the same. Then

$$k + m - s(k+m) = k - s(k) + m - s(m),$$

or

$$s(k+m) = s(k) + s(m).$$

This means that the binary addition of $k + m = 2012$ has no carrying.

Since $2012 = (11111011100)_2$, it consists of eight 1's and three 0's. If there is no carrying in the addition of $k + m = (11111011100)_2$, then on the bit of 0 in $(11111011100)_2$, $k$, $m$ are 0; on the bit of 1, one is 0, the other is 1, so there are two choices: $1 = 1 + 0$ and $1 = 0 + 1$. Thus, there are $2^8 = 256$ cases that the binary addition of two non-negative integers $k + m = 2012$ has no carrying. That is, there are 256 combination numbers that are odd, and the remaining $2013 - 256 = 1757$ combination numbers are even.

If the combination number

$$\binom{2012}{k} = \frac{2012!}{k! \times (2012 - k)!} = \frac{(k + m)!}{k! \times m!}$$

is even but not a multiple of 4, then it means that the power of 2 in the numerator is greater than that in the denominator by 1. Thus,

$$k + m - s(k + m) = k - s(k) + m - s(m) - 1,$$

or

$$s(k + m) = s(k) + s(m) - 1.$$

That is, there is only one carrying in the binary addition of $k + m = 2012$. The carrying happens at two bits as $01 + 01 = 10$. By

$$k + m = 2012 = (11111011100)_2,$$

we see that the carrying can only happen at the fifth (from the highest bit to the lowest bit) and sixth bits or at the ninth and tenth bits. So there exist $2^7 = 128$ cases. That is, there are 256 combinations whose values are even numbers but not the multiples of 4.

Thus, there are $2013 - 256 - 256 = 1501$ combinations whose values are multiples of 4. Now, we go back to consider the cases where $k$ is not the multiple of $p \doteq 503$.

We see that $\binom{4p}{0} = \binom{4p}{4p} = 1$ is not a multiple of 4. For $\binom{4p}{p} = \binom{4p}{3p} = \dfrac{(4p)!}{p!(3p)!}$, the power of 2 is $s(p) + s(3p) - s(4p) = s(3p) = 7$, and for $\binom{4p}{2p} = \dfrac{(4p)!}{(2p)!(2p)!}$, the power of 2 is $s(2p) + s(2p) - s(4p) = s(p) = 8$. Thus, there are three combination numbers which are multiples of 4 but not multiples of $p = 503$, that is, the number of $k$ such that the combinations $\binom{2012}{k}$ are multiples of 2012 is $1501 - 3 = 1498$. □

## $2013$ (Zhenhai, Zhejiang)

## First Day
(8:00 – 12:00 August 12, 2013)

**1** Let $A$ be the closed domain on the plane delimited by three lines $x = 1$, $y = 0$, and $y = t(2x - t)$, where $0 < t < 1$. Prove that the surface of any triangle inside the domain $A$ with $P(t, t^2)$ and $Q(1, 0)$ as two of its vertices cannot exceed $\dfrac{1}{4}$.

**Solution.** It is easy to observe that the domain is a closed triangle. Its three vertices are $B\left(\dfrac{t}{2}, 0\right)$, $Q(1,$

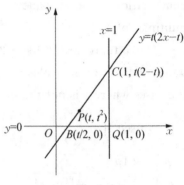

Fig. 1. 1

0) and $C(1, t(2-t))$. Pick a point $X$ inside the $\triangle BQC$, then the area of $\triangle PQX$ is equal to half of the product of $PQ$ with the distance from $X$ to $PQ$. So the area of $PQX$ takes its maximum value when the distance from $X$ to $PQ$ is maximized, i.e., when $X$ coincides with $B$ or $C$.

The area of $\triangle PQB$ is

$$\frac{1}{2}\left(1 - \frac{t}{2}\right)t^2 = \frac{1}{4}(2-t)t^2 \leqslant \frac{1}{4}(2-t)t$$

$$\leqslant \frac{1}{4}\left(\frac{2-t+t}{2}\right)^2 = \frac{1}{4};$$

the area of $\triangle PQC$ is

$$\frac{1}{2}(1-t)(2t-t^2) = \frac{1}{4}2t(1-t)(2-t)$$

$$\leqslant \frac{1}{4}\left(\frac{2t+1-t+2-t}{3}\right)^3 = \frac{1}{4}.$$

Hence, in the domain $A$, any triangle with $P$, $Q$ as two of its vertices cannot have an area that exceeds $\dfrac{1}{4}$. □

**②** As shown in Fig. 2. 1, In a trapezoid $ABCD$, $AB \parallel CD$, $\odot O_1$ is tangent to the segments $DA$, $AB$, $BC$, $\odot O_2$ is tangent to the segments $BC$, $CD$, $DA$. Let $P$ be the tangent point of $\odot O_1$ with $AB$, and $Q$ be the tangent point of $\odot O_2$ with $CD$. Prove that $AC$, $BD$, $PQ$ are concurrent.

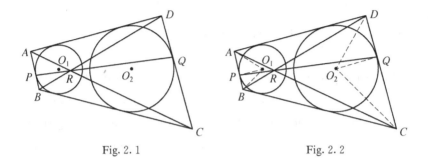

Fig. 2. 1          Fig. 2. 2

**Solution.** Let $R$ be the intersection of lines $AC$, $BD$, and join $O_1A$, $O_1B$, $O_1P$, $O_2C$, $O_2D$, $O_2Q$, $PR$, $QR$. As shown in Fig. 2. 2.

As $BA$ and $BC$ are tangent lines of $\odot O_1$,

$$\angle PBO_1 = \angle CBO_1 = \frac{1}{2}\angle ABC.$$

Similarly, we have $\angle QCO_2 = \frac{1}{2}\angle BCD$.

From $AB \parallel CD$, we know that $\angle ABC + \angle BCD = 180°$. Therefore, $\angle PBO_1 + \angle QCO_2 = 90°$, and as $\mathrm{Rt}\triangle O_1BP$ and $\mathrm{Rt}\triangle CO_2Q$ are similar, we have $\dfrac{O_1P}{BP} = \dfrac{CQ}{O_2Q}$.

Similarly, we have $\dfrac{AP}{O_1P} = \dfrac{O_2Q}{DQ}$. By multiplying these two identities, we obtain $\dfrac{AP}{BP} = \dfrac{CQ}{DQ}$, which in turn implies $\dfrac{AP}{AP+BP} = \dfrac{CQ}{CQ+DQ}$, i. e., $\dfrac{AP}{AB} = \dfrac{CQ}{CD}$.

Again by $AB \parallel CD$, we know that $\triangle ABR$ and $\triangle CDR$ are similar, so $\dfrac{AR}{AB} = \dfrac{CR}{CD}$. Comparing with $\dfrac{AP}{AB} = \dfrac{CQ}{CD}$, we have $\dfrac{AR}{AP} =$

$\dfrac{CR}{CQ}$. Meanwhile, $\triangle PAR$ is similar to $\triangle QCR$ as $\angle PAR =$

$\angle QCR$. Thus $\angle PRA = \angle QRC$. So $P$, $R$, $Q$ are collinear. Therefore, $AC$, $BD$, $PQ$ are concurrent. $\qquad\square$

**③** In a group of $m$ girls and $n$ boys, any two of them either know each other, or do not know each other. For any two boys and two girls, at least one boy and one girl do not know each other. Prove that the number of boy-girl pairs that know each other is at most $m + \dfrac{n(n-1)}{2}$.

**Solution.** From the hypothesis, for any two boys, there is at most one girl that knows both of them. Let $x_i$ be the number of girls that know exactly $i$ boys, $1 \leqslant i \leqslant n$. So $\sum_{i=1}^{n} x_i = m$. By counting the number of the above two boys-one girl combinations, we have

$$\sum_{i \geqslant 2} \dfrac{i(i-1)}{2} x_i \leqslant \dfrac{n(n-1)}{2}.$$

The number of boy-girl pairs that know each other is then

$$\sum_{i=1}^{n} i x_i = m + \sum_{i=2}^{n} (i-1) x_i \leqslant m + \sum_{i=2}^{n} \dfrac{i(i-1)}{2} x_i$$

$$\leqslant m + \dfrac{n(n-1)}{2}. \qquad\square$$

**④** Find the number of polynomials $f(x) = ax^3 + bx$ that satisfy the following conditions below:

(1) $a, b \in \{1, 2, \dots, 2013\}$;

(2) the difference of any two numbers among $f(1)$,

$f(2), \ldots, f(2013)$ is not a multiple of 2013.

**Solution.** 2013 is factorized as $2013 = 3 \times 11 \times 61$. Let $p_1 = 3$, $p_2 = 11$, $p_3 = 61$. We denote by $a_i$ the residue of $a$ modulo $p_i$, by $b_i$ the residue of $b$ modulo $p_i$ ($i = 1, 2, 3$), $a, b \in \{1, 2, \ldots, 2013\}$. By the Chinese Reminder Theorem, we have a bijection of $(a, b)$ with $(a_1, a_2, a_3, b_1, b_2, b_3)$.

Now, let $f_i(x) = a_i x^3 + b_i x$, $i = 1, 2, 3$. We call a polynomial " good modulo $n$ " if the residues of $f(0)$, $f(1), \ldots, f(n-1)$ modulo $n$ are all distinct.

If $f(x) = ax^3 + bx$ is not good modulo 2013, then there exists $x_1 \not\equiv x_2 \pmod{2013}$ such that $f(x_1) \equiv f(x_2) \pmod{2013}$. Suppose $x_1 \not\equiv x_2 \pmod{p_i}$. Let $u_1$ and $u_2$ be the residues of $x_1$ and $x_2$ modulo $p_i$, respectively. Then $u_1 \not\equiv u_2 \pmod{p_i}$ and $f_i(u_1) \equiv f_i(u_2) \pmod{p_i}$, so $f_i(x)$ is not good modulo $p_i$.

If $f(x) = ax^3 + bx$ is good modulo 2013, then for every $i$, $f_i(x)$ is good modulo $p_i$. The reason is as follows. For any distinct pair $r_1, r_2 \in \{0, 1, \ldots, p_i - 1\}$, there exist $x_1, x_2 \in \{1, 2, \ldots, 2013\}$ such that $x_1 \equiv r_1$, $x_2 \equiv r_2 \pmod{p_i}$ and $x_1 \equiv x_2 \left( \mod \dfrac{2013}{p_i} \right)$. Now $f(x_1) \equiv f(x_2) \left( \mod \dfrac{2013}{p_i} \right)$, but $f(x_1) \not\equiv f(x_2) \pmod{2013}$, so $f(r_1) \not\equiv f(r_2) \pmod{p_i}$.

Hence, we need to determine the number of good polynomials $f_i(x)$ modulo $p_i$.

For $p_1 = 3$, by Fermat's theorem, a good polynomial

$$f_1(x) \equiv a_1 x + b_1 x \equiv (a_1 + b_1) x \pmod{3}$$

is equivalent to say that $a_1 + b_1$ is not divisible by 3. There are in total six such $f_1(x)$.

For $i = 2, 3$, if $f_i(x)$ is good modulo $p_i$, then for any $u$ and $v \not\equiv 0 \pmod{p_i}$, $f_i(u + v) \not\equiv f_i(u - v) \pmod{p_i}$, i.e.,

$$f_i(u+v) - f_i(u-v) = 2v[a_i(3u^2 + v^2) + b_i]$$

is not divisible by $p_i$. If $a_i \neq 0$, the residues modulo $p_i$ of

elements in the sets $A = \left\{ 3a_i u^2 \mid u = 0, 1, \ldots, \dfrac{p_i - 1}{2} \right\}$ and

$B = \left\{ (-b_i - a_i v^2) \mid v = 1, 2, \ldots, \dfrac{p_i - 1}{2} \right\}$ do not coincide,

and $|A| + |B| = p_i$. So $A \cup B$ forms a complete residue system

modulo $p_i$. Their sum must be a multiple of $p_i$, i.e.,

$$\sum_{u=0}^{\frac{p_i-1}{2}} 3a_i u^2 + \sum_{v=1}^{\frac{p_i-1}{2}} (-b_i - a_i v^2) \equiv 0 \pmod{p_i}.$$

Now, $1^2 + 2^2 + \cdots + \left( \dfrac{p_i - 1}{2} \right)^2 = \dfrac{1}{6} \cdot \dfrac{p_i - 1}{2} \cdot \dfrac{p_i + 1}{2} \cdot p_i$ is·

a multiple of $p_i$, so $-\dfrac{p_i - 1}{2} \cdot b_i$ is also a multiple of $p_i$. Hence,

$b_i$ is divisible by $p_i$, i.e., exactly one of $a_i$, $b_i$ is 0.

If $a_i = 0$, $b_i \neq 0$, then $f_i(x) = b_i x$ is obviously good.
There are $p_i - 1$ such good polynomials.

If $a_i \neq 0$, $b_i = 0$, then $f_i(x) = a_i x^3$. For $p_2 = 11$, by
Fermat's theorem, $(x^3)^7 = x^{21} \equiv x \pmod{11}$, so for $x_1 \neq x_2 \pmod{11}$ and $x_1^3 \neq x_2^3 \pmod{11}$, $f_2(x) = a_2 x^3$ is good. There
are in total 10 such polynomials.

For $p_3 = 61$, as $4^3 = 64 \equiv 125 = 5^3 \pmod{61}$, $f_3(x) = a_3 x^3$
cannot be good.

Therefore, the total number that we are looking for is $6 \times (10 + 10) \times 60 = 7200$. $\qquad\square$

## Second Day
### 8:00 - 12:00, August 13, 2013

**5** Given positive real numbers $a_1$, $a_2$, $\ldots$, $a_n$. Prove that there exist positive real numbers $x_1$, $x_2$, ..., $x_n$ such that $\sum_{i=1}^{n} x_i = 1$, and that for any positive real numbers $y_1$, $y_2$, ..., $y_n$ satisfying $\sum_{i=1}^{n} y_i = 1$, one has

$$\sum_{i=1}^{n} \frac{a_i x_i}{x_i + y_i} \geqslant \frac{1}{2} \sum_{i=1}^{n} a_i.$$

**Solution.** Let $x_i = \dfrac{a_i}{\sum_{i=1}^{n} a_i}$. Then $\sum_{i=1}^{n} x_i = 1$. Moreover,

$$\sum_{i=1}^{n} \frac{a_i x_i}{x_i + y_i} = \sum_{i=1}^{n} a_i \sum_{i=1}^{n} \frac{x_i^2}{x_i + y_i}.$$

For any positive numbers $y_1$, $y_2$, ..., $y_n$ with $\sum_{i=1}^{n} y_i = 1$. By the Cauchy-Schwarz Inequality,

$$2 \sum_{i=1}^{n} \frac{x_i^2}{x_i + y_i} = \sum_{i=1}^{n} (x_i + y_i) \sum_{i=1}^{n} \frac{x_i^2}{x_i + y_i} \geqslant \left( \sum_{i=1}^{n} x_i \right)^2 = 1.$$

Hence,

$$\sum_{i=1}^{n} \frac{a_i x_i}{x_i + y_i} = \sum_{i=1}^{n} a_i \sum_{i=1}^{n} \frac{x_i^2}{x_i + y^i} \geqslant \frac{1}{2} \sum_{i=1}^{n} a_i. \qquad \square$$

**6** Let $S$ be a subset of $m$ elements of $\{0, 1, 2, \ldots, 98\}$, $m \geqslant 3$, such that for any $x$, $y \in S$, there exists $z \in S$ with $x + y \equiv 2z \pmod{99}$. Find all possible values of $m$.

**Solution.** Let $S = \{s_1, s_2, \ldots, s_m\}$. As $S' = \{0, s_2 - s_1, \ldots, s_m - s_1\}$ satisfies also the hypothesis, we may assume without loss of generality that $0 \in S$. For any $x$, $y \in S$, $50(x + y) \equiv z \pmod{99} \in S$. By taking $y = 0$, we have that for any $x \in S$, $50x \in S$. As 50 and 99 are coprime, there exists a positive integer $k$ such that $50^k \equiv 1 \pmod{99}$. Hence,

$$x + y \equiv 50^k (x + y) \pmod{99} \in S.$$

Let $d = \gcd(99, s_1, s_2, \ldots, s_m)$. The above argument then implies that $d \in S$, therefore $S = \{0, d, 2d, \ldots\}$. For any positive factor $d < 99$ of $99$, this $S$ satisfies all the requirements. Hence, all the possible values of $m$ are $3$, $9$, $11$, $33$, $99$. $\qquad\square$

**7** As shown in Fig. 7.1, $\odot O_1$ and $\odot O_2$ are tangent externally at point $T$. The quadrilateral $ABCD$ is inscribed in $\odot O_1$. The lines $DA$ and $CB$ are tangent to $\odot O_2$ at points $E$ and $F$, respectively. $BN$, the bisect of $\angle ABF$, intersects the segment $EF$ at point $N$. Line $FT$ intersects the arc $AT$ (which does not contain $B$) at point $M$. Prove that $M$ is the exocenter of $\triangle BCN$.

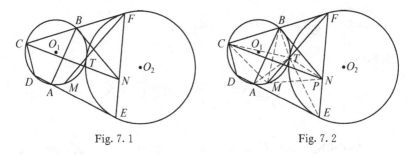

Fig. 7. 1          Fig. 7. 2

**Solution.** Let $P$ be the intersection of line $AM$ with $EF$. Join $AT$, $BM$, $BP$, $BT$, $CM$, $CT$, $ET$, $TP$. As shown in Fig. 7.2.

As $BF$ is tangent to $\odot O_2$ at $F$, we have $\angle BFT = \angle FET$. As $\odot O_1$ is tangent to $\odot O_2$ at $T$, we have $\angle MBT = \angle FET$. Hence, $\angle MBT = \angle BFM$. So $\triangle MBT$ and $\triangle MFB$ are similar, therefore $MB^2 = MT \cdot MF$. The same argument gives $MC^2 = MT \cdot MF$.

Now again from that $\odot O_1$ is tangent to $\odot O_2$ at $T$, we have $\angle MAT = \angle FET$. So $A$, $E$, $P$ and $T$ are concyclic, which implies that $\angle APT = \angle AET$. As $AE$ is tangent to $\odot O_2$ at $E$, we have $\angle AET = \angle EFT$. Thus, $\angle MPT = \angle PFM$, and $\triangle MPT$ is similar to $\triangle MF$. Therefore, $MP_2 = MT \cdot MF$.

From the above argument, we have $MC = MB = MP$, which means that $M$ is the excenter of $\triangle BCP$. So $\angle FBP = \frac{1}{2}\angle CMP$. Meanwhile, $\angle CMP = \angle CDA = \angle ABF$, and we have $\angle FBN = \frac{1}{2}\angle ABF$. Hence, $\angle FBN = \angle FBP$, i. e., $P$ and $N$ coincide. ☐

**8** Let $n \geqslant 4$ be an even number. At the vertices of a regular $n$ – gon we write in an arbitrary way $n$ distinct real numbers. Starting from one edge, we name all the edges in a clockwise way by $e_1$, $e_2$, ..., $e_n$. An edge is called "positive", if the difference of the numbers at its endpoint and its start point is positive. A set of two edges $\{e_i, e_j\}$ is called "crossing", if $2 \mid (i + j)$, and among the four, the numbers written at their vertices, the largest and the third largest ones belong to the same edge. Prove that the number of crossings and the number of positive edges have different parity.

**Solution 1.** Without loss of generality, we may assume that the numbers written on the vertices are $1, 2, \ldots, n$. Let $A$ be the number of crossings, and $B$ be the number of positive edges. We will prove that the parity of $S$ remains the same if we exchange numbers $i$ and $i + 1$. We distinguish two cases.

**Case I.** The numbers $i$ and $i + 1$ are written on adjacent vertices, i. e., the endpoints of edge $e_k$. Once we exchange $i$ and $i + 1$, the number of positive edges is modified by 1, thus the parity of $B$ is changed. On the other hand, the only two-edge subset that will become a new crossing (or change from a crossing to a non-crossing) is $\{e_{k-1}, e_{k+1}\}$ (the subindices are to be understood modulo $n$), all the other two-edge subsets will not be affected. So the parity of $A$ will change.

**Case II.** The numbers $i$ and $i + 1$ are written on non-adjacent vertices. Assume that they are written on the (common) endpoints of $e_j$, $e_{j+1}$ and of $e_k$, $e_{k+1}$, respectively. Once we exchange $i$ and $i + 1$, every positive edge will remain positive, so are non-positive ones. Therefore, $B$ is unchanged. Now, for number of crossings, if a two-edge subset does not involve at the same time $i$ and $i + 1$, then whether it is a crossing or not is not affected by the operation. So the only two-edge subsets to be considered are the two that have both $i$ and $i + 1$ written on their vertices and the sum of the edge number is even. They will both become crossing after the exchange if they are not before, and vice-versa. Hence, the parity of $A$ remains the same.

Now obviously, every pattern can be obtained from a finite number of such exchange if we start from writing $1, 2, \ldots, n$ consecutively in a clockwise way, and in the initial situation, $B = n - 1$, $A = 0$. So $A$ and $B$ have different parity.

**Solution 2.** Starting from one vertex, we denote the numbers written on the vertices by $x_1, x_2, \ldots, x_n$ in a clockwise way. We may assume that the numbers written on the endpoints of edge $e_i$ are $x_i$, $x_{i+1}$, $i = 1, 2, \ldots, n$, where $x_{n+1} = x_1$. Apparently $e_i$ is positive if and only if $x_{i+1} - x_i > 0$. Now, let $A$ be the number of positive edges, and $B$ be the number of crossings. Write

$$\beta = \prod_{i=1}^{n} (x_{i+1} - x_i).$$

As $n$ is even, the sign of $\beta$ is just $(-1)^A$.

On the other hand, $\{e_i, e_j\}$ is a crossing if and only if $2 \mid i + j$ and

$$(x_j - x_i)(x_j - x_{i+1})(x_{j+1} - x_i)(x_{j+1} - x_{i+1}) < 0.$$

We denote by $f(e_i, e_j)$ the left-hand side of above inequality, obviously $f(e_i, e_j) = f(e_j, e_i)$. This quantity is negative if and only if the two-edge set is a crossing. Now, define

$$\alpha = \prod_{\substack{1 \leqslant i < j \leqslant n \\ 2 \mid i+j}} (x_j - x_i)(x_j - x_{i+1})(x_{j+1} - x_i)(x_{j+1} - x_{i+1})$$

$$= \prod_{\substack{1 \leqslant i < j \leqslant n \\ 2 \mid i+j}} f(e_i, e_j),$$

then the sign of $\alpha$ is $(-1)^B$. Let us calculate the sign of $\alpha\beta$. For $1 \leqslant i < j \leqslant n$, consider the times of appearance, and the sign of $x_j - x_i$ in $\alpha$ and $\beta$, respectively. We distinguish several cases.

**Case I.** $j - i = 1$, $x_j - x_i$ appears once in $\beta$ with a positive sign, and once in $\alpha$. If $i > 1$, it appears in $f(e_{i-1}, e_{i+1})$ with a positive sign. If $i = 1$, $x_2 - x_1$ appears in $f(e_2, e_n)$ with a negative sign. The product of these numbers has a negative sign.

**Case II.** $2 \leqslant j - i < n - 1$, then $x_j - x_i$ does not appear in $\beta$, and appear in $\alpha$ twice. Among $(i - 1, j - 1)$ and $(i - 1, j)$, there is exactly one pair that is of the same parity, among $(i, j - 1)$ and $(i, j)$, there is also exactly one such pair. The sign of $x_j - x_i$ in each appearance is always positive. For $i = 1$, its appearance in $f(e_{j-1}, e_n)$ or $f(e_j, e_n)$ comes with a negative sign. Hence, there is a negative sign for each pair $(i, j)$ $(i = 1, j = 3, 4, \ldots, n - 1)$. This part of the product has the sign $(-1)^{n-3} = -1$.

**Case III.** $i = 1$, $j = n$, then $x_n - x_1$ appears once in $\beta$ with a negative sign, and once in $\alpha$ (in $f(e_{n-1}, e_1)$) with a positive sign. This part of the product has a negative sign.

In brief, the sign of $\alpha\beta$ is negative, i.e., $A + B$ is odd. $\square$

# China Western Mathematical Olympiad

$$2010 \quad \text{(Taiyuan Shanxi)}$$

## 28 – 29 October, 2010

**1** Suppose that $m$ and $k$ are non-negative integers, and $p = 2^{2^m} + 1$ is a prime number. Prove that

(a) $2^{2^{m+1} p^k} \equiv 1 \pmod{p^{k+1}}$;

(b) $2^{m+1} p^k$ is the smallest positive integer $n$ satisfying the congruence equation $2^n \equiv 1 \pmod{p^{k+1}}$.

**Solution.** We want to prove that $2^{2^{m+1} p^k} = p^{k+1} t_k + 1$ for some integer $t_k$ not divisible by $p$. We proceed by induction on $k$.

When $k = 0$, it follows from $2^{2^m} = p - 1$ that $2^{2^m} = (p-1)^2 = p(p-2) + 1$, in this case, $t_0 = p - 2$.

For inductive step, suppose that $2^{2^{m+1} p^k} = p^{k+1} t_k + 1$ where $k \geqslant 0$ and $p \nmid t_k$, then

$$2^{2^{m+1} p^{k+1}} = (2^{2^{m+1} p^k})^p = (p^{k+1} t_k + 1)^p$$

$$= \sum_{s=0}^{p} \binom{p}{s} (p^{k+1} t_k)^s$$

$$= 1 + p \cdot p^{k+1} t_k + \frac{p(p-1)}{2} (p^{k+1} t_k)^2 + \sum_{s=3}^{p} \binom{p}{s} (p^{k+1} t_k)^s.$$

As $k \geqslant 0$, then for any $s \geqslant 2$, we have $(k+1)s \geqslant 2(k+1) = 2k + 2 \geqslant k + 2$, so $2^{2^{m+1} p^{k+1}} = p^{k+2} t_{k+1} + 1$, where $t_{k+1} \in \mathbf{Z}_+$ and $p \nmid t_{k+1}$. It follows from mathematical induction that (a) holds.

Next, we prove (b). Write $2^{m+1} p^k = n\ell + r$, where $\ell, r \in \mathbf{Z}$ and $0 \leqslant r < n$. Then it follows from (a) that $1 \equiv 2^{2^{m+1} p^k} \equiv 2^{n\ell + r} \equiv (2^n)^\ell \cdot 2^r \equiv 2^r \pmod{p^{k+1}}$. As $0 \leqslant r < n$, it follows from the definition of $n$ that $r = 0$, i.e., $n \mid 2^{m+1} p^k$. By the Fundamental Theorem of Arithmetic, $n = 2^t p^s$. If $t \leqslant m$, then $2^{2^m p^k} = (2^{2^t p^k})^{2^{m-t}} \equiv 1 \pmod{p}$. On the other hand, $2^{2^m p^k} = (2^{2^m})^{p^k} \equiv (-1)^{p^k} \equiv -1 \pmod{p}$, which is a contradiction, and hence $t = m + 1$, i.e., $n = 2^{m+1} p^s$. $\qquad \square$

**2** As shown in Fig. 2.1, $AB$ is a diameter of a circle with center $O$. Let $C$ and $D$ be two different points on the circle on the same side of $AB$, and the lines tangent to the circle at points $C$ and $D$ meet at $E$. Segments $AD$ and $BC$ meet at $F$. Lines $EF$ and $AB$ meet

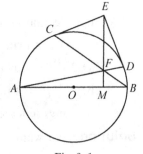

Fig. 2. 1

at $M$. Prove that $E$, $C$, $M$ and $D$ are concyclic.

**Solution 1.** As shown in Fig. 2. 2, join $OC$, $OD$ and $OE$. It follows from $\angle OCE + \angle EDO = 90° + 90° = 180°$ that $ECOD$ is cyclic. If $O = M$, then $ECMD$ is cyclic. We may assume that $O \neq M$ in the following. Let $G$ be the intersection of $BC$ and $AD$, and $H_1$ be the intersection of $GF$ and $AB$. Let $H_2$ be the foot of perpendicular of $E$ to $AB$. It follows from $AC \perp BG$ and $BD \perp AG$ that $F$ is the orthocenter of $\triangle BAG$, so $GH_2 \perp AB$. As $\angle CH_2D = \angle CBF + \angle DAF = 180° - 2\angle BGA$, and $\angle COD = 180° - \angle BOC - \angle AOD = 180° - (180° - 2\angle GBA) - (180° - 2\angle GAB) = 2(\angle GBA + \angle GAB) - 180° - 2\angle BGA$, then $\angle CH_2D = \angle COD$, so $C$, $O$, $H_2$ and $D$ are concyclic. It follows from $\angle ECO = \angle EDO = \angle EH_1O = 90°$ that $E$, $C$, $O$, $H_1$ and $D$ are concyclic. Hence, $H_1 = H_2$, i.e. $EF \perp AB$ and they meet at $M$. Consequently, $E$, $C$, $M$ and $D$ are concyclic.

**Solution 2.** As shown in Fig. 2. 2, join $EO$, $CO$, $DO$, $CA$, $CD$. It follows from $\angle COE = \angle CAF$ that $\mathrm{Rt}\triangle COE \backsim$ $\mathrm{Rt}\triangle COF$, and so $\dfrac{CE}{CF} = \dfrac{CO}{CA}$. As $\angle ECF = 90° - \angle BCO = \angle OCA$, so $\triangle ECF \backsim \triangle OCA$, and hence $\angle CFE = \angle CAO$. As $\angle BFM = \angle CFE$, so $\angle FBM + \angle BFM = \angle FBM + \angle CAO = 90°$. Hence, $EM \perp AB$, it follows that $O$, $M$, $D$, $E$, $C$ are concyclic. □

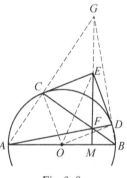

Fig. 2. 2

**3** Determine all possible values of positive integer $n$, such that there are $n$ different 3-element subsets $A_1$, $A_2$, ..., $A_n$ of the set $\{1, 2, \ldots, n\}$, with $|A_i \cap A_j| \neq 1$ for all

$i \neq j$.

**Solution.** The set of positive integers satisfying the given condition consists of all positive multiples of 4. We first prove that $n = 4k\, (k \in \mathbf{Z}_+)$ satisfying the condition. Define $A_1$, $A_2$, ... , $A_{4k}$ as follows: $A_{4i-j} = \{4i-3, 4i-2, 4i-1, 4i\} \backslash \{4i-j\}$, for all $1 \leqslant i \leqslant k$ and $0 \leqslant j \leqslant 3$.

Second, we want to prove that if $4 \nmid n$, then such $n$ subsets do not exist. Suppose to the contrary that $A_1$, ... , $A_n$ be different 3–element subsets fulfilling the condition stated in the problem. Let $A_1 = \{a, b, c\}$, consider all the given 3–element subsets with non-empty intersection, we may assume that they are $A_2$, ... , $A_m$ after relabeling. Let $U = A_1 \cup A_2 \cup \cdots \cup A_m$. We divide into the following different cases:

- If $|U| = 3$, then $m = 1 < |U|$.

- If $|U| = 4$, then $m \leqslant \dbinom{4}{3} = 4 = |U|$.

- If $|U| \geqslant 5$, then we prove that $m < |U|$ as follows.

Suppose that $m \geqslant |U|$, then for any $2 \leqslant i, j \leqslant m$, we have $|A_1 \cap A_i| = 2$, $|A_1 \cap A_j| = 2$. As $|A_1| = 3$, we have $A_i \cap A_j \neq \varnothing$. And it follows from $|A_i \cap A_j| \neq 1$ that $|A_i \cap A_j| = 2$. It means that any two distinct subsets of $A_1$, ... , $A_m$ have only two common elements.

Consider the four intersections $A_1 \cap A_2$, $A_1 \cap A_3$, $A_1 \cap A_4$, $A_1 \cap A_5$, it follows from the pigeon-hole principle that there are two intersections that are equal; by relabeling again, we may assume that they are $A_2 = \{a, b, d\}$, $A_3 = \{a, b, e\}$. Then for any $4 \leqslant i \leqslant m$, we have $a, b \in A_i$; otherwise, $A_i$ contains at least one of $a$ or $b$, and the other three elements $c$, $d$ and

$e$, in this case, $|A_i| \geqslant 4$, which is impossible. Hence $|U| = m + 2$, which contradicts to $|U| = m$.

From the argument above, one can divide the subsets $A_1, \ldots, A_n$ into several groups, and any two subsets in the same group have non-empty intersections. Moreover, the number of subsets in the same group is not more than the number of elements appeared in this group. As the number of subsets is $n$, which is equal the number of elements in $\{1, 2, \ldots, n\}$, so each group has exactly four subsets. It follows that $4 \mid n$, which contradicts the original assumption $4 \nmid n$.  □

**4** Let $a_1, a_2, \ldots, a_n, b_1, b_2, \ldots, b_n$ be non-negative numbers satisfying the following conditions simultaneously:

(1) $\sum_{i=1}^{n} (a_i + b_i) = 1$;

(2) $\sum_{i=1}^{n} i(a_i - b_i) = 0$;

(3) $\sum_{i=1}^{n} i^2(a_i + b_i) = 10$.

Prove that $\max\{a_k, b_k\} \leqslant \dfrac{10}{10 + k^2}$ for all $1 \leqslant k \leqslant n$.

**Solution.** For any $1 \leqslant k \leqslant n$, it follows from the given conditions and Cauchy's Inequality that

$$
\begin{aligned}
(ka_k)^2 &\leqslant \left( \sum_{i=1}^{n} ia_i \right)^2 = \left( \sum_{i=1}^{n} ib_i \right)^2 \\
&\leqslant \left( \sum_{i=1}^{n} i^2 b_i \right) \left( \sum_{i=1}^{n} b_i \right) \\
&= \left( 10 - \sum_{i=1}^{n} i^2 a_i \right) \left( 1 - \sum_{i=1}^{n} a_i \right) \\
&\leqslant (10 - ka^2 a_k)(1 - a_k) \\
&= 10 - (10 + k^2)a_k + k^2 a_k^2.
\end{aligned}
$$

It follows from that $a_k \leqslant \dfrac{10}{10 + k^2}$. Similarly, $b_k \leqslant \dfrac{10}{10 + k^2}$, and

hence the result follows. $\qquad\qquad\qquad\qquad\qquad\qquad\square$

**5** Let $k$ be an integer and $k > 1$. Define a sequence $\{a_n\}$ as follows: $a_0 = 0$, $a_1 = 1$ and $a_{n+1} = ka_n + a_{n-1}$ for $n = 1$, $2$, ....

Determine, with proof, all possible $k$ for which there exist non-negative integers $\ell$, $m$ ($\ell \neq m$) and positive integers $p$, $q$ such that $a_\ell + ka_p = a_m + ka_q$.

**Solution.** The answer is $k = 2$.

If $k = 2$, then $a_0 = 0$, $a_1 = 1$, $a_2 = 2$, so $a_0 + 2a_2 = a_2 + 2a_1 = 4$. Hence, $(\ell, m) = (0, 2)$ and $(p, q) = (2, 1)$.

For $k \geqslant 3$, it follows from the recurrent relation that the sequence $\{a_n\}$ is strictly increasing, and $k \mid a_{n+1} - a_{n-1}$ for all $n \geqslant 1$. In particular, for $n \geqslant 0$, we have

$$a_{2n} \equiv a_0 \equiv 0 (\mathrm{mod}\ k), \quad a_{2n+1} \equiv a_1 \equiv 1 (\mathrm{mod}\ k). \qquad (*)$$

Suppose there exist $\ell$, $m \in \mathbf{N}$, and $p$, $q \in \mathbf{Z}_+$ such that $\ell \neq m$ and $a_\ell + ka_p = a_m + ka_q$. We may assume that $\ell < m$, and divide into the following cases.

(a) $p < \ell < m$: Them $a_\ell + ka_p \leqslant a_\ell + ka_{\ell-1} < ka_\ell + a_{\ell-1} = a_{\ell+1} \leqslant a_m < a_m + ka_q$, which contradicts the original assumption.

(b) $\ell = p < m$: If $\ell = p = m - 1$, then we have $a_m + ka_q = a_\ell + ka_p = (k+1)a_{m-1}$, and modulo $k$ we have $a_m \equiv a_{m-1} \bmod k$, which contradicts ($*$).

If $\ell = p < m - 1$, then $a_\ell + ka_p < a_{m-2} + ka_{m-2} = a_m < a_m + ka_q$, which contradicts the original assumption.

(c) $\ell < p < m$: In this case, from $q$ we have $a_q > 0$, then

$a_\ell + ka_p \leqslant ka_p + a_{p-1} = a_{p+1} \leqslant a_m < a_m + ka_q$, which contradicts the original assumption.

(d) $\ell < m \leqslant p$: Then $a_p > \dfrac{a_\ell + ka_p - a_m}{k} = a_q$. It follows from

$$ka_q + a_m = ka_p + a_\ell \geqslant ka_p \qquad \text{①}$$

that

$$a_q \geqslant a_p - \frac{a_m}{k} \geqslant a_p - \frac{a_p}{k} = \frac{k-1}{k}a_p. \qquad \text{②}$$

Note that

$$a_p = ka_{p-1} + a_{p-2} \geqslant ka_{p-\ell}, \qquad \text{③}$$

hence

$$a_p > a_q \geqslant \frac{k-1}{k}a_p \geqslant (k-1)a_{p-1} \geqslant a_{p-1}. \qquad \text{④}$$

Then by increasing property of $\{a_n\}$ and ④, we have $a_q = a_{p-1}$, so inequalities ①–③ turn into equalities. It follows from ② that $m = p$, and from ③ that $p = 2$. Hence, $a_q = a_{p-1} = a_1 = 1$. And it follows from ① that $\ell = 0$. From $a_\ell + ka_p = a_m + ka_q$, we have $k^2 = k + k = 2k$, so $k = 2$, which is impossible.

So $k = 2$ is the only solution. □

**6** As shown in Fig. 6. 1, $\triangle ABC$ is a right-angled triangle, $\angle C = 90°$. Draw a circle centered at $B$ with radius $BC$. Let $D$ be a point on the side $AC$, and $DE$ be tangent to the

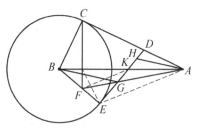

Fig. 6. 1

circle at $E$. The line through $C$ perpendicular to $AB$ meets line $BE$ at point $F$. Line $AF$ meets $DE$ at point $G$. The line through $A$ parallel to $BG$ meets $DE$ at $H$. Prove that $GE = GH$.

**Solution.** Let $K$ be the intersection of $AB$ and $DE$, and $M$ be the intersection of $AB$ and $CF$. Join $FK$, $AE$ and $ME$. In $\triangle ABC$, it follows from $CM \perp AB$ that $BM \cdot BA = BC^2 = BE^2$, so $\triangle BEM \backsim \triangle BAE$, and $\triangle BEM = \angle BAE$.

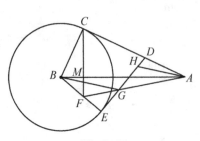

Fig. 6. 2

As $\angle FMK = \angle FEK = 90°$, so $MFEK$ is cyclic, and $\angle BEM = \angle FKM$. It follows from $\angle BAE = \angle BEM = \angle FKM$, so $FK \parallel AE$, and hence

$$\frac{KA}{KB} = \frac{EF}{BF}, \quad \text{i. e. ,} \quad \frac{KA}{KB} \cdot \frac{BF}{FE} = 1. \qquad ①$$

As the line $EGA$ intersects $\triangle EBK$, we have

$$\frac{EG}{GK} \cdot \frac{KA}{AB} \cdot \frac{BF}{FE} = 1 \qquad ②$$

As $BG \parallel AH$, we have $\dfrac{HK}{KG} = \dfrac{AK}{KB}$, so

$$\frac{HG}{GK} = \frac{AB}{BK}. \qquad ③$$

From ①-③, we have $\dfrac{EG}{HG} = 1$, and so $EG = HG$. $\qquad \square$

**7** There are $n (n \geqslant 3)$ players in a table tennis tournament, in which any two players have a match. Player $A$ is called not out-performed by player $B$, if at least one of player $A$'s losers is not a $B$'s loser.

Determine, with proof, all possible values of $n$, such that the following case could happen: after finishing all the matches, every player is not out-performed by any other player.

**Solution.** The answer is $n = 3$ or $n \geqslant 5$.

(1) For $n = 3$, suppose $A$, $B$ and $C$ are three players, and the result of three matches are as follows: $A$ wins $B$, $B$ wins $C$, and $C$ wins $A$. These results obviously satisfy the condition.

(2) If $n = 4$, suppose that the condition holds, i. e., in view of the results of all matches, every player is not out-performed by any other player. It is obvious that none of these four players wins in his three games, otherwise, the other three players will be not out-performed by this player. Similarly, none of these three players loses in his three games. It follows that each player wins one or two matches.

For the player $A$, assume that $A$ wins $B$ and $D$, but loses to $C$, then both $B$ and $D$ win $C$; otherwise, they would not out-performed $A$. For the loser in the match $B$ vs. $D$, he only wins $C$, and so the loser is impossible to be not out-performed by the winner. Consequently, for $n = 4$, the given condition could not happen.

(3) For $n = 6$, one can construct the tournament results by means of the following directed graph, in which each black dot represents a player, and $\cdot \rightarrow \circ$ represents a match with the result that player $\cdot$ wins player $\circ$.

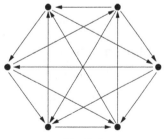

Fig. 7. 1

(4) If there exist tournament results such that each of $n$ players $A_i (1 \leqslant i \leqslant n)$ is not out-performed by any other player, we will prove that the same holds for $n + 2$ players as follows: Suppose that $M$ and $N$ are the additional players to the original $n$ players. Construct the game results of $M$ and $N$ as follows:

$$A_i \rightarrow M, \ M \rightarrow N, \ N \rightarrow A_i$$

for all $i = 1, 2, \ldots, n$, and the game results among $A_i$'s are still the original ones. Now we want to check that these $n + 2$ players satisfy the given condition. For any player $G \in \{A_1, A_2, \ldots, A_n\}$, then it suffices to consider the players $G$, $M$, $N$, and it reduces to the case $n = 3$.

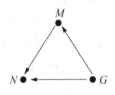

Fig. 7. 2

One can check that each of these three players $G$, $M$, $N$ is not out-performed by any one of the other two players. Hence, the tournament result of these $n + 2$ players satisfies the required condition.

In particular, it follows from (1) and the induction step of (4) that the required condition holds for any odd $n$ with $n \geqslant 3$. Moreover, it follows from (3) and the induction step of (4) that the required condition holds for any even $n$ with $n \geqslant 6$, and it completes the proof.    □

**8** Determine all possible values of integer $k$ for which there exist positive integers $a$ and $b$ such that $\dfrac{b+1}{a} + \dfrac{a+1}{b} = k$.

**Solution.** The answer is that $k = 3$ or $4$.

Fix a possible value $k$, among all the pairs $(A, B)$ of positive integers satisfying

$$\frac{B+1}{A} + \frac{A+1}{B} = k,$$

choose any $(a, b)$ such that $b$ is the smallest. Then the quadratic equation

$$x^2 + (1 - kb)x + b^2 + b = 0$$

has an integral root $x = a$. Let $x = a'$ be the second root, it follows from $a + a' = kb - 1$ that $a' \in \mathbf{Z}$, and from

$$a \cdot a' = b(b + 1)$$

that $a' > 0$. Hence, we have

$$\frac{b + 1}{a'} + \frac{a' + 1}{b} = k.$$

And it follows from the assumption on $b$ that

$$a \geqslant b, \ a' \geqslant b.$$

So one of $a$ and $a'$ is equal to $b$. Without loss of generality, we may assume $a = b$, so $k = 2 + \dfrac{2}{b}$, and so $b \mid 2$ i.e., $b = 1$ or $2$, and $k = 3$ or $4$, respectively.

If $a = b = 1$, then $k = 4$; if $a = b = 2$, then $k = 3$. Consequently, $k = 3$ or $4$ is the only solution. $\qquad\square$

# $2011$ (Yushan, Jiangxi)

## First Day

8:00 – 12:00, October 29, 2011

**1** Given that $0 < x, y < 1$, determine, with proof, the maximum value of $\dfrac{xy(1 - x - y)}{(x + y)(1 - x)(1 - y)}$. (posed by Liu

Shixiong)

**Solution.** When $x = y = \frac{1}{3}$, the value of expression is $\frac{1}{8}$.

We will prove that $\dfrac{xy(1-x-y)}{(x+y)(1-x)(1-y)} \leqslant \dfrac{1}{8}$ for any $0 < x, y < 1$ as follows.

If $x + y \geqslant 1$, then $\dfrac{xy(1-x-y)}{(x+y)(1-x)(1-y)} \leqslant 0 < \dfrac{1}{8}$.

If $x + y < 1$, then let $1 - x - y = z > 0$, it follows from AM-GM Inequality that

$$\frac{xy(1-x-y)}{(x+y)(1-x)(1-y)} = \frac{xyz}{(x+y)(y+z)(z+x)}$$

$$\leqslant \frac{xyz}{2\sqrt{xy} \cdot 2\sqrt{yz} \cdot 2\sqrt{zx}} = \frac{1}{8}.$$

In conclusion, the maximum value of the expression is $\frac{1}{8}$.

$\square$

**②** Let $M \subseteq \{1, 2, \ldots, 2011\}$ be a subset satisfying the following condition: For any three elements in $M$, there exist two of them $a$ and $b$, such that $a \mid b$ or $b \mid a$. Determine, with proof, the maximum value of $|M|$, where $|M|$ denotes the number of elements of $M$. (posed by Feng Zhigang)

**Solution.** One can check that $M = \{1, 2, 2^2, 2^3, \ldots, 2^{10}, 3, 3 \times 2, 3 \times 2^2, \ldots, 3 \times 2^9\}$ satisfies the condition, and $|M| = 21$.

Suppose that $|M| \geqslant 22$, and let $a_1 < a_2 < \cdots < a_k$ be the elements of $M$, where $|M| = k \geqslant 22$. We first prove that $a_{n+2} \geqslant 2a_n$ for all $n$; otherwise, we have $a_n < a_{n+1} < a_{n+2} < 2a_n$ for some $n < k + 2$, then any two of these three integers $a_n$, $a_{n+1}$, $a_{n+2}$ do not have any multiple relationship, which contradicts the assumption.

It follows from the inequality above that $a_4 \geqslant 2a_2 \geqslant 4$, $a_6 \geqslant 2a_4 \geqslant 8$, $\cdots a_{22} \geqslant 2a_{20} \geqslant 2^{11} > 2011$, which is a contradiction!

Hence, the maximum value of $| M |$ is 21. $\qquad\square$

**3** Let $n \geqslant 2$ be a given integer.

(1) Prove that one can arrange all the subsets of the set $\{1, 2, \ldots, n\}$ as a sequence of subsets $A_1$, $A_2$, $\ldots$, $A_{2^n}$, such that $| A_{i+1} | = | A_i | + 1$ or $| A_i | - 1$, where $i = 1, 2, \ldots, 2^n$ and $A_{2^n+1} = A_1$.

(2) Determine, with proof, all possible values of the sum $\sum_{i=1}^{2^n} (-1)^i S(A_i)$, where $S(A_i) = \sum_{x \in A_i} x$ and $S(\varnothing) = 0$, for any subset sequence $A_1$, $A_2$, $\ldots$, $A_{2^n}$ satisfying the condition in (1). (posed by Liang Yingde)

**Solution.** (1) We prove by mathematical induction that there exists a sequence $A_1$, $A_2$, $\ldots$, $A_{2^n}$ such that $A_1 = \{1\}$, $A_{2^n} = \varnothing$ and satisfies the condition in (1).

When $n = 2$, the sequence $\{1\}, \{1, 2\}, \{2\}, \varnothing$ of $\{1, 2\}$ works.

Assume that when $n = k$, there exists such a sequence $B_1$, $B_2$, $\ldots$, $B_{2^k}$ of subsets of $\{1, 2, \ldots, k\}$. As for $n = k + 1$, one can construct a sequence of subsets of $\{1, 2, \ldots, k+1\}$, as follows:

$$A_1 = B_1 = \{1\},$$
$$A_i = B_{i-1} \cup \{k+1\}, i = 2, 3, \ldots, 2^k + 1,$$
$$A_j = B_{j-2^k}, j = 2^k + 2, 2^k + 3, \ldots, 2^{k+1}.$$

One can easily check that the sequence fulfills the required conditions stated above. By induction we have proved (1) for $n \geqslant 2$.

(2) We will show that the sum is $0$, independent of the arrangement. Without loss of generality, we may assume that $A_1 = \{1\}$, otherwise shift the index cyclically. It follows from $|A_{i+1}| = |A_i| + 1$ or $|A_i| - 1$ that their parities are different, and hence the parities of the index label of any subset and its cardinality are the same.

It follows that $\sum_{i=1}^{2^n} (-1)^i S(A_i) = \sum_{A \in P} S(A) - \sum_{A \in Q} S(A)$, where $P$ consists of all subsets of $\{1, 2, \dots, n\}$ with even numbers of elements, and $Q$ consists of all subsets of $\{1, 2, \dots, n\}$ with odd numbers of elements.

For any $x \in \{1, 2, \dots, n\}$, among all $k$-element subsets, $x$ appears in exactly $C_{n-1}^{k-1}$ of them, hence it contributes to the sum $\sum_{A \in P} S(A) - \sum_{A \in Q} S(A)$ as

$$-C_{n-1}^0 + C_{n-1}^1 - C_{n-1}^2 + \cdots + (-1)^n C_{n-1}^{n-1} = -(1-1)^{n-1} = 0.$$

Therefore, $\sum_{i=1}^{2^n} (-1)^i S(A_i) = 0.$ □

**4** As shown in Fig. 4.1, $AB$ and $CD$ are two chords in the circle $\odot O$ meeting at point $E$, and $AB \neq CD$. $\odot I$ is tangent to $\odot O$ internally at point $F$, and is tangent to the chords $AB$ and $CD$ at points $G$ and $H$, respectively. $l$ is a line passing through $O$, meeting $AB$, $CD$ at points $P$, $Q$, respectively, such that $EP = EQ$. Line $EF$ meets the line $l$ at point $M$. Prove that the

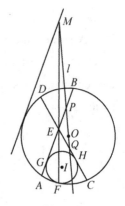

Fig. 4.1

line through $M$ and parallel to the line $AB$ is tangent to the circle $\odot O$. (posed by Li Qiusheng)

**Solution.** As shown in Fig. 4.2, draw a line parallel to $AB$ and tangent to circle $\odot O$ at point $L$, which meets the common tangent line to these two circles at a point $S$. Let $R$ be the intersection of lines $FS$ and $BA$, and join segments $LF$ and $GF$.

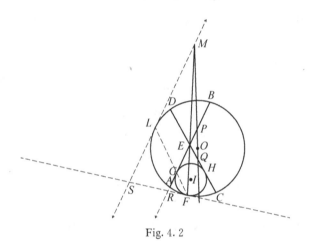

Fig. 4.2

First, we prove that the points $L$, $G$ and $F$ are collinear. As both $SL$ and $SF$ are tangent to $\odot O$, $SL = SF$; as both $RG$ and $RF$ are tangent to $\odot I$, $RG = RF$.

As $SL \parallel RG$, we have $\angle LSF = \angle GRF$, so

$$\angle LFS = \frac{180° - \angle LSF}{2} = \frac{180° - \angle GRF}{2}$$

$$= \angle GFR,$$

and hence $L$, $G$ and $F$ are collinear.

Similarly, as shown in Fig. 4.3, draw a line parallel to $CD$ and tangent to $\odot O$ at point $J$, then $F$, $H$ and $J$ are

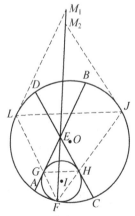

Fig. 4.3

collinear. Let tangent lines to the circle $\odot O$ at the points $L$ and $J$ meet $EF$ at points $M_1$ and $M_2$, respectively. In the following, we prove that the points $M_1$ and $M_2$ coincide. It follows from the homothety centered at $F$ mapping $\odot O$ to $\odot I$ that $LJ \parallel GH$, then $\dfrac{M_1 E}{EF} = \dfrac{LG}{GF}$

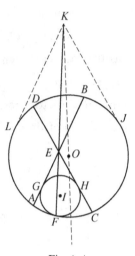

Fig. 4. 4

$= \dfrac{JH}{HF} = \dfrac{M_2 E}{EF}$, and hence $M_1$ and $M_2$ coincide. Denote this point by $K$.

Finally, we want to prove that points $M$ and $K$ also coincide. It suffices to show that $K$ lies on the line $l$.

As shown in Fig. 4. 4, join $KO$, as $KL$ and $KJ$ are tangent to $\odot O$, so $\angle LKO = \angle JKO$.

Note that $KO$ bisects $\angle LKJ$, it follows from $KL \parallel AB$ and $KJ \parallel CD$ that the line $KO$ meets lines $AB$ and $CD$ with the same angles of intersection, and hence $KO$ is just the line $l$, i. e., $K$ lies on the line $l$.

Hence, $K$ is just the intersection point $M$ of line $EF$ and $l$.

$\square$

## Second Day
### 8:00 – 12:00, October 30, 2011

**5** Determine, with proof, whether there is any odd integer $n \geqslant 3$ and $n$ distinct prime numbers $p_1$, $p_2$, ..., $p_n$ such that all $p_i + p_{i+1}$ ($i = 1, 2, \ldots, n$, and $p_{n+1} = p_1$) are perfect squares? (posed by Tao Pingsheng)

**Solution.** The answer is negative. Suppose that there exist odd integer $n \geqslant 3$ and $n$ distinct prime numbers $p_1$, $p_2$, $\ldots$, $p_n$ satisfying the given condition.

If all $p_1$, $p_2$, $\ldots$, $p_n$ are odd, then it follows from the given condition that all the sums $p_i + p_{i+1}$ are multiples of 4, so the prime numbers $p_1$, $p_2$, $\ldots$, $p_n$ modulo 4 appear to be 1 and 3 alternatively, and it contradicts the fact that $n$ is odd.

If one of $p_1$, $p_2$, $\ldots$, $p_n$ is 2, then without loss of generality, we may assume that $p_1 = 2$. As both $p_1 + p_2$ and $p_n + p_1$ are prefect squares and both are odd, it follows that $p_2$ and $p_n$ are congruent to 3 modulo 4. Similar to the discussion in the first case, we know that the primes $p_2$, $p_3$, $\ldots$, $p_n$ modulo 4 appear to be 1 and 3 alternatively, so $n - 1$ is odd, which is a contradiction.

Hence, there are no odd integer $n \geqslant 3$ and $n$ primes satisfying the given conditions. $\qquad\square$

**6** Let $a$, $b$, $c > 0$, prove that

$$\frac{(a-b)^2}{(c+a)(c+b)} + \frac{(b-c)^2}{(a+b)(a+c)} + \frac{(c-a)^2}{(b+c)(b+a)}$$
$$\geqslant \frac{(a-b)^2}{a^2+b^2+c^2}.$$

(posed by Li Shenghong)

**Solution 1.** It follows from $\dfrac{1}{2}(a - 2b)^2 + \dfrac{1}{2}(a - 2c)^2 + (b - c)^2 \geqslant 0$ that

$$3(a^2 + b^2 + c^2) \geqslant 2a^2 + 2ab + 2bc + 2ac = 2(a+b)(a+c),$$

so we have $(a + b)(a + c) \leqslant \dfrac{3}{2}(a^2 + b^2 + c^2)$. Similarly, we have

$$(b + a)(b + c) \leqslant \frac{3}{2}(a^2 + b^2 + c^2),$$

and $(c + a)(c + b) \leqslant \frac{3}{2}(a^2 + b^2 + c^2).$

Hence,

$$\frac{(a - b)^2}{(c + a)(c + b)} + \frac{(b - c)^2}{(a + b)(a + c)} + \frac{(c - a)^2}{(b + c)(b + a)}$$

$$\geqslant \frac{2}{3} \cdot \frac{(a - b)^2 + (b - c)^2 + (c - a)^2}{a^2 + b^2 + c^2}$$

$$\geqslant \frac{2}{3} \cdot \frac{(a - b)^2 + \frac{1}{2}(b - c + c - a)^2}{a^2 + b^2 + c^2}$$

$$= \frac{(a - b)^2}{a^2 + b^2 + c^2}.$$

**Solution 2.** It follows from Cauchy Inequality that

$$\left[ \frac{(a - b)^2}{(c + a)(c + b)} + \frac{(b - c)^2}{(a + b)(a + c)} + \frac{(c - a)^2}{(b + c)(b + a)} \right] \cdot$$

$$[(c + a)(c + b) + (a + b)(a + c) + (b + c)(b + a)]$$

$$\geqslant (|a - b| + |b - c| + |c - a|)^2$$

$$\geqslant (|a - b| + |b - c + c - a|)^2 = 4(a - b)^2.$$

And

$$(c + a)(c + b) + (a + b)(a + c) + (b + c)(b + a)$$

$$= (a^2 + b^2 + c^2) + 3(ab + bc + ca) \leqslant 4(a^2 + b^2 + c^2),$$

hence

$$\frac{(a - b)^2}{(c + a)(c + b)} + \frac{(b - c)^2}{(a + b)(a + c)} + \frac{(c - a)^2}{(b + c)(b + a)}$$

$$\geqslant \frac{4(a - b)^2}{(c + a)(c + b) + (a + b)(a + c) + (b + c)(b + a)}$$

$$\geqslant \frac{4(a - b)^2}{4(a^2 + b^2 + c^2)} = \frac{(a - b)^2}{a^2 + b^2 + c^2}. \qquad \square$$

**7** As shown in Fig. 7.1, $AB >$ $AC$, and the incircle $\odot I$ of $\triangle ABC$ is tangent to $BC$, $CA$ and $AB$ at points $D$, $E$ and $F$, respectively. Let $M$ be the midpoint of side $BC$, and $AH \perp BC$ at the point $H$. The bisector $AI$ of $\angle BAC$ intersects the lines $DE$, $DF$ at points $K$, $L$, respectively.

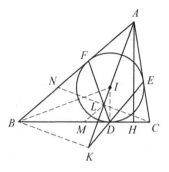

Fig. 7.1

Prove that $M$, $L$, $H$ and $K$ are concyclic. (posed by Bian Hongping)

**Solution.** Join $CL$, $BI$, $DI$, $BK$, $ML$ and $KH$. Extend $CL$ to meet $AB$ at point $N$.

As both $CD$ and $CE$ are the tangent to $\odot I$, so $CD = CE$.

As

$$\angle BIK = \angle BAI + \angle ABI = \frac{1}{2}(\angle BAC + \angle ABC)$$

$$= \frac{1}{2}(180° - \angle ACB) = \angle EDC = \angle BD,$$

so $B$, $K$, $D$ and $I$ are cyclic.

As $\angle BKI = \angle BDI = 90°$, i. e., $BK \perp AK$; similarly, $CL \perp AL$. As $AL$ is the bisector of $\angle BAC$, so $L$ is the midpoint of $CN$. As $M$ is the midpoint of $BC$, so $ML \parallel AB$.

Since $\angle BKA = \angle BHA = 90°$, it follows that points $B$, $K$, $H$ and $A$ are cyclic, so $\angle MHK = \angle BAK = \angle MLK$, hence $M$, $L$, $H$ and $K$ are cyclic. $\qquad \square$

**8** Determine, with proof, all pairs $(a, b)$ of integers, such that for any positive integer $n$, one has $n \mid (a^n + b^{n+1})$.

(posed by Chen Yonggao)

**Solution.** The solution pairs consist of $(0, 0)$ and $(-1, -1)$.

If one of $a$ and $b$ is $0$, it is obvious that the other is also $0$.

Now we assume that $ab \neq 0$, select a large prime $p$ such that $p > | a + b^2 |$, it follows from Fermat's Little Theorem that

$$a^p + b^{p+1} \equiv a + b^2 \pmod{p}.$$

As $p \mid (a^p + b^{p+1})$ and $p > | a + b^2 |$, we have $a + b^2 = 0$.

Then we select another prime $q$ such that $q > | b + 1 |$ and $(q, b) = 1$. Let $n = 2q$. Then we have

$$a^n + b^{n+1} = (-b^2)^{2q} + b^{2q+1} = b^{4q} + b^{2q+1} = b^{2q+1}(b^{2q-1} + 1).$$

It follows from $n \mid (a^n + b^{n+1})$ and $(q, b) = 1$ that

$$q \mid (b^{2q-1} + 1).$$

As $b^{2q-1} + 1 \equiv (b^{q-1})^2 \cdot b + 1 \equiv b + 1 \pmod{q}$, and $q > | b + 1 |$, it follows that $b + 1 = 0$, i.e., $b = -1$, and so $a = -b^2 = -1$.

In conclusion, there are only two solution pairs $(0, 0)$ and $(-1, -1)$. $\qquad \square$

## $2012$ (Hohhot, Inner Mongolia)

### First Day
8:00 – 12:00, September 28, 2012

**1** Find the least positive integer $m$, such that for every prime number $p > 3$,

$$105 \mid 9^{p^2} - 29^p + m.$$

(posed by Yang Hu)

**Solution.** As $105 = 3 \times 5 \times 7$, the original problem is equivalent to that of finding the least positive integer $m$ such that $9^{p^2} - 29^p + m$ can be divided simultaneously by 3, 5, 7.

Since $p^2$ and $p$ have the same odd-even character, we have

$$9^{p^2} - 29^p + m \equiv (-1)^{p^2} - (-1)^p + m \equiv m \,(\text{mod } 5).$$

Then we get $m \equiv 0 \pmod 5$.

As $p > 3$ is an odd number, we have

$$9^{p^2} - 29^p + m \equiv -(-1)^p + m \equiv m + 1 (\text{mod } 3).$$

Therefore, $m \equiv 2 \pmod 3$.

We have also $p^2 \equiv 1 (\text{mod } 3)$ for $p > 3$, or $p^2 = 3k + 1$. Then

$$9^{p^2} - 29^p + m \equiv 2^{3k+1} - 1 + m \equiv 8^k \cdot 2 - 1 + m$$
$$\equiv m + 1 (\text{mod } 7).$$

Therefore, $m \equiv 6 \pmod 7$.

In summary, we have

$$\begin{cases} m \equiv 0 (\text{mod } 5), \\ m \equiv 2 (\text{mod } 3), \\ m \equiv 6 (\text{mod } 7). \end{cases}$$

Then it is easy to find that the least positive integer $m$ is 20.

$\square$

**2** Prove that, among any $n$ vertices of a regular $2n - 1$ polygon ($n \geqslant 3$), there are three ones, which are the vertices of an isosceles triangle. (posed by Zou Jin)

**Solution.** Since it is easy to verify directly the assertion for the cases $n = 3$ (a pentagon) and $n = 4$ (a heptagon), we may assume that $n > 4$ in the following discussion.

By reduction to absurdity, assume we can select $n$ vertices in a regular $2n-1$ polygon $A_1A_2A_3\cdots A_{2n-1}$ such that no three of them constitutes an isosceles triangle. We mark these points with color red and the remainder $n-1$ ones with blue, respectively.

We may let $A_1$ be red, and divide the other $2n-2$ points into $n-1$ pairs (see the figure): $(A_2, A_{2n-1})$, $(A_3, A_{2n-2})$, $\ldots$, $(A_n, A_{n+1})$.

The two points in each pair cannot be both red as they plus $A_1$ constitute an isosceles triangle, and must be one red and one blue for there are exactly $n$ red points.

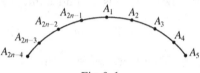

Fig. 2. 1

Assuming $A_2$ is red, then $A_{2n-1}$ is blue, and $A_3$ must be blue as $\triangle A_1A_2A_3$ is an isosceles triangle. Therefore $A_{2n-2}$ is red, from which we infer that $A_5$, $A_{2n-4}$ are both blue, as $\triangle A_{2n-2}A_2A_5$, $\triangle A_{2n-4}A_{2n-2}A_1$ are two isosceles triangles. But $A_5$, $A_{2n-4}$ are the two points in a pair, they must be one red and one blue. This is a contradiction! The assertion for $n > 4$ is then also true. The proof is complete. $\qquad\square$

**③** Let $E$ be a given set with $n$ elements. Suppose $A_1$, $A_2, \ldots, A_k$ are $k$ distinct non-empty subsets of $E$, with the property that, for any $1 \leqslant i < j \leqslant k$, either $A_i \cap A_j = \varnothing$ or one includes the other (i.e., $A_i \subset A_j$ or $A_j \subset A_i$).

Find the maximum value of $k$. (posed by Leng Gangsong)

**Solution.** We claim that the maximum value of $k$ is $2n-1$. To prove this, we at first give an example that satisfies $k = 2n-1$. We may let $E = \{1, 2, \ldots, n\}$, and

$$A_i = \begin{cases} \{i\} & \text{for } 1 \leqslant i \leqslant n, \\ \{1, 2, \ldots, i - n + 1\} & \text{for } n + 1 \leqslant i \leqslant 2n - 1. \end{cases}$$

It is easy to see that they possess the given property.

Now we will prove that $k \leqslant 2n - 1$ by induction.

When $n = 1$, it is obviously true.

Assume that it is true for $n \leqslant m - 1$. Then when $n = m$, we consider the set that contains the most elements among $A_1$, $A_2$, $\ldots$, $A_k$ excluding $E$. We may assume that it is $A_1$ containing $t (\leqslant m - 1)$ elements. Then we can divide $A_1$, $A_2$, $\ldots$, $A_k$ into three categories:

(1) the set $E$;

(2) sets that are included in $A_1$; and

(3) sets whose intersection with $A_1$ is empty.

(Categories (1) and (2) may be empty.)

Then the number of sets in category (1) is not greater than 1.

By induction, the number of sets in category (2) is not greater than $2t - 1$.

The number of elements contained in the union of the sets in category (3) is not greater than $m - t$. Then by induction, the number of sets in that category is not greater than $2(m - t) - 1$.

In total,

$$k \leqslant 1 + (2t - 1) + [2(m - t) - 1] = 2m - 1.$$

Therefore, the proposition is true for $n = m$.

By induction, we conclude that for any set $E$ of $n$ elements, we always have $k \leqslant 2n - 1$. This completes the proof. $\square$

**4** Let $P$ be any inner point of an acute $\triangle ABC$; $E$, $F$ be the projection points of $P$ onto lines $AC$, $AB$, respectively; and the extended lines of $BP$, $CP$ intersect the circumcircle of $\triangle ABC$ at points $B_1$, $C_1$ ($B_1 \neq B$, $C_1 \neq C$), *respectively.* Let $R$ and $r$ denote the radii of the circumcircle and incircle of $\triangle ABC$, respectively. Prove that $\dfrac{EF}{B_1 C_1} \geqslant \dfrac{r}{R}$, and, when the equality holds, determine completely the positions of $P$. (posed by Li Qiusheng)

**Solution.** As seen in Fig. 4. 1, we make $PD \perp BC$ with intersecting point $D$, extend line $AP$ to intersect with the circumcircle of $\triangle ABC$ at point $A_1$ and connect $DE$, $DF$, $A_1 B_1$, $A_1 C_1$.

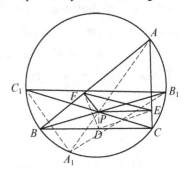

Fig. 4. 1

Since $P$, $D$, $B$, $F$ are concyclic, we have $\angle PDF = \angle PBF$; since $P$, $D$, $C$, $E$, we have $\angle PDE = \angle PCE$. Then

$$\angle FDE = \angle PDF + \angle PDE = \angle PBF + \angle PCE$$
$$= \angle AA_1 B_1 + \angle AA_1 C_1 = \angle C_1 A_1 B_1.$$

In the same way, we have $\angle DEF = \angle A_1 B_1 C_1$. Therefore, $\triangle DEF \sim \triangle A_1 B_1 C_1$.

The radius of the circumcircle of $\triangle A_1 B_1 C_1$ is also $R$, and let the radius of the circumcircle of $\triangle DEF$ be $R'$. We then have $\dfrac{EF}{B_1 C_1} = \dfrac{R'}{R}$.

Now let the incenter of $\triangle DEF$ be $O'$. Connect $AO'$, $BO'$, $CO'$ (see Fig. 4. 2). We have

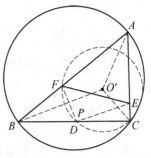

Fig. 4. 2

$$S_{\triangle ABC} = \frac{(AB + BC + CA) \cdot r}{2}$$

$$= S_{\triangle O'AB} + S_{\triangle O'BC} + S_{\triangle O'AC}$$

$$\leqslant \frac{AB \cdot O'F}{2} + \frac{BC \cdot O'D}{2} + \frac{CA \cdot O'E}{2}$$

$$= \frac{(AB + BC + CA) \cdot R'}{2}.$$

Therefore, $R' \geqslant r$. The equality holds if and only if

$$O'D \perp BC, \ O'E \perp CA, \ O'F \perp AB,$$

which implies $P = O'$ and $P$ is the incenter of $\triangle ABC$.

Therefore, $\dfrac{EF}{B_1 C_1} \geqslant \dfrac{r}{R}$ and the equality holds if and only $P$ is the incenter of $\triangle ABC$.

The proof is complete. □

## Second Day

8:00 – 12:00, September 29, 2012

**5** Let $H$ and $O$ be the orthocenter and circumcenter of acute triangle $\triangle ABC$, respectively ( $A$, $H$, $O$ are non-collinear). Suppose $D$ is the projection of $A$ onto line $BC$, and the perpendicular bisector of the segment $AO$ meets line $BC$ at $E$. Prove that the midpoint $N$ of $OH$ is on the circumcircle of $\triangle ADE$. ( posed by Feng Zhigang)

**Solution.** As seen in Fig. 5. 1, we extend $HD$ to let it intersect with the circumcircle of $\triangle ABC$ at point $H'$, link $FN$, $DN$, $BH$, $BH'$, $OH'$, and denote

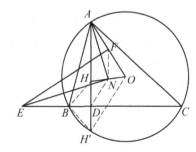

Fig. 5. 1

the midpoint of $AO$ as $F$.

As $H$ is the orthocenter, we have

$$\angle CBH' = \angle CAH' = \angle CBH.$$

Therefore, $D$ is the midpoint of $HH'$. On the other hand, $N$ is the midpoint of $HO$, so $DN$ is the median of $\triangle HOH'$. Therefore, $DN = \dfrac{1}{2}OH'$.

Since $OH' = OA$ and $F$ is the midpoint of $OA$, then

$$DN = \frac{1}{2}OH' = \frac{1}{2}OA = AF.$$

It is easy to see $FN \parallel AH$. Then $AFND$ is an isosceles trapezoid and $A$, $F$, $N$, $D$ are concyclic.

Furthermore, from $\angle ADE = 90° = \angle AFE$ we know $A$, $F$, $D$, $E$ are also concyclic. Then $A$, $F$, $N$, $D$, $E$ are concyclic, which implies that the circumcircle of $\triangle ADE$ crosses $N$ at the midpoint of $OH$. The proof is complete.  $\square$

**Remark.** By the facts that the radius of the nine-point circle of a triangle is half of that of the circumcircle and $N$ is the center of the nine-point circle of $\triangle ABC$, we can get also that $AFND$ is an isosceles trapezoid.

⑥ Sequence $\{a_n\}$ is defined by $a_0 = \dfrac{1}{2}$, $a_{n+1} = a_n + \dfrac{a_n^2}{2012}$, $n = 0, 1, 2, \ldots$. Find integer $k$, such that $a_k < 1 < a_{k+1}$.

(posed by Bian Hongping)

**Solution.** From the given condition, we have

$$\frac{1}{2} = a_0 < a_1 < \cdots < a_{2012}.$$

We note that

$$\frac{1}{a_{n+1}} = \frac{2012}{a_n(a_n + 2012)} = \frac{1}{a_n} - \frac{1}{a_n + 2012},$$

or

$$\frac{1}{a_n} - \frac{1}{a_{n+1}} = \frac{1}{a_n + 2012}.$$

By the telescoping sum, we have

$$\frac{1}{a_0} - \frac{1}{a_n} = \sum_{i=0}^{n-1} \frac{1}{a_i + 2012}.$$

Then

$$2 - \frac{1}{a_{2012}} = \sum_{i=0}^{2011} \frac{1}{a_i + 2012} < \sum_{i=0}^{2011} \frac{1}{2012} = 1.$$

Therefore,

$$a_0 < a_1 < \cdots < a_{2012} < 1.$$

Then we have

$$2 - \frac{1}{a_{2013}} = \sum_{i=0}^{2012} \frac{1}{a_i + 2012} > \sum_{i=0}^{2012} \frac{1}{1 + 2012} = 1,$$

which means $a_{2013} > 1$.

Consequently, we find that $k = 2012$. $\qquad\square$

**7** Given an $n \times n$ grid, we call two cells in it adjacent if they have a common side. At the beginning, each cell is assigned number $+1$. An operation on the grid is defined as follows: one chooses a cell, and then changes the signs of every number in its adjacent cells (but not change the sign of the number in itself). Find all the integers $n \geqslant 2$, such that after a finite of operations, all the numbers in the cells of the grid are changed to $-1$. (posed by Shen Huyue)

**Solution.** We will prove that $n$ meets the required condition if and only if it is an even number.

We denote the cell in the $i$-row and $j$-column of the grid as

$A_{ij}(i, j \in \{1, 2, \ldots, n\})$.

When $n = 2k$, $k \in \mathbf{N}^*$, we mark each $A_{ij}$ satisfying $i + j \equiv 0 \pmod 2$ with color red (presented by shaded areas), and that satisfying $j - i \equiv 3 \pmod 4$ and $j - i \not\equiv j + i \pmod 4$ with blue (presented by oblique line areas) (see Fig. 7.1).

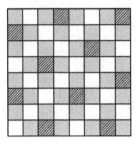

Fig. 7.1          Fig. 7.2

In this way, we can see that every cell adjacent to a blue one is red, and there is exactly one blue cell around each red one.

Now we do the operation on each blue cell. The number in each red cell is then changed from $+1$ to $-1$, while the numbers in the remaining cells are unchanged.

Since $n$ is an even number, we can rotate the grid around its center $O$ anticlockwise by $90°$. Then all the red cells of the rotated grid cover exactly all the cells that are not red in the original grid (see Fig. 7.2).

We do the operations again for the original grid on all the cells that are covered by blue cells of the rotated grid. Then we see that all the numbers with value $+1$ in the remaining cells of the original grid are changed to $-1$, while the numbers in the other cells are unchanged.

Therefore, when $n$ is an even number, all the numbers in the cells of the grid can be changed to $-1$ by a finite of operations.

When $n$ is odd, we denote the number in cell $A_{ii}$ as $M_i (i =$

$1, 2, \ldots, n$), and denote the number of operations on each of their adjacent cells as $x_1$, $x_2$, $\ldots$, $x_{n-1}$, $y_1$, $y_2$, $\ldots$, $y_{n-1}$, respectively (see Fig. 7. 3).

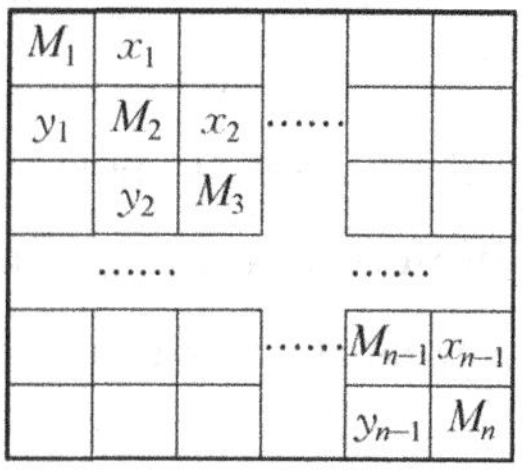

After finite operations, it is easy to see that:

Fig. 7. 3

$M_1$ is changed from $+1$ to $-1$ if and only $x_1 + y_1$ is odd;

$M_2$ is changed from $+1$ to $-1$ if and only $x_1 + y_1 + x_2 + y_2$ is odd;

$M_3$ is changed from $+1$ to $-1$ if and only $x_2 + y_2 + x_3 + y_3$ is odd;

$\vdots$

$M_{n-1}$ is changed from $+1$ to $-1$ if and only $x_{n-2} + y_{n-2} + x_{n-1} + y_{n-1}$ is odd, and

$M_n$ is changed from $+1$ to $-1$ if and only $x_{n-1} + y_{n-1}$ is odd.

Since $n$ is odd, the sum of $n$ odd numbers is still odd. Then,

$$(x_1 + y_1) + (x_1 + y_1 + x_2 + y_2) + (x_2 + y_2 + x_3 + y_3)$$
$$+ \cdots + (x_{n-2} + y_{n-2} + x_{n-1} + y_{n-1}) + (x_{n-1} + y_{n-1})$$
$$= 2(x_1 + x_2 + \cdots + x_{n-1} + y_1 + y_2 + \cdots + y_{n-1})$$

is odd. It is impossible!

Therefore, all the numbers in the cells of a given $n \times n$ grid can be changed from $+1$ to $-1$ after a finite of operations if and only if $n$ is an even number. $\square$

**8** Find all the prime numbers $p$, for which there are infinitely many positive integers $n$, such that $p \mid n^{n+1} + (n+1)^n$. ( posed by Chen Yonggao)

**Solution.** $n^{n+1} + (n+1)^n$ is always an odd number for any positive integer $n$, so the required $p$ will not be 2. Now we are going to prove that, for any prime number $p \geqslant 3$, there are infinite many positive integers that meet the condition. We present two proofs in the following.

**Proof 1.** For any prime number $p \geqslant 3$, let $n = pk - 2$ be an odd number. Then

$$n^{n+1} + (n+1)^n \equiv (-2)^{pk-1} + (-1)^{pk-2} \equiv 2^{pk-1} - 1$$
$$\equiv (2^{p-1})^k \cdot 2^{k-1} - 1 \equiv 2^{k-1} - 1 \pmod{p}.$$

Next, let $k - 1 = (p-1)t$.

Then $n^{n+1} + (n+1)^n \equiv 0 \pmod{p}$.

Therefore, when $n = p(p-1)t + p - 2$ (where $t$ is any positive integer), we have $p \mid n^{n+1} + (n+1)^n$.

**Proof 2.** Given any prime number $p \geqslant 3$, we have $(2, p) = 1$. By Fermat's Little Theorem, we know $2^{p-1} \equiv 1 \pmod{p}$. Let $n = p^t - 2$, where $t = 1, 2, 3, \ldots$. We have

$$n^{n+1} + (n+1)^n \equiv (-2)^{p^t-1} + (-1)^{p^t-2}$$
$$\equiv 2^{p^t-1} - 1 \equiv (2^{p-1})^{p^{t-1}+p^{t-2}+\cdots+p+1} - 1$$
$$\equiv 1 - 1 \equiv 0 \pmod{p}.$$

The proof is complete. $\square$

# 2013 (Lanzhou, Gansu)

## First Day
8:00 – 12:00, August 17, 2013

❶ Do there exist integers $a$, $b$ and $c$ such that $a^2bc + 2$,

$ab^2c + 2$, $abc^2 + 2$ are perfect squares. (posed by Li Qiusheng)

**Solution.** No. Suppose the contrary that there are such integers $a$, $b$ and $c$.

If one of them is even, say $a$, then $a^2bc + 2 \equiv 2 \pmod 4$, which contradicts the assumption that $a^2bc + 2$ is a perfect square. We may then assume that $a$, $b$ and $c$ are odd, so they are either 1 or 3 (mod 4). It follows from the Pigeon-hole Principle that two of them are congruent modulo 4. Relabel if necessary, we may assume that $a \equiv b \pmod 4$, so $abc^2 + 2 \equiv c^2 + 2 \equiv 1 + 2 \equiv 3 \pmod 4$, which violates the perfect square assumption. ☐

**2** Let $n$ be an integer, $n \geqslant 2$, and $x_1, x_2, \ldots, x_n \in [0, 1]$. Prove that

$$\sum_{1 \leqslant k < l \leqslant n} kx_kx_l \leqslant \frac{n-1}{3} \sum_{k=1}^{n} kx_k.$$

(posed by Leng Gangsong)

**Solution.** As $x_1, x_2, \ldots, x_n \in [0, 1]$, $x_ix_j \leqslant x_i$, so we have

$$3 \sum_{1 \leqslant k < l \leqslant n} kx_kx_l = \sum_{1 \leqslant k < l \leqslant n} 3kx_kx_l \leqslant \sum_{1 \leqslant k < l \leqslant n} (kx_k + 2kx_l).$$

For $1 \leqslant k \leqslant n$, the coefficient of $x_k$ in the last sum is

$$2[1 + 2 + \cdots + (k-1)] + k(n-k) = k(n-1),$$

so we have

$$3 \sum_{1 \leqslant k < l \leqslant n} kx_kx_l \leqslant \sum_{1 \leqslant k < l \leqslant n} (kx_k + 2kx_l) = \sum_{k=1}^{n} k(n-1)x_k$$

$$= (n-1) \sum_{k=1}^{n} kx_k,$$

and hence the desired inequality holds. ☐

**3** In $\triangle ABC$, point $B_2$ is the reflection of the center of B-excircle with respect to the midpoint of side $AC$, and point $C_2$ is the reflection of the center of C-excircle with respect to the midpoint of side $AB$. The A-excircle touches side $BC$ at point $D$. Prove that $AD \perp B_2C_2$.

(posed by Bian Hongping)

**Solution.** Let $A_1$ be the center of A-excircle. By the properties about centers of ex-circles, the following sets of three points are collinear: $\{B_1, A, C_1\}$, $\{A_1, C, B_1\}$ and $\{C_1, B, A_1\}$, and $A_1A \perp B_1C_1$.

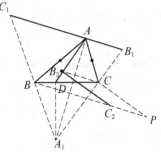

Fig. 3. 1

On the plane, choose point $P$ such that $\overrightarrow{C_2P} = \overrightarrow{B_2C}$, then it follows from $\overrightarrow{B_2C} = \overrightarrow{AB_1}$ that $\overrightarrow{C_2P} = \overrightarrow{AB_1}$. As $\overrightarrow{BC_2} = \overrightarrow{C_1A}$ and the points $C_1$, $B$, $A_1$ are collinear, the points $B$, $C_2$, $P$ are collinear, so $\overrightarrow{BP} = \overrightarrow{C_1B_1}$. It follows from

$$\angle AC_1B = 180° - \left(\frac{180° - \angle BAC}{2}\right) - \left(\frac{180° - \angle ABC}{2}\right)$$

$$= \frac{\angle BAC + \angle ABC}{2} = \frac{180° - \angle ACB}{2} = \angle BCA_1$$

that $\triangle A_1BC \backsim \triangle A_1B_1C_1$. Let $A_1D$ and $A_1A$ be the altitudes of the $\triangle A_1BC$ and $\triangle A_1B_1C_1$, respectively, with respect to the opposite sides, so $\dfrac{B_1C_1}{BC} = \dfrac{A_1A}{A_1D}$.

If $\overrightarrow{BP} = \overrightarrow{C_1B_1}$, then $A_1A \perp B_1C_1$, so $\dfrac{BP}{BC} = \dfrac{A_1A}{A_1D}$. If $BP \perp A_1A$, then $BC \perp A_1D$, so we have $\triangle BPC \backsim \triangle A_1AD$, and hence $CP \perp AD$. Again by $\overrightarrow{C_2P} = \overrightarrow{B_2C}$, one has $\overrightarrow{B_2C_2} = \overrightarrow{CP}$, and hence $AD \perp B_2 C_2$.                                    ☐

**④** There are $n(n \geqslant 2)$ coins in a row. If one of the coins is head, select an odd number of consecutive coins (or even 1 coin) with the one in head on the leftmost, and then flip all the selected coins upside down simultaneously. This is a *move*. No move is allowed if all $n$ coins are tails. Suppose $n$ coins are heads at the initial stage, determine if there is a way to carry out $\left\lfloor \dfrac{2^{n+1}}{3} \right\rfloor$ moves. (posed by Gu Bin )

**Solution.** The answer is possible.

For any configuration of the coins, we define a corresponding $01$ - sequence $c_1 c_2 \ldots c_n$ of length $n$ as follows: $c_i = 1$, if the $i$-th coin from the left is head, otherwise, $c_i = 0$. It is easy to see that the status of the $n$ coins as one-to-one correspondence to such $01$ - sequences, so in the following, we will consider this sequence model instead.

Initially, the sequence is $11 \cdots 11$, denoted by $1^n$ (with $n$ consecutive digits of 1). Similarly, $00 \cdots 00$, denoted by $0^n$ (with $n$ digits of 0). For any $01$ - sequence with at least a digit "1", consider the following move: locate the first digit "1" from right to left in the sequence, then take the $01$ - subsequence from left to right starting this "1" of maximal odd length, and change the $01$ - parity in this subsequence just like flipping the coins in a move. Denote by $a_n$ the total number of moves in the way stated above. We claim: $a_n = \left\lfloor \dfrac{2^{n+1}}{3} \right\rfloor$. When $n = 1$, it is easy to see that $a_1 = 1 = \left\lfloor \dfrac{2^2}{3} \right\rfloor$, proceed by induction. Assume formula $a_n = \left\lfloor \dfrac{2^{n+1}}{3} \right\rfloor$ holds for $n = k$, i.e., $a_k = \left\lfloor \dfrac{2^{k+1}}{3} \right\rfloor$.

Now, we discuss the case $n = k + 1$, i.e., we want to find the total number of moves as described above if the sequence is $1^{k+1}$, i.e, there are $k + 1$ coins in a row.

If $k$ is odd, then by induction hypothesis, the sequence $1^{k+1}$ (with $k + 1$ digits of 1) changes to $10^k$ (with $k$ digits of 0) after $a_k$ moves as stated above. After an additional move, it changes to $01^{k-1}0$ (with $k - 1$ digits of 1), then after applying $a_k - 1$ moves, the sequence changes to $0^{k+1}$ (with $k + 1$ digits of 0). In fact, recall that in sequence of $a_k$ moves from $1^k$ to $0^k$ by, the first move is from $1^k$ to $1^{k-1}0$. Therefore, we have

$$a_{k+1} = 2a_k = 2\left\lfloor \frac{2^{k+1}}{3} \right\rfloor = 2 \cdot \frac{2^{k+1} - 1}{3} = \frac{2^{k+2} - 2}{3} = \left\lfloor \frac{2^{k+2}}{3} \right\rfloor.$$

If $k$ is even, by the induction assumption it takes $a_k$ moves from $1^{k+1}$ to $10^k$, then apply an additional from $10^k$ to $01^k$, and finally by the induction assumption again, it takes $a_k$ moves from $01^k$ to $0^{k+1}$. Then we have

$$a_{k+1} = 2a_k + 1 = 2\left\lfloor \frac{2^{k+1}}{3} \right\rfloor + 1 = 2 \cdot \frac{2^{k+1} - 2}{3} + 1$$

$$= \frac{2^{k+2} - 1}{3} = \left\lfloor \frac{2^{k+2}}{3} \right\rfloor.$$

By induction that $a_n = \left\lfloor \dfrac{2^{n+1}}{3} \right\rfloor$, hence there exists a way to make the required number of moves. □

# Second Day
### 8:00 – 12:00, August 18, 2013

**5**  A non-empty set $A \subseteq \{1, 2, 3, \ldots, n\}$ is called a *good set* *of degree n* if $| A | \leqslant \min_{x \in A} x$. Denote by $a_n$ the number of good sets of 'degree $n$. Prove that $a_{n+2} = a_{n+1} + a_n + 1$ for any positive integer $n$. (posed by Li Weigu )

**Solution 1.** Let $A$ be a good set of degree $n$, and $| A | = k$, then $\min_{x \in A} x \geqslant k$, so $A \subseteq \{k, k + 1, \ldots, n\}$. Hence the number of good sets of degree $n$ with $k$ elements is $C_{n-k+1}^k$. It follows that $a_n = \sum_{k=1}^{\left[\frac{n+1}{2}\right]} C_{n-k+1}^k = C_n^1 + C_{n-1}^2 + C_{n-2}^3 + \cdots$.

If $n$ is even, $n = 2m$, then

$$a_{2m+2} = C_{2m+2}^1 + C_{2m+1}^2 + \cdots + C_{m+2}^{m+1}$$
$$= (C_{2m+1}^1 + C_{2m+1}^0) + (C_{2m}^2 + C_{2m}^1) + \cdots + (C_{m+1}^{m+1} + C_{m+1}^m)$$
$$= (C_{2m+1}^1 + C_{2m}^2 + \cdots + C_{m+1}^{m+1}) + (C_{2m}^1 + C_{2m-1}^2 + \cdots$$
$$+ C_{m+1}^m) + C_{2m+1}^0$$
$$= a_{2m+1} + a_{2m} + 1.$$

If $n$ is odd, $n = 2m - 1$, then

$$a_{2m+1} = C_{2m+1}^1 + C_{2m}^2 + \cdots + C_{m+2}^m + C_{m+1}^{m+1}$$
$$= (C_{2m}^1 + C_{2m}^0) + (C_{2m-1}^2 + C_{2m-1}^1) + \cdots + (C_{m+1}^m + C_{m+1}^{m-1}) + C_m^m$$
$$= (C_{2m}^1 + C_{2m-1}^2 + \cdots + C_{m+1}^m) + (C_{2m-1}^1 + C_{2m-2}^2 + \cdots$$
$$+ C_{m+1}^{m-1} + C_m^m) + C_{2m}^0$$
$$= a_{2m} + a_{2m-1} + 1.$$

In summary, the equality $a_{n+2} = a_{n+1} + a_n + 1$ holds for all positive integers $n$.

**Remark.** Let $F_n$ be the $n$-th term of Fibonacci sequence. From the combinatorial identity $\sum_{k=0}^{\left[\frac{n}{2}\right]} C_{n-k}^k = F_n$, one can derive that $a_n = F_{n+1} - 1$.

**Solution 2.** If $n = 1$, there is only one good set of degree 1, namely $\{1\}$, so $a_1 = 1$.

If $n = 2$, then there are only two good sets of degree 2:

$\{1\}$, $\{2\}$, so $a_2 = 2$.

Recall that $a_n$, $a_{n+1}$ are the numbers of the non-empty good subsets $A$ of $\{1, 2, \ldots, n\}$ and $\{1, 2, \ldots, n+1\}$, respectively, satisfying $|A| \leqslant \min_{x \in A} x$.

Consider the case $n + 2$: For any non-empty good set $A$ of degree $n + 2$, $A$ is a subset of $\{1, 2, \ldots, n + 2\}$, then we have the following three cases:

(a) $A$ does not contain the element $n + 2$;

(b) $A$ contains $n + 2$, and has at least 2 elements;

(c) $A = \{n + 2\}$.

In the following, we focus on the number of good sets of types (a) and (b). For any good set $A$ in (a), it follows from $|A| \leqslant \min_{x \in A} x$ and $\max_{x \in A} x \leqslant n + 1$ that $A$ is a good set of degree $n + 1$. Conversely, any good set of degree $n + 1$ is also a good set of degree $n + 2$, therefore, there are exactly $a_{n+1}$ good sets of type (a).

For any good set $A = \{a_1, a_2, \ldots, a_k, n + 2\}$ of degree $n + 2$ of type (b), where $a_1 < a_2 < \cdots < a_k < n + 2$. As $a_1 = \min_{x \in A} x \geqslant |A| \geqslant 2$, one can consider the non-empty set $A' = \{a_1 - 1, a_2 - 1, \ldots, a_k - 1\}$, where $1 \leqslant a_1 - 1 < a_2 - 1 < \cdots < a_k - 1 < n + 1$, and $A'$ satisfies $|A'| = |A| - 1 \leqslant a_1 - 1 = \min_{x \in A'} x$, hence $A'$ is a good set of degree $n$. Conversely, any good set of degree $n$ can be represented in the form $A' = \{a_1 - 1, a_2 - 1, \ldots, a_k - 1\}$, where $A = \{a_1, a_2, \ldots, a_k, n + 2\}$ is a good set of degree $n + 2$ of type (b), so the correspondence is one-to-one between $A$ and $A'$, and there are exactly $a_n$ good sets of degree $n + 2$ of type (b).

According to the discussion on the number of good sets of types (a), (b) and (c), we have $a_{n+2} = a_{n+1} + a_n + 1$. $\qquad \square$

**6** As shown in Fig. 6.1, $PA$, $PB$ are tangent to the circle with center $O$ at $A$ and $B$, point $C$ (different from $A$, $B$) is on minor arc $AB$. The line $l$ through point $C$ and perpendicular to $PC$ meets the angle bisectors $\angle AOC$ and $\angle BOC$ at points $D$ and $E$, respectively. Prove that $CD = CE$. (posed by He Yijie)

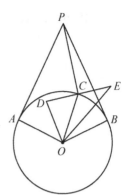

Fig. 6.1

**Solution 1.** As shown in Fig. 6.2, Line $PC$ meets the circle $\odot O$ at another point $F$. Join $BC$, $BE$, $BF$, $OF$, respectively. Since $B$, $C$ are on $\odot O$, $OE$ bisects $\angle BOC$, and hence $OE$ perpendicularly bisects $BC$, so $CE = BE$. By $FO = BO$ and $PC \perp DE$, we have

$$\angle ECB = 90° - \angle FCB = 90° - \frac{1}{2}\angle BOF = \angle OBF,$$ and so $\triangle CEB \backsim \triangle BOF$, hence

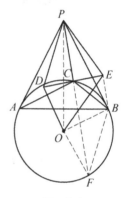

Fig. 6.2

$$\frac{CE}{BO} = \frac{CB}{BF}. \qquad ①$$

As $PB$ is tangent to $\odot O$, so $\angle PBC = \angle PFB$, and $\triangle PCB \backsim \triangle PBF$, and hence

$$\frac{CB}{BF} = \frac{PC}{PB}. \qquad ②$$

By ① and ②, we have $CE = BO \cdot \dfrac{PC}{PB}$. By symmetry, $CD = AO \cdot \dfrac{PC}{PA}$. It follows from $AO = BO$, $PA = PB$ that $CD = CE$.

**Solution 2.** As shown in Fig. 6. 3. Join $AB$, $AC$, $BC$. $OD$ meets $AC$ at point $M$. As $A$, $C$ are on $\odot O$, so $OD$ bisects $\angle AOC$, $\angle DMC = 90°$ and $MC = \frac{1}{2}AC$. As $PC \perp DE$, so $\cos\angle ACD = \sin\angle ACP$, and

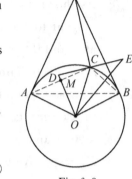

$$CD = \frac{MC}{\cos\angle ACD} = \frac{AC}{2\sin\angle ACP}. \quad ③$$

Fig. 6. 3

Let $R$ be the radius of $\odot O$. As $PA$ is tangent to $\odot O$, so $\angle ABC = \angle CAP$, and then it follows from ③ and the Sine Law that

$$CD = \frac{2R\sin\angle ABC}{2\sin\angle ACP} = R \cdot \frac{\sin\angle CAP}{\sin\angle ACP} = R \cdot \frac{CP}{AP}.$$

By symmetry, one has $CE = R \cdot \frac{CP}{BP}$, and it follows from $AP = BP$ that $CD = CE$.    □

**7** Label the sides of a regular $n$-gon in clockwise direction in order with $1, 2, \ldots, n$. Determine all integers $n$ $(n \geqslant 4)$ satisfying the following two conditions:

(1) $n - 3$ non-intersecting diagonals in the $n$-gon are selected, which subdivide the $n$-gon into $n - 2$ non-overlapping triangles, and

(2) each of the chosen $n - 3$ diagonals is labeled with an integer, such that the sum of labeled numbers on three sides of each triangles in (1) is equal to the other. (posed by Zou Jin)

**Solution.** The required integers are those satisfying both conditions: $n \geqslant 4$ and $n \not\equiv 2\,(\text{mod } 4)$. Suppose that $n$ satisfies the conditions (1) and (2), we first prove that $n \not\equiv 2(\bmod 4)$ as follows. Denote by $S$ the sum of the labels in three sides of any triangle, and by $m$ the sum of the labels in the $n - 3$ diagonals chosen. Adding up the sums of labels in three sides of all triangles in the subdivision in (a), every diagonal appears twice in those triangles, it follows that $(n - 2)S = (1 + 2 + \cdots + n) + 2m$. Suppose $n \equiv 2(\bmod 4)$, then $(n - 2)S$ is even but $(1 + 2 + \cdots + n) + 2m$ is odd, which is a contradiction, so $n \not\equiv 2(\bmod 4)$.

Let us remark that the condition on the regular $n$-gon in the problem can be relaxed to convex $n$-gon, so we can simplify the writing-up of the solution.

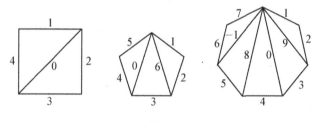

Fig. 7. 1

In the following, we prove that all integers $n$ with $n \geqslant 4$ and $n \not\equiv 2(\bmod 4)$ satisfy conditions (1) and (2).

The Fig. 7. 1 shows labeling for the case $n = 4$, 5 and 7, and one can verify directly that both conditions hold.

In the following, we show that if $n$ satisfies both conditions, so does $n + 4$.

As shown in Fig. 7. 2, label one diagonal which subdivides the given $(n + 4)$-gon into a convex $n$-gon and a convex hexagon (non-regular anymore). As $n$ satisfies conditions (1) and (2), one can subdivide the convex $n$-gon into $(n - 2)$ triangles such that the sum of the three labels in each triangle is equal to the same number $S$. One can also subdivide the hexagon with three diagonals with labels: $S - 2n - 1$, $n + 1$ and $S - 2n - 5$. One can check that the sum of the three labels in each of these four triangles is also $S$, hence $n + 4$ also satisfies conditions (1) and (2). So it follows that any integer $n$ such that $n \geqslant 4$ and $n \not\equiv 2 \pmod 4$ satisfies conditions (1) and (2).

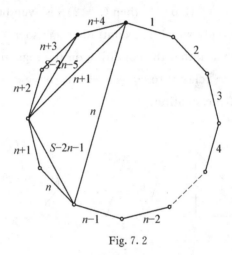

Fig. 7.2

In summary, those integers $n$ satisfying conditions (1) and (2) are exactly given by $n \geqslant 4$ and $n \not\equiv 2 \pmod 4$. □

**8** Find all positive integers $a$ such that $(2^n - n^2) \mid (a^n - n^a)$
for all positive integers $n \geqslant 5$. (posed by Yang Mingliang)

**Solution.** The only answers for $a$ are 2 and 4.

First, we prove that $a$ is even. It follows from the given condition by choosing an even integer $n \geqslant 6$.

Next, we prove that $a$ has no odd prime factor. Suppose the contrary, let $p$ be an odd prime factor of $a$. If $p = 3$, let $n = 8$, then $2^n - n^2 = 192$ has a factor 3, but $a^n - n^a$ is not divisible by 3, contradicting to $(2^n - n^2) \mid (a^n - n^a)$, hence $p$ is not 3.

If $p = 5$, let $n = 16$, then $2^n - n^2 = 64\,110$ has a factor 5, but $a^n - n^a$ is not divisible by 5, contradicting $(2^n - n^2) \mid (a^n - n^a)$, hence $p$ is not 5.

If $p \geq 7$, let $n = p - 1$, it follows from Fermat's Little Theorem that $2^{p-1} \equiv 1 \pmod{p}$.

As $(p - 1)^2 \equiv 1 \pmod{p}$, so $p \mid (2^n - n^2)$. Moreover, since $a$ is even and $p \mid a$, so $n^a \equiv (p - 1)^a \equiv (-1)^a \equiv 1 \pmod{p}$ and $p \mid a^n$, and hence $p$ does not divide $(a^n - n^a)$, contradicting $(2^n - n^2) \mid (a^n - n^a)$.

Finally, we prove that $a$ is 2 or 4. For this, let $a = 2^t$ where $t$ is a positive integer, then it follows from $(2^n - n^2) \mid (2^{tn} - n^{2^t})$ and $(2^n - n^2) \mid (2^{tn} - n^{2t})$ that $(2^n - n^2) \mid (n^{2^t} - n^{2t})$.

If we choose $n$ to be sufficiently large, it follows from the fact $\lim\limits_{n \to \infty} \dfrac{n^{2^t}}{2^n} = 0$ that $n^{2^t} - n^{2t} = 0$, hence $2^t = 2t$. $t = 1$ and $t = 2$ are obvious solutions.

If $t \geq 3$, then by the Binomial Theorem, we have $t = 2^{t-1} = (1 + 1)^{t-1} > 1 + (t - 1) = t$, which is impossible. At last, one can easily check that $a = 2$ and $a = 4$ satisfy the condition in the problem, so the solutions for $a$ are 2 and 4. ☐

# China Southeastern Mathematical Olympiad

<u>2010</u> **(Lukang, Changhua, Taiwan)**

### First Day

8: 00 – 12: 00, Auguest 17, 2010

**1** Let $a$, $b$, $c \in \{0, 1, 2, \ldots, 9\}$. The quadratic equation $ax^2 + bx + c = 0$ has a rational root. Prove that the three-digit number $\overline{abc}$ is not a prime number.

**Solution.** We prove by contradiction. If $\overline{abc} = p$ is a prime number, the rational root of quadratic equation $f(x) = ax^2 + bx + c = 0$ is $x_1$, $x_2 = \dfrac{-b \pm \sqrt{b^2 - 4ac}}{2a}$. Obviously, $b^2 - 4ac$ is a perfect square number, and $x_1$, $x_2$ are all negative, and

$$f(x) = a(x - x_1)(x - x_2).$$

Thus,

$$p = f(10) = a(10 - x_1)(10 - x_2).$$

So,

$$4ap = (20a - 2ax_1)(20a - 2ax_2).$$

It is easy to see that $(20a - 2ax_1)$ and $(20a - 2ax_2)$ are all positive integers. Consequently, $p \mid (20a - 2ax_1)$ or $p \mid (20a - 2ax_2)$. If $p \mid (20a - 2ax_1)$, then $p \leqslant 20a - 2ax_1$, so, $80 - 8x_1 - 10x_2 + x_1x_2 \leqslant 0$, which contradicts to $x_1$, $x_2 < 0$. Similarly, $p \mid (20a - 2ax_1)$ is not true. $\qquad\square$

**②** For any set $A = \{a_1, a_2, \ldots, a_m\}$, denote $P(A) = a_1a_2\ldots a_m$. Let $A_1$, $A_2$, $\ldots$, and $A_n$ be all $99$ - element subsets of $\{1, 2, \ldots, 2010\}$, $n = C_{2010}^{99}$. Prove that $2010 \bigm| \sum_{i=1}^{n} P(A_i)$.

**Solution 1.** For each 99-elements subset, $A_i = \{a_1, a_2, \ldots, a_{99}\}$ of $\{1, 2, \ldots, 2010\}$ uniquely corresponds to $99$ - elements subset $B_i = \{b_1, b_2, \ldots, b_{99}\}$ of $\{1, 2, \ldots, 2010\}$ by $b_k = 2011 - a_k$, $k = 1, 2, \ldots, 99$.

Since $\sum_{k=1}^{99} (a_k + b_k) = 99 \times 2011$ is odd, we see that $A_i$, $B_i$ are different subsets of $\{1, 2, \ldots, 2010\}$. When $A_i$ take all $99$ - elements subset of $\{1, 2, \ldots, 2010\}$, so are $B_i$. Moreover

$$P(A_i) + P(B_i)$$
$$= a_1a_2\cdots a_{99} + (2011 - a_1)(2011 - a_2)\cdots(2011 - a_{99})$$
$$\equiv a_1a_2\cdots a_{99} + (-a_1)(-a_2)\cdots(-a_{99})(\mathrm{mod}\ 2011)$$
$$\equiv 0(\mathrm{mod}\ 2011).$$

Thus,

$$2\sum_{i=1}^{n} P(A_i) = \sum_{i=1}^{n} P(A_i) + \sum_{i=1}^{n} P(B_i) \equiv 0 \pmod{2011},$$

hence $2011 \Big| \sum_{i=1}^{n} P(A_i)$.

**Solution 2.** Let $f(n) = (n-1)(n-2)\cdots(n-2010) - n^{2010} - 2010!$, where $n \in \mathbf{Z}$.

Since 2011 is prime, by Fermat's Little Theorem, $n^{2010} \equiv 1 \pmod{2011}$. By Wilson's Theorem, we have $2010! \equiv -1 \pmod{2011}$. Thus,

(i) If $2011 \nmid n$, then

$$f(n) \equiv (n-1)(n-2)\cdots(n-2010) \equiv 0 \pmod{2011}.$$

(ii) If $2011 \mid n$, then

$$f(n) \equiv (2011-1)(2011-2)\cdots(2011-2010) - 2011^{2010} - 2010!$$
$$\equiv 2010! - 2010! \pmod{2011}$$
$$\equiv 0 \pmod{2011}.$$

So $f(n) \equiv 0 \pmod{2011}$ has 2011 solutions in the sense of mod 2011.

Since $f(n)$ is a polynomial of order 2009, and for all $n \in \mathbf{Z}$, $2011 \mid f(n)$, we see that each coefficient of $f(n)$ can be divided by 2011.

Turn to the original problem, $\sum_{i=1}^{n} P(A_i)$ is the coefficient of term with order 1911 of $f(n)$, thus $2011 \Big| \sum_{i=1}^{n} P(A_i)$.                    $\square$

**3** As shown in Fig. 3. 1. Let the inscribed circle $I$ of $\triangle ABC$ touch $BC$ and $AB$ at $D$ and $F$, respectively. Let $I$ intersect the segments $AD$ and $CF$ at $H$ and $K$, respectively. Prove that

$$\frac{FD \times HK}{FH \times DK} = 3.$$

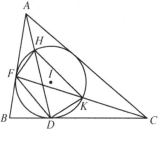

Fig. 3. 1

**Solution 1.** Suppose that the lengths of segments are $AF = x$, $BF = y$, $CD = z$, then by Stewart's Theorem, we have

$$AD^2 = \frac{BD}{BC} \cdot AC^2 + \frac{CD}{BC} \cdot AB^2 - BD \cdot DC$$

$$= \frac{y(x+z)^2 + z(x+y)^2}{y+z} - yz$$

$$= x^2 + \frac{4xyz}{y+z}.$$

By Tangent-Secant Theorem, we have $AH = \dfrac{AF^2}{AD} = \dfrac{x^2}{AD}.$

Thus,

$$HD = AD - AH = \frac{AD^2 - x^2}{AD} = \frac{4xyz}{AD(y+z)}.$$

Similarly, we have

$$KF = \frac{4xyz}{CF(x+y)}.$$

Since $\triangle CDK \backsim \triangle CFD$, we see that

$$DK = \frac{DF \times CD}{CF} = \frac{DF}{CF}z.$$

By the fact of $\triangle AFH \backsim \triangle ADF$, we have

$$FH = \frac{DF \times AF}{AD} = \frac{DF}{AD}x.$$

By the Cosine Law,

$$DF^2 = BD^2 + BF^2 - 2BD \cdot BF \cos B$$

$$= 2y^2 \left( 1 - \frac{(y+z)^2 + (x+y)^2 - (x+z)^2}{2(x+y)(y+z)} \right)$$

$$= \frac{4xy^2z}{(x+y)(y+z)}.$$

Thus

$$\frac{KF \times HD}{FH \times DK} = \frac{\dfrac{4xyz}{CF(x+y)} \cdot \dfrac{4xyz}{AD(y+z)}}{\dfrac{DF}{AD}x \cdot \dfrac{DF}{CF}z}$$

$$= \frac{16xy^2z}{DF^2(x+y)(y+z)} = 4.$$

Appling Ptolemy's Theorem to cyclic quadrilateral $DKHF$, we have

$$KF \cdot HD = DF \cdot HK + FH \cdot DK.$$

Combining with $\dfrac{KF \times HD}{FH \times DK} = 4$, we obtain $\dfrac{FD \times HK}{FH \times DK} = 3$.

**Solution 2.** First we prove a lemma.

**Lemma.** As shown in Fig. 3. 2. If the circle touches $AB$ and $AC$ at $B$ and $C$, respectively, $Q$ is a point on the circle. $AQ$ intersects the circle at point $P$, then we have

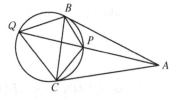

Fig. 3. 2

$$PQ \cdot BC = 2BP \cdot QC = 2BQ \cdot PC.$$

**Proof of the lemma.** By Ptolemy's Theorem, we see that

$$PQ \cdot BC = BP \cdot QC + BQ \cdot PC.$$

Since $AB$ tangents to the circle, we see that $\angle ABP = \angle AQB$.

Further by $\angle BAP = \angle QAB$, we have $\triangle ABP \backsim \triangle AQB$. Consequently,

$$\frac{BP}{BQ} = \frac{AP}{AB} = \frac{AB}{AQ}.$$

Thus,

$$\left(\frac{BP}{BQ}\right)^2 = \frac{AP}{AB} \times \frac{AB}{AQ} = \frac{AP}{AQ}.$$

Similarly,

$$\left(\frac{CP}{CQ}\right)^2 = \frac{AP}{AQ}.$$

Thus, $\left(\dfrac{CP}{CQ}\right)^2 = \left(\dfrac{BP}{BQ}\right)^2$, that is $BP \cdot QC = BQ \cdot PC$. So

$$PQ \cdot BC = 2BP \cdot QC = 2BQ \cdot PC.$$

The lemma is proved.

Turn to the original problem. Let circle $I$ intersect $AC$ at point $O$ and draw segments $HO$, $OK$, $OD$ and $FO$. As shown in Fig. 3. 3.

By Ptolemy's Theorem, we have

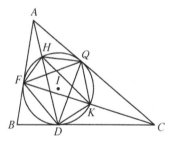

Fig. 3. 3

$$KF \cdot HD = DF \cdot HK + FH \cdot DK.$$

Thus,

$$\frac{FD \times HK}{FH \times DK} = 3 \Leftrightarrow \frac{KF \times HD}{FH \times DK} = 4.$$

Since $CQ$ and $CD$ are tangent to circle $I$, by Lemma, we see that

$$KF \cdot DO = 2DK \cdot FO,$$
$$HD \cdot FO = 2FH \cdot DO.$$

Multiply above two equations by each side, we have

$$KF \cdot HD \cdot DO \cdot FO = 4DK \cdot FH \cdot DO \cdot FO,$$

that is, $\dfrac{KF \times HD}{FH \times DK} = 4.$  □

**4**　Let $a$ and $b$ be positive integers such that $1 \leqslant a < b \leqslant 100$. If there exists a positive integer $k$ such that $ab \mid (a^k + b^k)$, then we say that the pair $(a, b)$ is *good*. Determine the number of good pairs.

**Solution.** Let $(a, b) = d$, $a = sd$, $b = td$, $(s, t) = 1$, $t > 1$, then $std^2 \mid d^k(s^k + t^k)$. So $k \geqslant 2$ and $st \mid d^{k-2}(s^k + t^k)$. Since $(st, s^k + t^k) = 1$, we have $st \mid d^{k-2}$. Therefore, any prime factor of $st$ can be divided by $d$.

If there is a prime factor $p$ of $s$ or $t$ no less than 11, then $p$ divides $d$. So $p^2 \mid a$ or $p^2 \mid b$, but $p^2 > 100$, which is a contradiction.

So the prime factor of $st$ may be 2, 3, 5 or 7.

If there are at least three prime factors of $st$ among 2, 3, 5, 7, then there is a prime factor of $s$ or $t$ no less than 5. And $d > 2 \times 3 \times 5 = 30$, so that $a$ or $b \geqslant 5d > 100$, which is a contradiction. The prime factor set of $st$ cannot be $\{3, 7\}$, otherwise, $a$ or $b \geqslant 7 \times 3 \times 7 > 100$, which is a contradiction.

Similarly, the prime factor set of $st$ cannot be $\{5, 7\}$.

Therefore, the prime factor set of $st$ can only be $\{2\}$, $\{3\}$, $\{5\}$, $\{7\}$, $\{2, 3\}$, $\{2, 5\}$, $\{2, 7\}$ or $\{3, 5\}$.

( i ) If the prime factor set of $st$ is $\{3, 5\}$, then $d$ can only be 15. Then, $s = 3$, $t = 5$. So there is one good pair $(a, b) = (45, 75)$.

( ii ) If the prime factor set of $st$ is $\{2, 7\}$, then $d$ can only be 14. Then $s = 2$, $t = 7$ or $s = 4$, $t = 7$. So there are two good pairs $(a, b) = (28, 98)$ and $(56, 98)$.

(iii) If the prime factor of $st$ is $\{2, 5\}$, then $d$ can only be 10 or 20.

For $d = 10$, then $s = 2$, $t = 5$; $s = 1$, $t = 10$; $s = 4$, $t = 5$; $s = 5$, $t = 8$.

For $d = 20$, then $s = 2$, $t = 5$; $s = 4$, $t = 5$.

There are six good pairs.

(iv) If the prime factor of $st$ is $\{2, 3\}$, then $d$ can only be 6, 12, 18, 24 or 30.

For $d = 6$, $s = 1$, $t = 6$; $s = 1$, $t = 12$; $s = 2$, $t = 3$; $s = 2$, $t = 9$; $s = 3$, $t = 4$; $s = 3$, $t = 8$; $s = 3$, $t = 16$; $s = 4$, $t = 9$; $s = 8$, $t = 9$; $s = 9$, $t = 16$.

For $d = 12$, $s = 1$, $t = 6$; $s = 2$, $t = 3$; $s = 3$, $t = 4$; $s = 3$, $t = 8$.

For $d = 18$, $s = 2$, $t = 3$; $s = 3$, $t = 4$.

For $d = 24$, $s = 2$, $t = 3$; $s = 3$, $t = 4$. $d = 30$, $s = 2$, $t = 3$.

There are 19 good pairs.

(v) If the prime factor set of $st$ is $\{7\}$, then $s = 1$, $t = 7$, $d$ can only be 7 or 14.

So, there are two good pairs.

(vi) If the prime factor set of $st$ is $\{5\}$, then $s = 1$, $t = 5$, $d$ can only be 5, 10, 15 or 20. So, there are four good pairs.

(vii) If the prime factor set of $st$ is $\{3\}$ then we have the following:

when $s = 1$, $t = 3$, $d$ can only be 3, 6, ..., or 33;

when $s = 1$, $t = 9$, $d$ can only be 3, 6 or 9;

when $s = 1$, $t = 27$, $d$ can only be 3.

There are 15 good pairs.

(viii) If the prime factor set of $st$ is $\{2\}$, then we have the following:

when $s = 1$, $t = 2$, $d$ can only be 2, 4, ..., or 50;

when $s = 1$, $t = 4$, then $d$ can only be 2, 4, ..., or 24;

when $s = 1$, $t = 8$, then $d$ can only be 2, 4, ..., or 12;

when $s = 1$, $t = 16$, then $d$ can only be 2, 4 or 6;

when $s = 1$, $t = 32$, then $d$ can only be 2.

There are 47 good pairs.

Therefore, there are all together $1 + 2 + 6 + 19 + 2 + 4 + 15 + 47 = 96$ good pairs. □

## Second Day

### 8: 00 – 12: 00, Auguest 18, 2010

**5** As shown in Fig. 5. 1. Let $C$ be the right angle of $\triangle ABC$. $M_1$ and $M_2$ are two arbitrary points inside $\triangle ABC$, and $M$ is the midpoint of $M_1 M_2$. The extensions of $BM_1$, $BM$ and $BM_2$ intersect $AC$ at $N_1$, $N$ and $N_2$ respectively. Prove that $\dfrac{M_1 N_1}{BM_1} + \dfrac{M_2 N_2}{BM_2} \geqslant 2 \dfrac{MN}{BM}$.

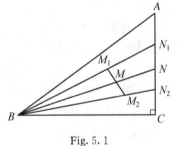

Fig. 5. 1

**Solution.** As shown in Fig. 5. 2. Let $H_1$, $H_2$ and $H$ be the projection points of $M_1$, $M_2$ and $M$ on line $BC$, respectively. Then

$$\frac{M_1 N_1}{BM_1} = \frac{H_1 C}{BH_1},$$

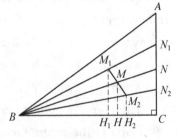

Fig. 5. 2

$$\frac{M_2 N_2}{BM_2} = \frac{H_2 C}{BH_2},$$

$$\frac{MN}{BM} = \frac{HC}{BH} = \frac{H_1 C + H_2 C}{BH_1 + BH_2}.$$

Suppose that $BC = 1$, $BH_1 = x$ and $BH_2 = y$. We have

$$\frac{M_1 N_1}{BM_1} = \frac{H_1 C}{BH_1} = \frac{1 - x}{x},$$

$$\frac{M_2 N_2}{BM_2} = \frac{H_2 C}{BH_2} = \frac{1 - y}{y},$$

$$\frac{MN}{BM} = \frac{HC}{BH} = \frac{1 - x + 1 - y}{x + y}.$$

Thus, the inequality we are to prove is equivalent to

$$\frac{1 - x}{x} + \frac{1 - y}{y} \geq 2 \frac{1 - x + 1 - y}{x + y},$$

which is equivalent to $\dfrac{1}{x} + \dfrac{1}{y} \geq \dfrac{4}{x + y}$, that is, $(x - y)^2 \geq 0$

which is obviously true. $\qquad\qquad\qquad\qquad\qquad\qquad\qquad$ $\square$

**6** Let $\mathbf{N}^*$ be the set of positive integers. Define $a_1 = 2$, and for $n = 1, 2, \ldots,$

$$a_{n+1} = \min\left\{\lambda \,\middle|\, \frac{1}{a_1} + \frac{1}{a_2} + \cdots + \frac{1}{a_n} + \frac{1}{\lambda} < 1, \lambda \in \mathbf{N}^*\right\}.$$

Prove that $a_{n+1} = a_n^2 - a_n + 1$ for $n = 1, 2, \ldots.$

**Solution.** By $a_1 = 2$, $a_2 = \min\left\{\lambda \,\middle|\, \frac{1}{a_1} + \frac{1}{\lambda} < 1, \lambda \in \mathbf{N}^*\right\}$,

consider $\dfrac{1}{a_1} + \dfrac{1}{\lambda} < 1$, then $\dfrac{1}{\lambda} < 1 - \dfrac{1}{2} = \dfrac{1}{2}$, $\lambda > 2$, hence $a_2 =$

3. So the conclusion is true for $n = 1$.

Suppose that the conclusion is true for all integer $n \leq k -$

$1 (k \geq 2)$. If $n = k$, then $a_{k+1} = \min\left\{\lambda \,\middle|\, \frac{1}{a_1} + \frac{1}{a_2} + \cdots + \frac{1}{a_k} + \frac{1}{\lambda}\right.$

$< 1, \lambda \in \mathbf{N}^{*} \Big\}$. Considering

$$\frac{1}{a_1} + \frac{1}{a_2} + \cdots + \frac{1}{a_k} + \frac{1}{\lambda} < 1,$$

that is, $0 < \dfrac{1}{\lambda} < 1 - \left( \dfrac{1}{a_1} + \dfrac{1}{a_2} + \cdots + \dfrac{1}{a_k} \right)$, we have

$$\lambda > \frac{1}{1 - \dfrac{1}{a_1} - \dfrac{1}{a_2} - \cdots - \dfrac{1}{a_k}}.$$

In the following, we show that

$$\frac{1}{1 - \dfrac{1}{a_1} - \dfrac{1}{a_2} - \cdots - \dfrac{1}{a_k}} = a_k(a_k - 1).$$

By the induction hypotheses, for $2 \leqslant n \leqslant k$, $a_n = a_{n-1}$ $(a_{n-1} - 1) + 1$, we have

$$\frac{1}{a_n - 1} = \frac{1}{a_{n-1}(a_{n-1} - 1)} = \frac{1}{a_{n-1} - 1} - \frac{1}{a_{n-1}},$$

therefore $\dfrac{1}{a_{n-1}} = \dfrac{1}{a_{n-1} - 1} - \dfrac{1}{a_n - 1}$. By taking the sum, we have

$\displaystyle\sum_{i=2}^{k} \frac{1}{a_{i-1}} = 1 - \frac{1}{a_k - 1}$, that is

$$\sum_{i=1}^{k} \frac{1}{a_i} = 1 - \frac{1}{a_k - 1} + \frac{1}{a_k} = 1 - \frac{1}{a_k(a_k - 1)}.$$

Consequently, $\dfrac{1}{1 - \dfrac{1}{a_1} - \dfrac{1}{a_2} - \cdots - \dfrac{1}{a_k}} = a_k(a_k - 1).$

Therefore,

$$a_{k+1} = \min\left\{ \lambda \mid \frac{1}{a_1} + \frac{1}{a_2} + \cdots + \frac{1}{a_k} + \frac{1}{\lambda} < 1, \lambda \in \mathbf{N}^{*} \right\}$$

$$= a_k(a_k - 1) + 1.$$

By induction on $n$, for all positive integer $n$, we have, $a_{n+1} = a_n^2 - a_n + 1$. $\quad\square$

**7** There are $2n$ real numbers $a_1$, $a_2$, $\ldots$, $a_n$, $r_1$, $r_2$, $\ldots$, and $r_n$ satisfying $a_1 \leqslant a_2 \leqslant \cdots \leqslant a_n$ and $0 \leqslant r_1 \leqslant r_2 \leqslant \cdots \leqslant r_n$. Prove that $\sum_{i=1}^{n} \sum_{j=1}^{n} a_i a_j \min(r_i, r_j) \geqslant 0$.

**Solution.** Write a matrix of $n \times n$ elements as follows:

$$A_1 = \begin{pmatrix} a_1 a_1 r_1 & a_1 a_2 r_1 & a_1 a_3 r_1 & \cdots & a_1 a_n r_1 \\ a_2 a_1 r_1 & a_2 a_2 r_2 & a_2 a_3 r_2 & \cdots & a_2 a_n r_2 \\ a_3 a_1 r_1 & a_3 a_2 r_2 & a_3 a_3 r_3 & \cdots & a_3 a_n r_3 \\ \vdots & \vdots & \vdots & \ddots & \vdots \\ a_n a_1 r_1 & a_n a_2 r_2 & a_n a_3 r_3 & \cdots & a_n a_n r_n \end{pmatrix}.$$

Since

$$\sum_{i=1}^{n} \sum_{j=1}^{n} a_i a_j \min(r_i, r_j) = \sum_{j=1}^{n} a_1 a_j \min(r_1, r_j) + \sum_{j=1}^{n} a_2 a_j \min(r_2, r_j)$$

$$+ \cdots + \sum_{j=1}^{n} a_k a_j \min(r_k, r_j) + \cdots$$

$$+ \sum_{j=1}^{n} a_n a_j \min(r_n, r_j),$$

its $k$th term is

$$\sum_{j=1}^{n} a_k a_j \min(r_k, r_j) = a_k a_1 r_1 + a_k a_2 r_2 + \cdots + a_k a_k r_k$$

$$+ a_k a_{k+1} r_k + \cdots + a_k a_n r_k$$

which is the sum of elements of the $k$th row of $A_1$, $k = 1$, $2$, $\ldots$, $n$.

Therefore, $\sum_{i=1}^{n} \sum_{j=1}^{n} a_i a_j \min(r_i, r_j)$ is the sum of all elements of $A_1$.

On the other hand, the summation can also be done as follows: Take the elements of first column and the first row of $A_1$, sum up; denote the rest $(n-1) \times (n-1)$ element by matrix $A_2$, then take the first column and the first row of $A_2$, sum up, denote the rest by matrix $A_3, \ldots$, so we get

$$\sum_{i=1}^{n}\sum_{j=1}^{n} a_i a_j \min(r_i, r_j) = \sum_{k=1}^{n} r_k (a_k^2 + 2a_k(a_{k+1} + a_{k+2} + \cdots + a_n))$$

$$= \sum_{k=1}^{n} r_k \left( \left( a_k + \sum_{i=k+1}^{n} a_i \right)^2 - \left( \sum_{i=k+1}^{n} a_i \right)^2 \right)$$

$$= \sum_{k=1}^{n} r_k \left( \left( \sum_{i=k}^{n} a_i \right)^2 - \left( \sum_{i=k+1}^{n} a_i \right)^2 \right)$$

$$= r_1 \left( \sum_{i=1}^{n} a_i \right)^2 + r_2 \left( \sum_{i=2}^{n} a_i \right)^2 + r_3 \left( \sum_{i=3}^{n} a_i \right)^2 + \cdots$$

$$+ r_n \left( \sum_{i=n}^{n} a_i \right)^2 - r_1 \left( \sum_{i=2}^{n} a_i \right)^2 - r_2 \left( \sum_{i=3}^{n} a_i \right)^2$$

$$- \cdots - r_{n-1} \left( \sum_{i=n}^{n} a_i \right)^2$$

$$= \sum_{k=1}^{n} (r_k - r_{k-1}) \left( \sum_{i=k}^{n} a_i \right)^2 \geqslant 0$$

(where $r_0 = 0$).  □

**8** Given eight points $A_1, A_2, \ldots, A_8$ on a circle, determine the smallest positive integer $n$ such that among any $n$ triangles with vertices in these eight points, there are two which have a common side.

**Solution.** First, we consider the maximal number of triangles with no common side.

Consider the maximal number of triangles with no common side pairwise. There are $C_8^2 = 28$ chords by connecting eight points. If each chord only belongs to one triangle, then these chords can only form $r \leqslant \left[\frac{28}{3}\right] = 9$ triangles with no common side pairwise. But if there are nine such triangles, then there are 27 vertices. So, there is one point in eight points, which is the common vertex of four triangles.

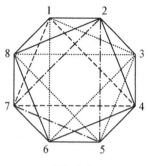

Fig. 8. 1

Suppose that point is $A_8$, then eight edges are connected to seven points $A_1$, $A_2$, $\ldots$, $A_7$. So, there must exist an edge $A_8 A_k$, which is the common side of two triangles, which is a contradiction. So $r \leqslant 8$.

On the other hand, when $r = 8$, we can make such eight triangles, see figure. Denote the triangles by three vertices as: (1, 2, 8), (1, 3, 6), (1, 4, 7), (2, 3, 4), (2, 5, 7), (3, 5, 8), (4, 5, 6) and (6, 7, 8). So the minimal number of $n$ is 9.

□

# $2011$ (Ningbo, Zhejiang)

## First Day

### 8: 00 – 12: 00, July 27, 2011

**1** If $\min_{x \in \mathbf{R}} \dfrac{ax^2 + b}{\sqrt{x^2 + 1}} = 3$, find

(1) the range of $b$;

(2) the value of $a$ for given $b$. (posed by Lu Xingjiang)

**Solution 1.** Denote $f(x) = \dfrac{ax^2 + b}{\sqrt{x^2 + 1}}$. It is easy to see that $a > 0$. By $f(0) = b$, we see that $b \geqslant 3$.

(i) If $b - 2a \geqslant 0$,

$$f(x) = \frac{ax^2 + b}{\sqrt{x^2 + 1}} = a\sqrt{x^2 + 1} + \frac{b - a}{\sqrt{x^2 + 1}} \geqslant 2\sqrt{a(b - a)} = 3,$$

equality holds if $a\sqrt{x^2 + 1} = \dfrac{b - a}{\sqrt{x^2 + 1}}$, that is, if $x = \pm\sqrt{\dfrac{b - 2a}{a}}$.

The value of $a$ for given $b$ is $a = \dfrac{b - \sqrt{b^2 - 9}}{2}$, especially

when $b = 3$, $a = \dfrac{3}{2}$.

(ii) If $b - 2a < 0$, let $\sqrt{x^2 + 1} = t(t \geqslant 1)$. $f(x) = g(t) =$

$at + \dfrac{b - a}{t}$ monotonically increasing when $t \geqslant 1$, so,

$$\min_{x \in \mathbf{R}} f(x) = g(1) = a + b - a = b = 3, \text{ when } a > \frac{3}{2}.$$

Summing up, we get (1) the range of $b$ is $[3, +\infty)$.

(2) If $b = 3$, then $a \geqslant \dfrac{3}{2}$; if $b > 3$, then $a = \dfrac{b - \sqrt{b^2 - 9}}{2}$.

**Solution 2.** Let $f(x) = \dfrac{ax^2 + b}{\sqrt{x^2 + 1}}$. It is easy to see that $a > 0$.

Since $\min_{x \in \mathbf{R}} \dfrac{ax^2 + b}{\sqrt{x^2 + 1}} = 3$, and $f(0) = b$, we see that $b \geqslant 3$.

$$f'(x) = \frac{ax\left(x^2 - \dfrac{b - 2a}{a}\right)}{(x^2 + 1)^{3/2}}.$$

(i) If $b - 2a \leqslant 0$, let $f'(x) = 0$, we have the solution $x_0 = 0$, and if $x < 0$, then $f'(x) < 0$; and if $x > 0$, then $f'(x) > 0$.

$f(0) = b$ is the minimal value. Hence $b = 3$, and $a \geqslant \dfrac{b}{2}$.

(ii) If $b - 2a > 0$, let $f'(x) = 0$, we have the solutions $x_0 = 0$, $x_{1,2} = \pm \sqrt{\dfrac{b - 2a}{a}}$.

It is easy to see $f(0) = b$ is not the minimal value, which implies $b > 3$; and $f(x_{1,2})$ is the minimal value

$$f(x_{1,2}) = \dfrac{a \cdot \dfrac{b - 2a}{a} + b}{\sqrt{\dfrac{b - 2a}{a}} + 1} = 3 \Rightarrow 2\sqrt{a(b - a)} = 3$$

$$\Rightarrow a^2 - ab + \dfrac{9}{4} = 0 \Rightarrow a = \dfrac{b - \sqrt{b^2 - 9}}{2},$$

that is, $b > 3$ and $a = \dfrac{b - \sqrt{b^2 - 9}}{2}$.

Summing up, we get

(1) the range of $b$ is $[3, +\infty)$.

(2) If $b = 3$, then $a \geqslant \dfrac{3}{2}$; if $b > 3$, then $a = \dfrac{b - \sqrt{b^2 - 9}}{2}$.

$\square$

**2** Let $a$, $b$ and $c$ be coprime positive integers so that $a^2 \mid (b^3 + c^3)$, $b^2 \mid (a^3 + c^3)$ and $c^2 \mid (a^3 + b^3)$. Find the values of $a$, $b$ and $c$. (posed by Yang Xiaoming)

**Solution.** By the condition of the problem, we have $a^2 \mid (a^3 + b^3 + c^3)$, $b^2 \mid (a^3 + b^3 + c^3)$ and $c^2 \mid (a^3 + b^3 + c^3)$. Since $a$, $b$ and $c$ are coprime, we see that $a^2 b^2 c^2 \mid (a^3 + b^3 + c^3)$.

Without loss of generality, suppose that $a \geqslant b \geqslant c$, so

$$3a^3 \geqslant a^3 + b^3 + c^3 \geqslant a^2 b^2 c^2 \Rightarrow a \geqslant \dfrac{b^2 c^2}{3},$$

and

$$2b^3 \geqslant b^3 + c^3 \geqslant a^2 \Rightarrow 2b^3 \geqslant \frac{b^4 c^4}{9} \Rightarrow b \leqslant \frac{18}{c^4}.$$

We see that if $c \geqslant 2 \Rightarrow b \leqslant 1$, the result contradicts $b \geqslant c$. Thus, $c = 1$.

If $c = 1$ and $b = 1$, then $a = 1$, so $(a, b, c) = (1, 1, 1)$ is a solution.

If $c = 1$, $b \geqslant 2$ and $a = b$, then $b^2 \mid b^3 + 1$, which is a contradiction!

If $b \geqslant 2$ and $a > b > c = 1$, then

$$a^2 b^2 \mid (a^3 + b^3 + 1) \Rightarrow 2a^3 \geqslant a^3 + b^3 + 1 \geqslant a^2 b^2 \Rightarrow a \geqslant \frac{b^2}{2},$$

and by $c = 1$,

$$a^2 \mid (b^3 + 1) \Rightarrow b^3 + 1 \geqslant a^2 \geqslant \frac{b^4}{4} \Rightarrow 4b^3 + 4 \geqslant b^4.$$

For $b > 5$, the inequality has no solution. Take $b = 2, 3, 4, 5$, we see that the solutions are $c = 1$, $b = 2$, $a = 3$.

Therefore, all solutions are $(a, b, c) = (1, 1, 1)$, $(1, 2, 3)$, $(1, 3, 2)$, $(2, 1, 3)$, $(2, 3, 1)$, $(3, 2, 1)$ and $(3, 1, 2)$. $\quad\square$

**3**  Let set $M = \{1, 2, 3, \ldots, 50\}$. Find all positive integer $n$, such that there are at least two different elements $a$ and $b$ in any subset with 35 elements of $M$, such that $a + b = n$ or $a - b = n$. (posed by Li Shenghong)

**Solution.** Take $A = \{1, 2, 3, \ldots, 35\}$, then for any $a, b \in A$,

$$a - b \leqslant 34, \quad a + b \leqslant 34 + 35 = 69.$$

In the following, we show that $1 \leqslant n \leqslant 69$. Let $A = \{a_1,$

$a_2, \ldots, a_{35}\}$, without loss of generality, suppose that $a_1 < a_2 < \cdots < a_{35}$.

(i) If $1 \leqslant n \leqslant 19$, by

$$1 \leqslant a_1 < a_2 < \cdots < a_{35} \leqslant 50,$$
$$2 \leqslant a_1 + n < a_2 + n < \cdots < a_{35} + n \leqslant 50 + 19 = 69,$$

and by Dirichlet's Drawer Theorem, there exist $1 \leqslant i, j \leqslant 35$ $(i \neq j)$ such that $a_i + n = a_j$, that is, $a_i - a_j = n$.

(ii) If $51 \leqslant n \leqslant 69$, by

$$1 \leqslant a_1 < a_2 < \cdots < a_{35} \leqslant 50,$$
$$1 \leqslant n - a_{35} < n - a_{34} < \cdots < n - a_1 \leqslant 68,$$

and by Dirichlet's Drawer Theorem, there exist at least $1 \leqslant i,$ $j \leqslant 35 (i \neq j)$ such that $n - a_i = a_j$, that is, $a_i + a_j = n$. ❦

(iii) If $20 \leqslant n \leqslant 24$, since

$$50 - (2n + 1) + 1 = 50 - 2n \leqslant 50 - 40 = 10,$$

we see that there are at least 25 elements in $a_1, a_2, \ldots, a_{35}$ that belong to $[1, 2n]$.

There are at most 24 elements in $\{1, n + 1\}, \{2, n + 2\}, \ldots,$ $\{n, 2n\}$ such that $\{a_i, a_j\} = \{i, n + i\}$. Hence, $a_j - a_i = n$.

(iv) If $25 \leqslant n \leqslant 34$, since $\{1, n + 1\}, \{2, n + 2\}, \ldots, \{n, 2n\}$ have at most 34 elements, by Dirichlet Drawer Theorem, there exist $1 \leqslant i, j \leqslant 35 (i \neq j)$ such that $a_i = i, a_j = n + i,$ that is $a_j - a_i = n$.

(v) If $n = 35$, there are 33 elements $\{1, 34\}, \{2, 33\}, \ldots,$ $\{17, 18\}, \{35\}, \{36\}, \ldots, \{50\}$. Hence, there exist $1 \leqslant i, j \leqslant 35 (i \neq j)$ such that $a_i + a_j = 35$.

(vi) If $36 \leqslant n \leqslant 50$,

if $n = 2k + 1$, $\{1, 2k\}, \{2, 2k - 1\}, \ldots, \{k, k + 1\},$ $\{2k + 1\}, \ldots, \{50\};$

if $18 \leqslant k \leqslant 20$, $50 - (2k + 1) + 1 = 50 - 2k \leqslant 50 - 36 = 14;$

if $21 \leqslant k \leqslant 24$, $50 - (2k + 1) + 1 = 50 - 2k \leqslant 50 - 42 = 8$,
there exist $1 \leqslant i, j \leqslant 35 (i \neq j)$ such that $a_i + a_j = 2k + 1 = n$.

If $n = 2k$, $\{1, 2k - 1\}$, $\{2, 2k - 2\}$, $\ldots$, $\{k - 1, k + 1\}$,
$\{k\}$, $\{2k\}$, $\{2k + 1\}$, $\ldots$, $\{50\}$;

if $18 \leqslant k \leqslant 19$, $50 - (2k + 1) + 3 \leqslant 16k - 1 \leqslant 19 - 1 = 18$;

if $20 \leqslant k \leqslant 23$, $50 - (2k + 1) + 3 \leqslant 50 - 2k + 2 \leqslant 12k - 1$
$\leqslant 23 - 1 = 22$;

if $24 \leqslant k \leqslant 25$, $50 - (2k + 1) + 3 \leqslant 50 - 2k + 2 \leqslant 4k - 1 \leqslant$
$25 - 1 = 24$,
there exist $1 \leqslant i, j \leqslant 35 (i \neq j)$ such that $a_i + a_j = 2k$.   $\square$

**4** Suppose that a line passing the circumcentre $O$ of $\triangle ABC$
intersects $AB$ and $AC$ at points $M$ and $N$, respectively, and
$E$ and $F$ are the midpoints of $BN$ and $CM$, respectively.
Prove that $\angle EOF = \angle A$. (posed by Tao Pingsheng)

**Solution.** We show that the above conclusion is true for any triangle.

If $\triangle ABC$ is right-angled. The conclusion is obvious. In
fact, see Fig. 4.1, where $\angle ABC = 90°$. So, the circumcentre $O$
is the midpoint of $AC$, $OA = OB$ and $N = O$. Since $F$ is the
midpoint of $CM$, we see that the median line $OF \parallel AM$. Hence
$\angle EOF = \angle OBA = \angle OAB = \angle A$.

If $\triangle ABC$ is not right-angled, see Fig. 4.2 and Fig. 4.3.

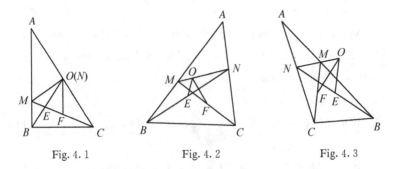

Fig. 4.1          Fig. 4.2          Fig. 4.3

First we give a lemma.

**Lemma.** Let $A$ and $B$ be two points on the diameter $KL$ of circle $\odot O$ with radius $R$, and $OA = OB = a$. See figure.

Let $CD$ and $EF$ be two chords passing $A$ and $B$, respectively. Suppose $CE$ and $DF$ intersect $KL$ at $M$ and $N$, respectively. Then $MA = NB$.

**Proof of the lemma.** As shown in Fig. 4. 4. Suppose that $CD \cap EF = P$. Think of that lines $CE$ and $DF$ intersect $\triangle PAB$. By Menelaus' Theorem, we have

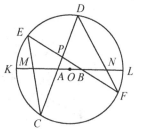

Fig. 4. 4

$$\frac{AC}{CP} \cdot \frac{PE}{EB} \cdot \frac{BM}{MA} = 1,$$

$$\frac{BF}{FP} \cdot \frac{PD}{DA} \cdot \frac{AN}{NB} = 1.$$

Then

$$\frac{MA}{NB} = \frac{AC}{BE} \cdot \frac{AD}{BF} \cdot \frac{PE}{PC} \cdot \frac{PF}{PD} \cdot \frac{BM}{AN}. \qquad ①$$

By the Intersecting Chord Theorem, we get

$$PC \cdot PD = PE \cdot PF. \qquad ②$$

So

$$AC \cdot AD = AK \cdot AL = R^2 - a^2 = BK \cdot BL = BE \cdot BF. \qquad ③$$

By ①, ③, we have $\dfrac{MA}{NB} = \dfrac{MB}{NA}$, that is

$$\frac{MA}{NB} = \frac{MA + AB}{NB + AB} = \frac{AB}{AB} = 1.$$

Thus, $MA = NB$.

Now, return to the original problem, see Fig. 4. 5 and

Fig. 4. 6. Extend $MN$ to diameter $KK_1$, take point $M_1$ on $KK_1$ such that $OM_1 = OM$. Let $CM_1 \cap \odot O = A_1$, and let $A_1B$ intersect $KK_1$ at $N_1$. By the lemma, $MN_1 = M_1N$ (or $M_1N_1 = MN$ on the right figure). So, $O$ is the midpoint of $NN_1$. Thus, $OE$ and $OF$ are the median line of $\triangle NBN_1$ and $\triangle MCM_1$, respectively. Thus, we have $\angle EOF = \angle BA_1C = \angle A$. $\square$

Fig. 4. 5

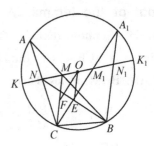

Fig. 4. 6

## Second Day

8: 00 – 12: 00, July 28, 2011

**5** Let $AA_0$, $BB_0$ and $CC_0$ be angular bisectors of $\triangle ABC$. Let $A_0A_1 \parallel BB_0$ and $A_0A_2 \parallel CC_0$, where $A_1$ and $A_2$ lie on $AC$ and $AB$, respectively, and let line $A_1A_2$ intersect $BC$ at $A_3$. The points $B_3$ and $C_3$ are obtained similarly. Prove that points $A_3$, $B_3$, $C_3$ are collinear. (posed by Tao Pingsheng)

**Solution.** By the Menelaus

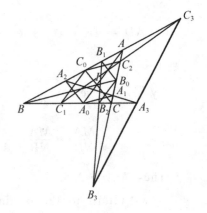

Fig. 5. 1

Inverse Theorem, we need only to show that

$$\frac{AB_3}{B_3C} \cdot \frac{CA_3}{A_3B} \cdot \frac{BC_3}{C_3A} = 1. \qquad \textcircled{1}$$

Since line $A_1A_2A_3$ intersects $\triangle ABC$, by Menelaus' Theorem, we have $\dfrac{CA_3}{A_3B} \cdot \dfrac{BA_2}{A_2A} \cdot \dfrac{AA_1}{A_1C} = 1$. So

$$\frac{CA_3}{A_3B} = \frac{A_2A}{BA_2} \cdot \frac{A_1C}{AA_1}. \qquad \textcircled{2}$$

Similarly, we have

$$\frac{AB_3}{B_3C} = \frac{B_2B}{CB_2} \cdot \frac{B_1A}{BB_1}, \qquad \textcircled{3}$$

$$\frac{BC_3}{C_3A} = \frac{C_2C}{AC_2} \cdot \frac{C_1B}{CC_1}. \qquad \textcircled{4}$$

By $BA_2 = \dfrac{BC_0}{BC} \cdot BA_0$ and $AA_2 = \dfrac{AA_0}{AI} \cdot AC_0$, we have

$$\frac{AA_2}{BA_2} = \frac{AA_0 \cdot AC_0}{BA_0 \cdot BC_0} \cdot \frac{BC}{AI}. \qquad \textcircled{5}$$

Moreover, by $AA_1 = \dfrac{AA_0}{AI} \cdot AB_0$ and $CA_1 = \dfrac{CA_0}{CB} \cdot CB_0$, we have

$$\frac{A_1C}{AA_1} = \frac{CA_0 \cdot CB_0}{AA_0 \cdot AB_0} \cdot \frac{AI}{BC}. \qquad \textcircled{6}$$

Then by $\textcircled{2}$, $\textcircled{5}$ and $\textcircled{6}$, we have

$$\frac{CA_3}{A_3B} = \frac{CA_0}{BA_0} \cdot \frac{AC_0}{BC_0} \cdot \frac{CB_0}{AB_0} = \left(\frac{CA_0}{A_0B}\right)^2.$$

Similarly, we have

$$\frac{CB_3}{B_3A} = \left(\frac{CB_0}{B_0A}\right)^2,$$

$$\frac{AC_3}{C_3B} = \left(\frac{AC_0}{C_0B}\right)^2. \tag{7}$$

Since three angle bisectors $AA_0$, $BB_0$, $CC_0$ of $\triangle ABC$ are concurrent, by Ceva's Theorem, we have

$$\frac{AB_0}{B_0C} \cdot \frac{CA_0}{A_0B} \cdot \frac{BC_0}{C_0A} = 1. \tag{8}$$

Thus, by ⑦ and ⑧, we have

$$\frac{AB_3}{B_3C} \cdot \frac{CA_3}{A_3B} \cdot \frac{BC_3}{C_3A} = \left(\frac{AB_0}{B_0C} \cdot \frac{CA_0}{A_0B} \cdot \frac{BC_0}{C_0A}\right)^2 = 1,$$

that is ①. $\qquad\square$

**6** Given $n$ points $P_1$, $P_2$, ..., $P_n$ on a plane, let $M$ be any point on segment $AB$ on the plane. Denote by $|P_iM|$ the distance between $P_i$ and $M$, $i = 1, 2, 3, \ldots, n$. Prove that

$$\sum_{i=1}^{n} |P_iM| \leqslant \max\left\{\sum_{i=1}^{n} |P_iA|, \sum_{i=1}^{n} |P_iB|\right\}.$$

(posed by Jin Mengwei)

**Solution.** Let $O$ be the origin. Then we have $\overrightarrow{OM} = t\overrightarrow{OA} + (1-t)\overrightarrow{OB}$, $t \in (0, 1)$.

$$
\begin{aligned}
|P_iM| &= |\overrightarrow{OM} - \overrightarrow{OP_i}| \\
&= |t\overrightarrow{OA} + (1-t)\overrightarrow{OB} - t\overrightarrow{OP_i} - (1-t)\overrightarrow{OP_i}| \\
&\leqslant t|\overrightarrow{OA} - \overrightarrow{OP_i}| + (1-t)|\overrightarrow{OB} - \overrightarrow{OP_i}| \\
&= t|P_iA| + (1-t)|P_iB|.
\end{aligned}
$$

Hence,

$$\sum_{i=1}^{n} |P_iM| \leqslant t\sum_{i=1}^{n} |\overrightarrow{P_iA}| + (1-t)\sum_{i=1}^{n} |\overrightarrow{P_iB}|$$

$$\leqslant \max\left\{\sum_{i=1}^{n} |P_iA|, \sum_{i=1}^{n} |P_iB|\right\}. \qquad\square$$

**7** Suppose that the sequence $\{a_n\}$ defined by $a_1 = a_2 = 1$, $a_n = 7a_{n-1} - a_{n-2}$, $n \geq 3$. Prove that $a_n + a_{n+1} + 2$ is a perfect square for any positive integer $n$. (posed by Tao Pingsheng)

**Solution.** It is well known that the solution of the sequence can be obtained by solving two geometric sequence. We get $a_n = C_1\lambda_1^n + C_2\lambda_2^n$, where

$$\lambda_1 = \frac{7 + \sqrt{45}}{2} = \left(\frac{3 + \sqrt{5}}{2}\right)^2, \quad \lambda_2 = \frac{7 - \sqrt{45}}{2} = \left(\frac{3 - \sqrt{5}}{2}\right)^2,$$

$\lambda_1$ and $\lambda_2$ are solutions of the equation $\lambda^2 - 7\lambda + 1 = 0$, $a_1 = 1 = C_1\lambda_1 + C_2\lambda_2$, $a_2 = 1 = C_1\lambda_1^2 + C_2\lambda_2^2$.

Therefore,

$$\begin{aligned}
a_n + a_{n+1} + 2 &= \lambda_1^{n-1}C_1(\lambda_1 + \lambda_1^2) + \lambda_2^{n-1}C_2(\lambda_2 + \lambda_2^2) + 2 \\
&= \lambda_1^{n-1} + \lambda_2^{n-1} + 2 \\
&= \left(\left(\frac{3 + \sqrt{5}}{2}\right)^{n-1}\right)^2 + \left(\left(\frac{3 - \sqrt{5}}{2}\right)^{n-1}\right)^2 + 2 \\
&= \left[\left(\frac{3 + \sqrt{5}}{2}\right)^{n-1} + \left(\frac{3 - \sqrt{5}}{2}\right)^{n-1}\right]^2 = x_n^2,
\end{aligned}$$

where $x_n = \left(\frac{3 + \sqrt{5}}{2}\right)^{n-1} + \left(\frac{3 - \sqrt{5}}{2}\right)^{n-1}$ is the solution of the sequence $\{x_n\}$ of positive integers $x_1 = 2$, $x_2 = 3$, $x_n = 3x_{n-1} - x_{n-2}$, $n \geq 3$. □

**8** Consider 12 figures on the clock face as 12 points. Color them in four colors: red, yellow, blue and green. Each color is used for three points. Configure $n$ convex quadrilaterals with vertices in these points, such that

(1) there are no same color of vertices for each quadrilateral.

(2) among any three of these quadrilaterals, there is a

color of vertices such that the vertices of that color are different.

Find the largest number of $n$. (posed by Tao Pingsheng)

**Solution.** We use $A$, $B$, $C$, $D$ to represent these four colors, respectively, and the points in the same color by lower letters as $a_1$, $a_2$, $a_3$; $b_1$, $b_2$, $b_3$; $c_1$, $c_2$, $c_3$ and $d_1$, $d_2$, $d_3$ respectively.

Now consider color $A$. If, in $n$ quadrilaterals, the number of points $a_1$, $a_2$, $a_3$ in color $A$ are $n_1$, $n_2$, $n_3$ respectively, then $n_1 + n_2 + n_3 = n$. Suppose that $n_1 \geqslant n_2 \geqslant n_3$. If $n \geqslant 10$, then $n_1 + n_2 \geqslant 7$. Consider these seven quadrilaterals (its $A$ color vertex is either $a_1$ or $a_2$), if the numbers of points $b_1$, $b_2$, $b_3$ in color $B$ are $m_1$, $m_2$, $m_3$, respectively, then $m_1 + m_2 + m_3 = 7$. By symmetricity, we may suppose that $m_1 \geqslant m_2 \geqslant m_3$, then $m_3 \leqslant 2$, that is, $m_1 + m_2 \geqslant 5$.

Consider these five quadrilaterals (its $A$ color point is either $a_1$ or $a_2$, and its $B$ color point is either $b_1$ or $b_2$), if the numbers of points $c_1$, $c_2$, $c_3$ in color $C$ are $k_1$, $k_2$, $k_3$, respectively, then $k_1 + k_2 + k_3 = 5$. By symmetricity, we may suppose that $k_1 \geqslant k_2 \geqslant k_3$, then $k_3 \leqslant 1$, that is, $k_1 + k_2 \geqslant 4$.

Consider these four quadrilaterals, denoted as $T_1$, $T_2$, $T_3$, $T_4$ (its color $A$ point is either $a_1$ or $a_2$, its color $B$ point is either $b_1$ or $b_2$, and its color $C$ point is either $c_1$ or $c_2$). Since there are only three points in color $D$, there are two quadrilaterals that have the same color $D$ point. Suppose that the same color $D$ point of $T_1$, $T_2$ is $d_1$.

Then, in three quadrilaterals $T_1$, $T_2$, $T_3$, whatever be the color of the vertex, there are repeated points, which contradicts condition (2). Hence, $n \leqslant 9$.

We show the maximal number $n = 9$ by construction of these nine quadrilaterals.

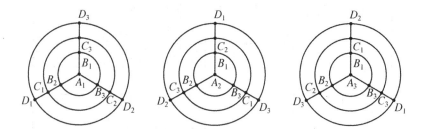

We draw three "concentric annulus" with four points on each radius representing four vertices and the color. So nine radii represent nine quadrilaterals, which satisfy condition (1).

Next, we show that they also satisfy condition (2). Take any three radii (or three quadrilaterals).

If these three radii come from a concentric annulus, for each color except $A$, there are three points.

If these three radii come from three concentric annuli, then for color $A$, there are three points.

If these three radii come from two concentric annuli, call these three figures Fig. 1, Fig. 2 and Fig. 3. The radius directions are called "up radius", "left radius" and "right radius", and denoted, respectively, by $S$, $Z$ and $Y$. If three radii have three directions, then there are three points of color $B$ in three quadrilaterals. If the three radii have only two directions, then there are all cases as shown in the tables below, where 1, 2 and 3 stand for Fig. 1, Fig. 2 and Fig. 3, respectively.

Here, the color in figure means that the three quadrilaterals have color with different figures.

| $S$ | 1, 2 | 1, 2 | 2 | | 1 | |
|---|---|---|---|---|---|---|
| $Z$ | 1 | | 1, 2 | 1, 2 | | 2 |
| $Y$ | | 2 | | 1 | 1, 2 | 1, 2 |

C

| $S$ | 1, 2 | 1, 2 | 1 | | 2 | |
|---|---|---|---|---|---|---|
| $Z$ | 2 | | 1, 2 | 1, 2 | | 1 |
| $Y$ | | 1 | | 2 | 1, 2 | 1, 2 |

D

| S | 1, 3 | 1, 3 | 1 | | 3 | |
|---|---|---|---|---|---|---|
| Z | 3 | | 1, 3 | 1, 3 | | 1 |
| Y | | 1 | | 3 | 1, 3 | 1, 3 |

C

| S | 1, 3 | 1, 3 | 3 | | 1 | |
|---|---|---|---|---|---|---|
| Z | 1 | | 1, 3 | 1, 3 | | 3 |
| Y | | 3 | | 1 | 1, 3 | 1, 3 |

D

| S | 2, 3 | 2, 3 | 3 | | 2 | |
|---|---|---|---|---|---|---|
| Z | 2 | | 2, 3 | 2, 3 | | 3 |
| Y | | 3 | | 2 | 2, 3 | 2, 3 |

C

| S | 2, 3 | 2, 3 | 2 | | 3 | |
|---|---|---|---|---|---|---|
| Z | 3 | | 2, 3 | 2, 3 | | 2 |
| Y | | 2 | | 3 | 2, 3 | 2, 3 |

D

Thus, the maximal number of $n$ is 9.  ▢

# 2012  (Putian, Fujian)

## First Day

8: 00 – 12: 00, July 27, 2012

**1** Find a triple $(l, m, n)(1 < l < m < n)$ of positive integers such that $\sum_{k=1}^{l} k$, $\sum_{k=l+1}^{m} k$, $\sum_{k=m+1}^{n} k$ form a geometric sequence in order. (posed by Tao Pingsheng)

**Solution.** For $t \in \mathbf{N}^*$, denote $S_t = \sum_{k=1}^{t} k = \dfrac{t(t+1)}{2}$. Let

$$\sum_{k=1}^{l} k = S_l, \sum_{k=l+1}^{m} k = S_m - S_l, \sum_{k=m+1}^{n} k = S_n - S_m,$$

form a geometric sequence in order. Then

$$S_l(S_n - S_m) = (S_m - S_l)^2, \qquad ①$$

that is $S_l(S_n + S_m - S_l) = S_m^2$. Thus, $S_l \mid S_m^2$, that is

$$2l(l+1) \mid m^2(m+1)^2.$$

Let $m + 1 = l(l + 1)$ and take $l = 3$. Then $m = 11$ and $S_l = S_3 = 6$, $S_m = S_{11} = 66$. Substitute it into ①, we have $S_n = 666$, that is $\dfrac{n(n+1)}{2} = 666$. So $n = 36$.

Therefore, $(l, m, n) = (3, 11, 36)$ is a solution satisfying the condition. $\qquad\square$

**Remark.** The solution is not unique. For example, there are other solutions as $(8, 11, 13)$, $(5, 9, 14)$, $(2, 12, 62)$, $(3, 24, 171)$. (We may show that the number of solutions is infinite.)

**2** Let $\odot I$ be the incircle of $\triangle ABC$. The circle $\odot I$ intersects sides $AB$, $BC$ and $CA$ at points $D$, $E$ and $F$, respectively. Line $EF$ intersects lines $AI$, $BI$ and $DI$ at points $M$, $N$ and $K$, respectively. Prove that $DM \cdot KE = DN \cdot KF$. (posed by Zhang Pengcheng)

**Solution.** It is easy to see that points $I$, $D$, $E$ and $B$ are concyclic and

$$\angle AID = 90° - \angle IAD,$$
$$\angle MED = \angle FDA = 90° - \angle IAD.$$

So $\angle AID = \angle MED$, thus points $I$, $D$, $E$ and $M$ are concyclic.

Hence, five points $I$, $D$, $B$, $E$, $M$ are concyclic and $\angle IMB = \angle IEB = 90°$, that is $AM \perp BM$.

Similarly, points $I$, $D$, $A$, $N$ and $F$ are concyclic and $BN \perp AN$.

Let lines $AN$ and $BM$ intersect at point $G$. We see point $I$ is the

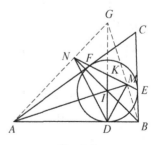

Fig. 2.1

orthocenter of $\triangle GAB$ and $ID \perp AB$, so points $G$, $I$ and $D$ are collinear.

Since points $G$, $N$, $D$ and $B$ are concyclic, we see that $\angle ADN = \angle G$.

Similarly, $\angle BDM = \angle G$. So $DK$ bisects $\angle MDN$, thus

$$\frac{DM}{DN} = \frac{KM}{KN}. \qquad \text{①}$$

Since points $I$, $D$, $E$ and $M$ are concyclic, and points $I$, $D$, $N$ and $F$ are concyclic, we see that

$$KM \cdot KE = KI \cdot KD = KF \cdot KN.$$

Therefore,

$$\frac{KM}{KN} = \frac{KF}{KE}. \qquad \text{②}$$

By ① and②, we see that $\dfrac{DM}{DN} = \dfrac{KF}{KE}$, that is $DM \cdot KE = DN \cdot KF$. $\qquad \square$

**3** For positive composite number $n$, denote by $f(n)$ and $g(n)$ the sum of the smallest three positive divisors of $n$ and the largest two positive divisors of $n$, respectively. Find all $n$ such that $g(n)$ equals $f(n)$ to some power of positive integers. (posed by He Yijie)

**Solution.** If $n$ is odd, then all factors of $n$ are odd. So $f(n)$ is odd and $g(n)$ is even. $g(n)$ cannot be $f(n)$ to some power of positive integer. Therefore $n$ is even. The smallest two divisors of $n$ are 1 and 2, and the largest two divisors of $n$ are $n$ and $n/2$.

Let $d$ be the third smallest divisor of $n$. If there exists $k \in \mathbf{N}^*$ such that $g(n) = f^k(n)$, then

$$\frac{3n}{2} = (1 + 2 + d)^k = (3 + d)^k \equiv d^k \,(\mathrm{mod}\ 3).$$

Since $3 \left| \dfrac{3n}{2} \right.$, we see that $3 \mid d^k$. So $3 \mid d$, and since $d$ is the third smallest, we see that $d = 3$.

Thus, $\dfrac{3}{2}n = 6^k$, we get $n = 4 \times 6^{k-1}$. Since $3 \mid n$, we see that $k \geqslant 2$.

Summing up, $n = 4 \times 6^l\,(l \in \mathbf{N}^*)$. $\qquad\qquad\square$

**④** Let real numbers $a$, $b$, $c$ and $d$ satisfy

$$f(x) = a\cos x + b\cos 2x + c\cos 3x + d\cos 4x \leqslant 1$$

for any real number $x$. Find the values of $a$, $b$, $c$ and $d$ such that $a + b - c + d$ takes the maximum number.

(posed by Li Shenghong)

**Solution.** Since

$$f(0) = a + b + c + d,$$
$$f(\pi) = -a + b - c + d,$$
$$f\left(\frac{\pi}{3}\right) = \frac{a}{2} - \frac{b}{2} - c - \frac{d}{2},$$

then

$$a + b - c + d = f(0) + \frac{2}{3}f(\pi) + \frac{4}{3}f\left(\frac{\pi}{3}\right) \leqslant 3$$

if and only if $f(0) = f(\pi) = f\left(\dfrac{\pi}{3}\right) = 1$, that is, if $a = 1$, $b + d = 1$ and $c = -1$, then the equality holds. Let $t = \cos x$, $-1 \leqslant t \leqslant 1$. Then

$$f(x) - 1 = \cos x + b\cos 2x - \cos 3x + d\cos 4x - 1$$
$$= t + (1 - d)(2t^2 - 1) - (4t^3 - 3t) + d(8t^4 - 8t^2 + 1) - 1$$
$$= 2(1 - t^2)[-4dt^2 + 2t + (d - 1)] \leqslant 0, \ \forall t \in [-1, 1],$$

that is

$$4dt^2 - 2t + (1 - d) \geqslant 0, \ \forall t \in (-1, 1).$$

Taking $t = 1/2 + \varepsilon$, $|\varepsilon| < 1/2$, then $\varepsilon[(2d - 1) + 4d\varepsilon] \geqslant 0$,

$|\varepsilon| < 1/2$. So we see that $d = \dfrac{1}{2}$. If $d = \dfrac{1}{2}$, then

$$4dt^2 - 2t + (1 - d) = 2t^2 - 2t + 1/2 = 2(t - 1/2)^2 \geqslant 0.$$

So, the maximal number of $a + b - c + d$ is 3, and $(a, b, c,$

$d) = \left(1, \dfrac{1}{2}, -1, \dfrac{1}{2}\right)$.                    □

## Second Day

### 8: 00 – 12: 00, July 28, 2012

**⑤** A non-negative number $m$ is called a *six match number*. If $m$ and the sum of its digits are both multiples of 6, find the number of the six match numbers less than 2012. (posed by Tao Pingsheng)

**Solution.** Let $n = \overline{d_1 d_2 d_3 d_4} = 1000d_1 + 100d_2 + 10d_3 + d_4$, $d_1, d_2, d_3, d_4 \in [0, 1, 2, \ldots, 9]$, and $S(n) = d_1 + d_2 + d_3 + d_4$.

Match the non-negative multiples of 6 less than 2000 into 167 pairs $(x, y)$, $x + y = 1998$, such that

$$(0, 1998), (6, 1992), (12, 1986), \ldots, (996, 1002).$$

For each pair $(x, y)$, let $x = \overline{a_1 a_2 a_3 a_4}$, $y = \overline{b_1 b_2 b_3 b_4}$, then

$$1000(a_1 + b_1) + 100(a_2 + b_2) + 10(a_3 + b_3) + (a_4 + b_4)$$
$$= x + y = 1998.$$

Since $x$, $y$ are even, $a_4$, $b_4 \leqslant 8$. So $a_4 + b_4 \leqslant 16 < 18$. Thus, $a_4 + b_4 = 8$. Since $a_3 + b_3 \leqslant 18 < 19$, $a_3 + b_3 = 9$.

Similarly, we can obtain $a_2 + b_2 = 9$ and $a_1 + b_1 = 1$. Thus,

$$S(x) + S(y) = (a_1 + b_1) + (a_2 + b_2) + (a_3 + b_3) + (a_4 + b_4)$$
$$= 1 + 9 + 9 + 8 = 27.$$

Consequently, there is only one of $S(x)$ and $S(y)$ that is the multiple of 6. (This is because that $x$, $y$ are all multiples of 3, so are $S(x)$ and $S(y)$. ) That is, there is only one of $x$ and $y$ which is a six match number.

Therefore, there are 167 six match numbers less than 2000, and there is just one six match number between 2000 and 2011. Therefore, the answer is $167 + 1 = 168$.  □

**6** Find the minimum positive integer $n$ such that

$$\sqrt{\frac{n - 2011}{2012}} - \sqrt{\frac{n - 2012}{2011}} < \sqrt[3]{\frac{n - 2013}{2011}} - \sqrt[3]{\frac{n - 2011}{2013}}.$$

(posed by Liu Guimei)

**Solution.** We see that if $2012 \leqslant n \leqslant 4023$, then $\sqrt{\dfrac{n - 2011}{2012}} - \sqrt{\dfrac{n - 2012}{2011}} \geqslant 0$ and $\sqrt[3]{\dfrac{n - 2013}{2011}} - \sqrt[3]{\dfrac{n - 2011}{2013}} < 0$.

Otherwise,

$$\sqrt{\frac{n - 2011}{2012}} \leqslant \sqrt{\frac{n - 2012}{2011}} \Leftrightarrow n > 4023$$

$$\sqrt[3]{\frac{n - 2013}{2011}} \geqslant \sqrt[3]{\frac{n - 2011}{2013}} \Leftrightarrow n \geqslant 4024.$$

Thus, if $n \geqslant 4024$, then

$$\sqrt{\frac{n - 2011}{2012}} - \sqrt{\frac{n - 2012}{2011}} < 0 \leqslant \sqrt[3]{\frac{n - 2013}{2011}} - \sqrt[3]{\frac{n - 2011}{2013}}.$$

So the minimum of $n$ is 4024.  □

**7** In $\triangle ABC$ with $AB = 1$, let $D$ be a point on $AC$ such that $\angle ABD = \angle C$ and let $E$ be a point on $AB$ such that $BE = DE$. Let $H$ be a point on $DE$ such that $AH \perp DE$ and $M$ be the midpoint of $CD$. If $AH =$

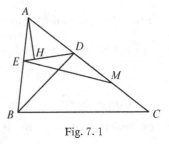

Fig. 7. 1

$2 - \sqrt{3}$, find the size of $\angle AME$. (posed by Xiong Bin)

**Solution.** Let $\angle ABD = \angle C = \alpha$ and $\angle DBC = \beta$. It is easy to see that $\angle BDE = \alpha$, $\angle AED = 2\alpha$,

$$\angle ADE = \angle ADB - \angle BDE = (\alpha + \beta) - \alpha = \beta,$$
$$AB = AE + EB = AE + EH + HD.$$

Hence,

$$\frac{AB}{AH} = \frac{AE + EH}{AH} + \frac{HD}{AH} = \frac{1 + \cos 2\alpha}{\sin 2\alpha} + \cot \beta$$
$$= \cot \alpha + \cot \beta. \qquad ①$$

Draw lines $EK \perp AC$ and $EL \perp BD$ with pedals $K$ and $L$, respectively. Then, $L$ is the midpoint of $BD$. Combining with the Sine Theorem, we obtain

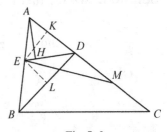

$$\frac{EL}{EK} = \frac{DE \sin \angle EDL}{DE \sin \angle EDK} = \frac{\sin \alpha}{\sin \beta}$$
$$= \frac{BD}{CD} = \frac{LD}{MD}.$$

Fig. 7. 2

Thus,

$$\cot \alpha = \frac{LD}{EL} = \frac{MD}{EK} = \frac{MK}{EK} - \frac{DK}{EK} = \cot \angle AME - \cot \beta. \qquad ②$$

By ①, ② and known conditions, we have

$$\cot \angle AME = \frac{AB}{AH} = \frac{1}{2 - \sqrt{3}} = 2 + \sqrt{3}.$$

Therefore, $\angle AME = 15°$. ☐

**8** Let $m$ be a positive integer, $n = 2^m - 1$, and $P_n = \{1, 2, \ldots, n\}$ be the set of $n$ points on number axis. A grasshopper jumps between adjacent points on $P_n$. Find the maximal number of $m$ such that for any $x$, $y \in P_n$, the number of ways that a grasshopper jumping from $x$ to $y$ by 2012 steps is even (passing $x$ or $y$ on the way is permitted) . (posed by Zhang Sihui)

**Solution.** If $m \geqslant 11$, then $n = 2^m - 1 > 2013$. Since there is only one way a grasshopper jumps from point 1 to point 2013 by 2012 steps, we see that $m \leqslant 10$.

In the following, we show that the answer is $m = 10$. To show this, we will prove a stronger proposition by induction on $m$ : for any $k \geqslant n = 2^m - 1$ and any $x$, $y \in P_n$, the number of ways the grasshopper jumps from point $x$ to $y$ by $k$ steps is even.

If $m = 1$, the number of ways is 0, where 0 is even.

If $m = l$, the number of ways is even. Then, for $k \geqslant n = 2^{l+1} - 1$, there are three kind of routes from point $x$ to point $y$ by $k$ steps. We show that the number of ways is even for each kind of route.

(1) The route does not pass point $2^l$. So points $x$ and $y$ both are on one side of point $2^l$. By the induction hypotheses, there are even routes.

(2) The route passes point $2^l$ just once.

Suppose that the grasshopper is at point $2^l$ at the $i$-th steps.

($i \in \{0, 1, \ldots, k\}$, $i = 0$ means $x = 2^l$, $i = k$ means $y = 2^l$ ). We show that, for any $i$, the number of routes is even.

Suppose that the route is $x$, $a_1$, $\ldots$, $a_{i-1}$, $2^l$, $a_{i+1}$, $\ldots$, $a_{k-1}$, $y$. Divided it into two sub-routes: from point $x$ to point $a_{i-1}$ of $i - 1$ steps and from point $a_{i+1}$ to point $y$ of $k - i - 1$ steps (for $i = 0$ or $k$, only one sub-route of $k - 1$ steps).

If $i - 1 < 2^l - 1$ and $k - i - 1 < 2^l - 1$, then $k \leqslant 2^{l+1} - 2$, which contradicts $k \geqslant n = 2^{l+1} - 1$. So, we must have $i - 1 \geqslant 2^l - 1$ or $k - i - 1 \geqslant 2^l - 1$. By the induction hypotheses, there are even ways for a sub-route. So, by the Multiplication Principle, the number of ways is even.

(3) The route passes point $2^l$ no less than two times.

Consider the sub-routes from $2^l$ to $2^l$, the number of ways is even, since we can consider the routes symmetric to $2^l$. So by the Multiplication Principle, the number of ways is even.

Summing up, the maximal $m$ is 10. $\qquad\square$

# $2013$ (Yingtan, Jiangxi)

## First Day

8: 00 – 12: 00, July 27, 2013

**1** Let $a$, $b$ be real numbers such that the equation $x^3 - ax^2 + bx - a = 0$ has only real roots. Find the minimum of $\dfrac{2a^3 - 3ab + 3a}{b + 1}$.

**Solution.** Let $x_1$, $x_2$ and $x_3$ be the real roots of the equation

$x^3 - ax^2 + bx - a = 0$. By Vieta's Formula, we have

$$x_1 + x_2 + x_3 = a, \ x_1 x_2 + x_2 x_3 + x_1 x_3 = b, \ x_1 x_2 x_3 = a.$$

By $(x_1 + x_2 + x_3)^2 \geqslant 3(x_1 x_2 + x_2 x_3 + x_1 x_3)$, we have $a^2 \geqslant 3b$, and by $a = x_1 + x_2 + x_3 \geqslant 3\sqrt[3]{x_1 x_2 x_3} = 3\sqrt[3]{a}$, we have $a \geqslant 3\sqrt{3}$. Thus,

$$\frac{2a^3 - 3ab + 3a}{b+1} = \frac{a(a^2 - 3b) + a^3 + 3a}{b+1}$$

$$\geqslant \frac{a^3 + 3a}{b+1} \geqslant \frac{a^3 + 3a}{\dfrac{a^2}{3} + 1}$$

$$= 3a \geqslant 9\sqrt{3}.$$

If $a = 3\sqrt{3}$, $b = 9$, then the equality holds when each root is equal to $\sqrt{3}$.

Summing up, the answer is $9\sqrt{3}$.  □

**2** Let $\odot I$ be the incircle of $\triangle ABC$ with $AB > AC$. $\odot I$ tangent to $BC$ and $AD$ at $D$ and $E$, respectively. The tangent line $EP$ of $\odot I$ intersects the extended line of $BC$ at $P$. Segment $CF$ is parallel to $PE$ and intersects $AD$ at point $F$. Line $BF$ intersects $\odot I$ at points $M$ and $N$ such that $M$ is on segment $BF$. Segment $PM$ intersects $\odot I$ at the other point $Q$. Prove that $\angle ENP = \angle ENQ$.

**Solution.** Suppose that $\odot I$ touches $AC$ and $AB$ at $S$ and $T$, respectively. Suppose that $ST$ intersects $AI$ at point $G$, we see that $IT \perp AT$ and $TG \perp AI$. So we

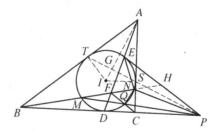

have $AG \cdot AI = AT^2 = AD \cdot AE$, thus points $I$, $G$, $E$ and $D$ are concyclic.

Since $IE \perp PE$ and $ID \perp PD$, we see that points $I$, $E$, $P$ and $D$ are concyclic. Hence, points $I$, $G$, $E$, $P$ and $D$ are concyclic.

Therefore, $\angle IGP = \angle IEP = 90°$, that is, $IG \perp PG$. Hence, points $P$, $S$ and $T$ are collinear.

Line $PST$ intersects $\triangle ABC$. By Menelaus' Theorem, we have

$$\frac{AS}{SC} \cdot \frac{CP}{PB} \cdot \frac{BT}{TA} = 1.$$

Since $AS = AT$, $CS = CD$ and $BT = BD$, we have

$$\frac{PC}{PB} \cdot \frac{BD}{CD} = 1. \qquad\qquad ①$$

Let the extension of $BN$ intersect $PE$ at point $H$. Then line $BFH$ intersects $\triangle PDE$. By Menelaus' Theorem,

$$\frac{PH}{HE} \cdot \frac{EF}{FD} \cdot \frac{DB}{BP} = 1.$$

Since $CF$ parallel to $BE$, $\dfrac{EF}{FD} = \dfrac{PC}{CD}$, we have

$$\frac{PH}{HE} \cdot \frac{PC}{CD} \cdot \frac{DB}{BP} = 1. \qquad\qquad ②$$

By ① and ②, we have $PH = HE$. Hence, $PH^2 = HE^2 = HM \cdot HN$. Thus, we have $\dfrac{PH}{HM} = \dfrac{HN}{PH}$, $\triangle PHN \backsim \triangle MHP$ and $\angle HPN = \angle HMP = \angle NEQ$. Further, since $\angle PEN = \angle EQN$, therefore $\angle ENP = \angle ENQ$. $\qquad\square$

**3** Let the sequence $\{a_n\}$ be defined by

$$a_1 = 1, \ a_2 = 2, \ a_{n+1} = \frac{a_n^2 + (-1)^n}{a_{n-1}} (n = 2, 3, \dots ).$$

Prove that the sum of squares of any two adjacent terms of the sequence is also in the sequence.

**Solution.** By $a_{n+1} = \dfrac{a_n^2 + (-1)^n}{a_{n-1}}$, we have $a_{n+1}a_{n-1} = a_n^2 + (-1)^n (n = 2, 3, \dots , )$, so

$$\frac{a_n - a_{n-2}}{a_{n-1}} = \frac{a_n a_{n-2} - a_{n-2}^2}{a_{n-1} a_{n-2}} = \frac{a_{n-1}^2 + (-1)^{n-1} - a_{n-2}^2}{a_{n-1} a_{n-2}}$$

$$= \frac{a_{n-1}^2 - a_{n-1} a_{n-3}}{a_{n-1} a_{n-2}} = \frac{a_{n-1} - a_{n-3}}{a_{n-2}}$$

$$= \dots = \frac{a_3 - a_1}{a_2} = 2,$$

that is, $a_n = 2a_{n-1} + a_{n-2} (n \geqslant 3)$, $a_1 = 1$, $a_2 = 2$.

Therefore, $a_n = C_1 \lambda_1^n + C_2 \lambda_2^n$, $\lambda_1 + \lambda_2 = 2$, $\lambda_1 \lambda_2 = -1$, $a_1 = 1$, $a_2 = 2 \ n \in \mathbf{N}^+$.

Then, since $\lambda_1 \lambda_2 = -1$ and $\lambda_2 = 2 - \lambda_1$ , we have

$$\begin{cases} 1 = C_1 \lambda_1 + C_2 \lambda_2 \\ 2 = C_1 \lambda_1^2 + C_2 \lambda_2^2 \end{cases} \Rightarrow \begin{cases} \lambda_2 = C_1 \lambda_1 \lambda_2 + C_2 \lambda_2^2 \\ 2 = C_1 \lambda_1^2 + C_2 \lambda_2^2 \end{cases}$$

$$\Rightarrow \begin{cases} 2 - \lambda_1 = -C_1 + C_2 \lambda_2^2 \\ 2 = C_1 \lambda_1^2 + C_2 \lambda_2^2 \end{cases}$$

$$\Rightarrow C_1 (1 + \lambda_1^2) = \lambda_1.$$

Thus, by symmetric condition, we have $C_2 (1 + \lambda_2^2) = \lambda_2$. So, since $1 + \lambda_1 \lambda_2 = 0$, we have

$$a_n^2 + a_{n+1}^2 = C_1^2 (1 + \lambda_1^2) \lambda_1^{2n} + C_2^2 (1 + \lambda_2^2) \lambda_2^{2n} + 2C_1 C_2 (\lambda_1 \lambda_2)^n (1 + \lambda_1 \lambda_2)$$

$$C_1 \lambda_1^{2n+1} + C_2 \lambda_2^{2n+1} = a_{2n+1}. \qquad \Box$$

**4** Suppose that 12 acrobats labeled 1 – 12 divided into two circles $A$ and $B$, with six persons in each. Let each acrobat in $B$ stand on the shoulders of two adjacent acrobats of $A$. We call it a *tower* if the label of each acrobat of $B$ is equal to the sum of the labels of the acrobats under his feet. How many different towers can they make?

(**Remark.** We treat two towers as the same if one can be obtained by rotation or reflection of the other. For example, the following towers are the same, where the labels inside the circle refer to the bottom acrobat, the labels outside the circle refers to the upper acrobat.)

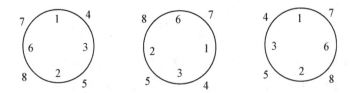

**Solution.** Denote the sum of labels of $A$ and $B$ by $x$ and $y$, respectively. Then $y = 2x$. Thus, we have

$$3x = x + y = 1 + 2 + \cdots + 12 = 78, \ x = 26.$$

Obviously, $1, 2 \in A$ and $11, 12 \in B$. Denote $A = \{1, 2, a, b, c, d\}$, where $a < b < c < d$. Then $a + b + c + d = 23$, and $a \geqslant 3$, $8 \leqslant d \leqslant 10$ (if $d \leqslant 7$, then $a + b + c + d \leqslant 4 + 5 + 6 + 7 = 22$, which is a contradiction.)

(1) If $d = 8$, then $A = \{1, 2, a, b, c, 8\}$, $c \leqslant 7$, $a + b + c = 15$. Thus, $(a, b, c) = (3, 5, 7)$ or $(4, 5, 6)$, that is, $A = \{1, 2, 3, 5, 7, 8\}$ or $A = \{1, 2, 4, 5, 6, 8\}$.

If $A = \{1, 2, 3, 5, 7, 8\}$, then $B = \{4, 6, 9, 10, 11, 12\}$. Since $B$ contains 11, 4, 6 and 12, there is only one tower that, in $A$, 8 and 3, 3 and 1, 1 and 5, 5 and 7 are adjacent.

If $A = \{1, 2, 4, 5, 6, 8\}$, then $B = \{3, 7, 9, 10, 11, 12\}$. Similarly, we see that, in $A$, 1 and 2, 5 and 6, 4 and 8 are adjacent, respectively. There are two arrangements, that is, there are two towers.

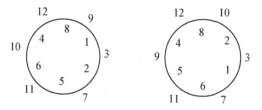

(2) If $d = 9$, then $A = \{1, 2, a, b, c, 9\}$, $c \leqslant 8$, $a + b + c = 14$, where $(a, b, c) = (3, 5, 6)$ or $(3, 4, 7)$, that is, $A = \{1, 2, 3, 5, 6, 9\}$ or $A = \{1, 2, 3, 4, 7, 9\}$.

If $A = \{1, 2, 3, 5, 6, 9\}$, then $B = \{4, 7, 8, 10, 11, 12\}$. To obtain 4, 10 and 12 in $B$, 1, 3, and 9 in $A$ must be adjacent pairwise, it is impossible!

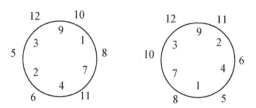

If $A = \{1, 2, 3, 4, 7, 9\}$, then $B = \{5, 6, 8, 10, 11, 12\}$. To obtain 6, 8 and 12 in $B$, 2 and 4, 1 and 7, 9 and 3 must be adjacent in $A$, respectively. There are two arrangements, that is, there are two towers.

(3) If $d = 10$, then $A = \{1, 2, a, b, c, 10\}$, where $c \leqslant 9$, $a + b + c = 13$. Thus, $(a, b, c) = (3, 4, 6)$, that is, $A = \{1, 2, 3, 4, 6, 10\}$ and $B = \{5, 7, 8, 9, 11, 12\}$. To

obtain 8, 9, 11 and 12 in $B$, 6 and 2, 6 and 3, 10 and 1, 10 and 2 must be adjacent, respectively. There is only one tower.

Summing up, there are six different towers all together. □

## Second Day
### 8: 00 – 12: 00, July 28, 2013

**5** Let $f(x) = \left[\frac{x}{1!}\right] + \left[\frac{x}{2!}\right] + \cdots + \left[\frac{x}{2013!}\right]$, where $[x]$ is the greatest integer no greater than $x$. Call an integer $n$ a *good number* if the equation $f(x) = n$ has a real solution $x$. Find the number of good numbers in the set $\{1, 3, 5, \ldots, 2013\}$.

**Solution.** First, we point out two obvious facts:

(a) If $m$ is a positive integer and $x$ is real, then

$$\left[\frac{x}{m}\right] = \left[\frac{[x]}{m}\right].$$

(b) For any integer $l$ and positive even number $m$, we have

$$\left[\frac{2l+1}{m}\right] = \left[\frac{2l}{m}\right].$$

Let $m = k!(k = 1, 2, \ldots, 2013)$ in (a) and summing up, we have

$$f(x) = \sum_{k=1}^{2013}\left[\frac{x}{k!}\right] = \sum_{k=1}^{2013}\left[\frac{[x]}{k!}\right] = f([x]),$$

that is, $f(x) = n$ has a real solution if and only if $f(x) = n$ has an integer solution. So, we only consider $x$ as an integer. Since

$$f(x+1) - f(x) = [x+1] - [x] + \sum_{k=2}^{2013}\left(\left[\frac{x+1}{k!}\right] - \left[\frac{x}{k!}\right]\right) \geqslant 1,$$

①

we see that $f(x)(x \in \mathbf{Z})$ is monotonously increasing. Now, we find integers $a$ and $b$, such that

$$f(a-1) < 0 \leqslant f(a) < f(a+1) < \cdots$$
$$< f(b-1) < f(b) \leqslant 2013 < f(b+1).$$

Note that $f(-1) < 0 = f(0)$, so $a = 0$. Since

$$f(1173) = \sum_{k=1}^{6} \left[\frac{1173}{k!}\right] = 1173 + 586 + 195 + 48 + 9 + 1$$
$$= 2012 \leqslant 2013,$$

$$f(1174) = \sum_{k=1}^{6} \left[\frac{1174}{k!}\right] = 1174 + 587 + 195 + 48 + 9 + 1$$
$$= 2014 > 2013,$$

we see that $b = 1173$.

So the good numbers in $\{1, 3, 5, \ldots, 2013\}$ are the odd numbers in

$$\{f(0), f(1), \ldots, f(1173)\}.$$

Let $x = 2l(l = 0, 1, \ldots, 586)$ in ①. By (b), we have

$$\left[\frac{2l+1}{k!}\right] = \left[\frac{2l}{k!}\right](2 \leqslant k \leqslant 2013).$$

Thus,

$$f(2l+1) - f(2l) = 1 + \sum_{k=2}^{2013} \left(\left[\frac{2l+1}{k!}\right] - \left[\frac{2l}{k!}\right]\right) = 1,$$

that is, there is exactly one odd number in $f(2l)$ and $f(2l+1)$. Therefore, there are $\frac{1174}{2} = 587$ odd numbers in $\{f(0), f(1), \ldots, f(1173)\}$, that is, there are 587 good numbers in the set $\{1, 3, 5, \ldots, 2013\}$. $\square$

**6** Let $n$ be an integer greater than 1. Denote the first $n$ primes in increasing order by $p_1$, $p_2$, ..., $p_n$ ( i. e., $p_1 = 2$, $p_2 = 3$, ...). Let $A = p_1^{\rho_1} p_2^{\rho_2} \cdots p_n^{\rho_n}$. Find all positive integers $x$ such that $\dfrac{A}{x}$ is even and has exactly $x$ distinct positive divisors.

**Solution.** By $2x \mid A$, note that $A = 4 \cdot p_2^{\rho_2} \cdots p_n^{\rho_n}$. We may suppose that $x = 2^{\alpha_1} p_2^{\alpha_2} \cdots p_n^{\alpha_n}$, where $0 \leqslant \alpha_1 \leqslant 1$, $0 \leqslant \alpha_i \leqslant p_i$ ($i = 2, 3, \ldots, n$). Then, we have

$$\frac{A}{x} = 2^{2-\alpha_1} p_2^{\rho_2-\alpha_2} \cdots p_n^{\rho_n-\alpha_n}.$$

Hence, the number of different divisors of $\dfrac{A}{x}$ is

$$(3 - \alpha_1)(p_2 - \alpha_2 + 1)\cdots(p_n - \alpha_n + 1).$$

We know that

$$(3 - \alpha_1)(p_2 - \alpha_2 + 1)\cdots(p_n - \alpha_n + 1) = x = 2^{\alpha_1} p_2^{\alpha_2} \cdots p_n^{\alpha_n}.$$
$$\textcircled{1}$$

By induction on $n$, we shall prove that the array $(\alpha_1, \alpha_2, \ldots, \alpha_n)$ satisfying $\textcircled{1}$ is $(1, 1, \ldots, 1)(n \geqslant 2)$.

(1) If $n = 2$, then $\textcircled{1}$ becomes $(3 - \alpha_1)(4 - \alpha_2) = 2^{\alpha_1} 3^{\alpha_2}$, where $\alpha_1 \in \{0, 1\}$. If $\alpha_1 = 0$, then $3(4 - \alpha_2) = 3^{\alpha_2}$ which has no integer solution $\alpha_2$. If $\alpha_1 = 1$, then $2(4 - \alpha_2) = 2 \cdot 3^{\alpha_2}$. We have $\alpha_2 = 1$. Thus, $(\alpha_1, \alpha_2) = (1, 1)$. That is, the conclusion is true for $n = 2$.

(2) Suppose that the conclusion is true for $n = k - 1$. ($k \geqslant 3$), then when $n = k$, $\textcircled{1}$ becomes

$$(3 - \alpha_1)(p_2 - \alpha_2 + 1)\cdots(p_{k-1} - \alpha_{k-1} + 1)(p_k - \alpha_k + 1)$$
$$= 2^{\alpha_1} p_2^{\alpha_2} \cdots p_{k-1}^{\alpha_{k-1}} p_k^{\alpha_k}. \qquad \textcircled{2}$$

If $\alpha_k \geqslant 2$, considering

$$0 < p_k - \alpha_k + 1 < p_k,$$
$$0 < p_i - \alpha_i + 1 \leqslant p_i + 1 < p_k(1 \leqslant i \leqslant k - 1),$$

we see that the left-hand side of ② cannot be divided by $p_k$, but the right-hand side of ② is a multiple of $p_k$, which is a contradiction.

If $\alpha_k = 0$, then ② becomes

$$(3 - \alpha_1)(p_2 - \alpha_2 + 1)\cdots(p_{k-1} - \alpha_{k-1} + 1)(p_k + 1)$$
$$= 2^{\alpha_1} p_2^{\alpha_2} \cdots p_{k-1}^{\alpha_{k-1}}. \qquad ③$$

Note that $p_2, p_3, \ldots, p_k$ are odd prime, thus, on the one hand, $p_k + 1$ is even. So the left-hand side of ③ is even. On the other hand, the right side of ③ is odd. So $\alpha_1 = 1$. But then $3 - \alpha_1 = 2$, so the left-hand side of ③ is a multiple of 4, but the right-hand side of ③ is not, which is a contradiction.

By the above argument, we must have $\alpha_k = 1$, and in ②,

$$p_k - \alpha_k + 1 = p_k^{\alpha_k} = p_k.$$

Thus,

$$(3 - \alpha_1)(p_2 - \alpha_2 + 1)\cdots(p_{k-1} - \alpha_{k-1} + 1) = 2^{\alpha_1} p_2^{\alpha_2} \cdots p_{k-1}^{\alpha_{k-1}}.$$

By the induction hypotheses, $\alpha_1 = \alpha_2 = \cdots = \alpha_{k-1} = 1$.

Thus, $\alpha_1 = \alpha_2 = \cdots = \alpha_{k-1} = \alpha_k = 1$, that is, the conclusion is true for $n = k$.

By (1) and (2), we conclude that $(\alpha_1, \alpha_2, \ldots, \alpha_n) = (1, 1, \ldots, 1)$, so the integer required is $x = 2p_2 \cdots p_n = p_1 p_2 \cdots p_n$.

$\square$

**7**   Cut off a corner of $2 \times 2$ unit squares from a $3 \times 3$ unit squares, the remaining figure is called a *horn* (Fig. 7.1 is a horn). Now, put some horns without overlapping on a board of $10 \times 10$ unit squares (Fig. 7.2) such that the boundaries of the horn coincide with the grid of the board. Find the maximum of $k$ such that whatever the $k$

horns put on the board, one can always put another horn on the board. (posed by He Yijie)

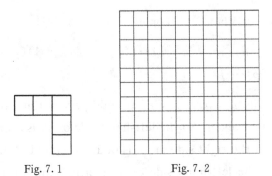

Fig. 7. 1              Fig. 7. 2

**Solution.** First, we have $k_{max} < 8$, this is because of that if we put eight horns as in Fig. 7. 3, then no more horn can be put.

Next, we show that, after putting any seven horns, there is always room for another horn.

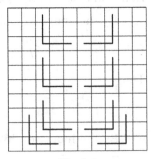

Fig. 7. 3

Consider four $4 \times 4$ squares in four corners of $10 \times 10$ squares. Then any horn can only intersect one of these $4 \times 4$ squares. So, seven horns are placed on $10 \times 10$ squares. By the Dirichlet Drawer Theorem, there exists a $4 \times 4$ square $S$ such that there is at most one horn $H$ that intersects $S$, and $H$ can be contained by a $3 \times 3$ squares, so $S \cap H$ is contained by a $3 \times 3$ squares on a conner of $S$.

We can put a horn on $S$ as shown in Fig. 7. 4.

Summing up, we have $k_{max} = 7$.

Fig. 7. 4

**8** Let $n \geqslant 3$ be integer. Suppose that $\alpha$, $\beta$, $\gamma \in (0, 1)$ and $a_k$, $b_k$, $c_k \geqslant 0$ $(k = 1, 2, \ldots, n)$ satisfy $\sum_{k=1}^{n}(k + \alpha)a_k \leqslant \alpha$, $\sum_{k=1}^{n}(k + \beta)b_k \leqslant \beta$ and $\sum_{k=1}^{n}(k + \gamma)c_k \leqslant \gamma$. Find the minimum of $\lambda$ such that $\sum_{k=1}^{n}(k + \lambda)a_kb_kc_k \leqslant \lambda$.

**Solution.** Let $a_1 = \dfrac{\alpha}{1 + \alpha}$, $b_1 = \dfrac{\beta}{1 + \beta}$, $c_1 = \dfrac{\gamma}{1 + \gamma}$, $a_i$, $b_i$, $c_i = 0(i = 2, 3, \ldots, n)$. We see that all conditions are satisfied. So, we must have

$$(1 + \lambda) \frac{\alpha}{1 + \alpha} \cdot \frac{\beta}{1 + \beta} \cdot \frac{\gamma}{1 + \gamma} \leqslant \lambda,$$

that is,

$$\lambda \geqslant \frac{\alpha\beta\gamma}{(1 + \alpha)(1 + \beta)(1 + \gamma) - \alpha\beta\gamma}.$$

Denote $\dfrac{\alpha\beta\gamma}{(1 + \alpha)(1 + \beta)(1 + \gamma) - \alpha\beta\gamma} = \lambda_0$. We will show that, for any $a_k$, $b_k$, $c_k$, $k = 1, 2, \ldots, n$, we have

$$\sum_{k=1}^{n}(k + \lambda_0)a_kb_kc_k \leqslant \lambda_0, \qquad \qquad ①$$

which satisfies all the conditions.

By the conditions of the problem,

$$\sum_{k=1}^{n}\left(\frac{k + \alpha}{\alpha}a_k \cdot \frac{k + \beta}{\beta}b_k \cdot \frac{k + \gamma}{\gamma}c_k\right)^{\frac{1}{3}}$$

$$\leqslant \left(\sum_{k=1}^{n}\frac{k + \alpha}{\alpha}a_k\right)^{\frac{1}{3}} \cdot \left(\sum_{k=1}^{n}\frac{k + \beta}{\beta}b_k\right)^{\frac{1}{3}} \cdot \left(\sum_{k=1}^{n}\frac{k + \gamma}{\gamma}c_k\right)^{\frac{1}{3}} \leqslant 1,$$

where we use the Hölder's Inequality: if $x_i$, $y_i$, $z_i \geqslant 0(i = 1, 2, \ldots, n)$, then

$$\left(\sum_{i=1}^{n}x_iy_iz_i\right)^3 \leqslant \left(\sum_{i=1}^{n}x_i^3\right)\left(\sum_{i=1}^{n}y_i^3\right)\left(\sum_{i=1}^{n}z_i^3\right). \qquad ②$$

Therefore, to prove ①, it suffices to show that, for $k = 1$, $2, \ldots, n$, we have

$$\frac{k + \lambda_0}{\lambda_0} a_k b_k c_k \leqslant \left( \frac{k + \alpha}{\alpha} \cdot \frac{k + \beta}{\beta} \cdot \frac{k + \gamma}{\gamma} \cdot a_k b_k c_k \right)^{\frac{1}{3}},$$

that is

$$\frac{k + \lambda_0}{\lambda_0} (a_k b_k c_k)^{\frac{2}{3}} \leqslant \left( \frac{(k + \alpha)(k + \beta)(k + \gamma)}{\alpha \beta \gamma} \right)^{\frac{1}{3}}. \qquad ③$$

In fact,

$$\lambda_0 = \frac{\alpha \beta \gamma}{1 + (\alpha + \beta + \gamma) + (\alpha\beta + \beta\gamma + \gamma\alpha)}$$

$$\geqslant \frac{\alpha \beta \gamma}{k^2 + (\alpha + \beta + \gamma)k + (\alpha\beta + \beta\gamma + \gamma\alpha)}$$

$$= \frac{k \alpha \beta \gamma}{(k + \alpha)(k + \beta)(k + \gamma) - \alpha \beta \gamma}.$$

Thus,

$$\frac{k + \lambda_0}{\lambda_0} \leqslant \frac{(k + \alpha)(k + \beta)(k + \gamma)}{\alpha \beta \gamma}. \qquad ④$$

And since $(k + \alpha)a_k \leqslant \alpha$, $(k + \beta)b_k \leqslant \beta$, $(k + \gamma)c_k \leqslant \gamma$, we have

$$(a_k b_k c_k)^{\frac{2}{3}} \leqslant \left( \frac{\alpha \beta \gamma}{(k + \alpha)(k + \beta)(k + \gamma)} \right)^{\frac{2}{3}}. \qquad ⑤$$

By ④ and ⑤, we see that equation ③ holds, thus ① holds.

Summing up, we have $\lambda_{\min} = \lambda_0 = \dfrac{\alpha \beta \gamma}{(1 + \alpha)(1 + \beta)(1 + \gamma) - \alpha \beta \gamma}$.

$\square$

# International Mathematical Olympiad

$2011$ (Amsterdam, Holland)

## First Day

9: 00~13: 30, July 18, 2011

**1** Given any set $A = \{a_1, a_2, a_3, a_4\}$ of four distinct positive integers, we denote the sum $a_1 + a_2 + a_3 + a_4 = s_A$. Let $n_A$ denote the number of pares $(i, j)$ with $1 \leqslant i < j \leqslant 4$ for which $a_i + a_j$ divides $s_A$. Finds all sets $A$ of four distinct positive integers which achieve the largest possible value of $n_A$. (posed by Mexico)

**Solution.** (Based on the solution by Jin Zhaorong) Let set $A = \{a_1, a_2, a_3, a_4\}$ of four positive integers with $a_1 < a_2 < a_3 < a_4$. Since

$$\frac{1}{2}s_A = \frac{1}{2}(a_1 + a_2 + a_3 + a_4) < a_2 + a_4 < a_3 + a_4 < s_A,$$

$a_2 + a_4$, $a_3 + a_4$ do not divide $s_A$. Therefore,

$$n_A \leqslant C_4^2 - 2 = 4.$$

On the other hand, if $A = \{1, 5, 7, 11\}$, $n_A = 4$, the largest possible value of $n_A$ is 4.

Next, we will find all sets $A$ of four positive integers with $n_A = 4$.

First, we see that $a_2 + a_4$, $a_3 + a_4$ do not divide $s_A$, and

$$\frac{1}{2}s_A \leqslant \max\{a_1 + a_4, a_2 + a_3\} < s_A.$$

Thus,        $\dfrac{1}{2}s_A = \max\{a_1 + a_4, a_2 + a_3\}$,

and then                $a_1 + a_4 = a_2 + a_3$.

By $a_1 + a_3 \mid s_A$, let $s_A = k(a_1 + a_3)$, where $k$ is a positive integer. And by $a_1 + a_3 < a_2 + a_3$ we know that $k > 2$.

As $2(a_2 + a_3) = s_A = k(a_1 + a_3)$, $a_2 = \dfrac{1}{2}(ka_1 + (k-2)a_3)$, and from

$$a_2 = \frac{1}{2}(ka_1 + (k-2)a_3) < a_3,$$

we have $k < 4$. Thus, $k = 3$. Consequently, we have

$$2(a_2 + a_3) = 2(a_1 + a_4) = 3(a_1 + a_3) = s_A,$$

from which we can derive $a_2 = \dfrac{1}{2}(3a_1 + a_3)$, $a_4 = \dfrac{1}{2}(a_1 +$

$3a_3$ ).

And by $a_1 + a_2 \mid s_A$, let $s_A = l(a_1 + a_2)$, we have

$$3(a_1 + a_3) = l\left(a_1 + \frac{1}{2}(3a_1 + a_3)\right),$$

that is, $(6 - l) a_3 = (5l - 6)a_1.$

Since $a_1$ and $a_3$ are positive integers and $a_1 < a_3$, $l = 3$, 4 or 5.

If $l = 3$, we have $a_2 = a_3$, which is a contradiction.

If $l = 4$, we have $a_3 = 7a_1$, consequently, it follows that $a_2 = 5a_1$, $a_4 = 11a_1$.

If $l = 5$, we have $a_3 = 19a_1$, consequently, we have $a_2 = 11a_1$, $a_4 = 29a_1$.

It is easy to verify that, when $l = 4$, 5, each of $a_1 + a_2$, $a_1 + a_3$, $a_1 + a_4$ and $a_2 + a_3$ divides $s_A$.

To sum up, all sets $A$ of four distinct positive integers which achieve the largest possible value of $n_A = 4$ are $A = \{a, 5a, 7a, 11a\}$, and $A = \{a, 11a, 19a, 29a\}$, where $a$ is any positive integer. ☐

**2** Let $S$ be a finite set of at least two points in the plane. Assume that no three points of $S$ are collinear. A *windmill* is a process that starts with a line $l$ going through a single point $P \in S$. The line rotates clockwise about the point $P$ until the first time that the line meets some other point belonging to $S$. This process continues indefinitely.
Show that we can choose a point $P$ in $S$ and a line $l$ going through $P$ such that the resulting windmill uses each point of $S$ as a pivot infinitely many times. (posed by British)

**Solution.** (Based on the solution by Zhou Tianyou) Consider each line $\ell$ has a direction, which varies continuously when the line $\ell$ rotates. First, consider the case when $\mid S \mid = 2k + 1$ is old.

In the process of a windmill, we say a line $\ell$ is on a *good* position, if there are $k$ points of $S$ on each side of the line $\ell$ when $\ell$ just leaves one point of $S$.

We consider two good positions are equal if $\ell$ passes the same two points with the same direction, otherwise they are different good positions. It is easy to see that the number of good positions is finite in all windmills. Given a positive direction such as $x$-axis, denote all good positions in the order of clockwise angles in $[0, 2\pi)$ by $\ell_1, \ell_2, \ldots, \ell_m$. We proceed to prove in three steps.

Firstly, given any good position, we can see there is at most one good position. Since if there were two good positions $\ell_i$ and $\ell_j$ that had the same direction, then they were parallel and did not coincide, the number on the right side to the lines $\ell_i$ and $\ell_j$ should be $k$ or $k - 1$, it could not be possible to both lines $\ell_i$ and $\ell_j$.

Secondly, for any point in $S$, there exists some $\ell_i$ passing through $P$ as follows.

Take any point $P \in S$, and line $\ell$ passing through $P$ but not through the other point of $S$. If the number of points to the right side of line $\ell$ is $s$, then the number of points to the left side of line $\ell$ is $2k - s$. The difference of two numbers is $2k - 2s$. Now, rotate $\ell$ about $P$ clockwise, with $s$ increasing or decreasing by 1 when $\ell$ is passing a point, so that $2k - 2s$ changes 2. When $\ell$ rotates $180°$, $2k - 2s$ becomes its opposite number. Therefore, there exists a moment that an $\ell$ of which the numbers of points on two sides are equal. Denote the last passing point by $Q$, then there is a good position $\ell_i$ passing $P$ and $Q$.

Thirdly, a windmill starting from a good position meets the requirement of the problem.

Without loss of generality, we may start with $\ell_1$. And we show that the next meeting point is just $\ell_2$, so the windmill meets all $\ell_1$ and continues infinitely many times. By step two, we know that each point of $S$ is used as a pivot infinitely many times.

Suppose that $\ell = \ell_1$ passing $P$ and $Q$ with $P$ as a pivot and dividing $S$ equally when $\ell_1$ is just leaving $Q$. Suppose that $\ell$ met point $R$ next and use $R$ as a pivot, then $\ell$ remain divides $S$ equally when $\ell$ leaving $P$ which is valid for $R$ on any side of $P$.

This shows that $\ell$ is still a good position when it met $R$ and denotes this good position as $\ell'$. It remains to be shown that there is no good position between $\ell$ and $\ell'$, so $\ell' = \ell_2$. Since if $\ell_2$ was a good position between $\ell$ and $\ell'$, take directed line $\ell''$ passing $P$ and parallel to $\ell_2$, then $\ell''$ divides $S$ equally, so that there are at least $k + 1$ points on the left or right side of $\ell_2$, which contradicts the fact that $\ell_2$ is a good position.

Now, consider the case when $|S| = 2k$ is even. The argument above is still valid with suitable revision. We say a line $\ell$ is on a *good* position if there are $k$ points of $S$ on the right side of the line $\ell$ when $\ell$ just leaves one point of $S$. Then the first and second steps can be proven similarly. For the third step, starting from a *good* position, we can show similarly that the next meeting point by $\ell$ is still a good position. The only slight difference to the odd case is to show that there is no good position between $\ell_1$ and $\ell_2$.

Take directed line $\ell''$ passing $P$ and parallel to $\ell_2$, then there are $k$ points of $S$ to the right of $\ell''$. If $\ell_2$ were on the right side to $\ell''$, then there would be no more than $k - 2$ points, so $\ell''$ would not be a good position. If $\ell_2$ were on the left side to $\ell''$, then there would be at least $k + 1$ to the right side to $\ell''$, so $\ell''$ would not be a good position. Thus, we proved for the even case.                                                                    □

**3** Let $f : \mathbf{R} \to \mathbf{R}$ be a real-valued function defined on the set of real numbers that satisfies

$$f(x + y) \leqslant yf(x) + f(f(x)) \qquad \text{①}$$

for all real numbers $x$ and $y$. Prove that $f(x) = 0$ for all $x \leqslant 0$. (posed by Belarus)

**Solution.** (Based on the solution by Wu Mengxi)

Let $y = f(x) - x$ in inequality ①, then we have

$$f(f(x)) \leqslant (f(x) - x)f(x) + f(f(x)),$$

thus,

$$(f(x) - x)f(x) \geqslant 0. \qquad \text{②}$$

Consequently, for any real number $x$, we have

$$(f(f(x)) - f(x))f(f(x)) \geqslant 0.$$

Note that by taking $y = 0$, ① implies $f(x) \leqslant f(f(x))$. Therefore,

$$f(f(x)) \geqslant 0, \text{ or } f(f(x)) = f(x) < 0. \qquad \text{③}$$

We first show that $f(x) \leqslant 0$ for any real number $x$ by contradiction. If there is a real number $x_0$ such that $f(x_0) > 0$, then for any real number $y$, we have $f(x_0 + y) \leqslant yf(x_0) + f(f(x_0))$. Thus, for any $y < -\dfrac{f(f(x_0))}{f(x_0)}$, we have $f(x_0 + y)$

$<0$. So, for any real number $z < x_0 - \dfrac{f(f(x_0))}{f(x_0)}$, we have $f(z)$

$<0$. Therefore, for $z < \min\left\{0, \; x_0 - \dfrac{f(f(x_0))}{f(x_0)}\right\}$, we see that

$$z < 0 \text{ and } f(z) < 0. \qquad\qquad ④$$

Thus, by ②, we have

$$f(z) \leqslant z < \min\left\{0, \; x_0 - \frac{f(f(x_0))}{f(x_0)}\right\} \leqslant x_0 - \frac{f(f(x_0))}{f(x_0)}.$$

Consequently, $f(f(z)) = f(x_0 + (f(z) - x_0)) < 0$. Thus, by ② and ④, we have

$$f(f(z)) = f(z) < 0. \qquad\qquad ⑤$$

Hence, for any real number $y$, by ① and ⑤, we have

$$f(z + y) \leqslant yf(z) + f(f(z)) = (y + 1)f(z). \qquad ⑥$$

Let $y = x_0 - z$ in ⑥, then

$$f(x_0) \leqslant (1 + x_0 - z)f(z). \qquad\qquad ⑦$$

Taking $z$ to be sufficiently negative such that $1 + x_0 - z > 0$, then by ⑤ and ⑦, we have $f(x_0) < 0$, which is a contradiction.

Therefore, for any real number $x$,

$$f(x) \leqslant 0 \text{ and } f(f(x)) \leqslant 0. \qquad\qquad ⑧$$

Let $y = -x$ in ①, then $f(0) \leqslant -xf(x) + f(f(x)) \leqslant -xf(x)$ by ⑧.

Thus, we only need to show that $f(0) = 0$, then for $x < 0$, $f(x) \geqslant 0$, and by ⑧, we proved $f(x) = 0$ for all $x \leqslant 0$.

In fact, if $f(t) = t$ has no negative solution, then for any real number $x$, we have

$$f(f(x)) \neq f(x) \text{ and } f(f(f(x))) \neq f(f(x)). \qquad ⑨$$

So $f(f(x)) \geqslant 0$ and $f(f(f(x))) \geqslant 0$ by ③. Consequently, $f(f(x)) = 0$ and $f(f(f(x))) = 0$ by ⑧. That is, $f(0) = 0$.

If $f(t) = t$ has a negative solution, we show that $t$ is unique. Since if there are two solutions $t_1 < t_2 < 0$, then by ①, we have

$$t_2 = f(t_2) \leqslant (t_2 - t_1)f(t_1) + f(f(t_1)) = (t_2 - t_1)t_1 + t_1,$$

thus $(t_2 - t_1)(1 - t_1) \leqslant 0$, which is a contradiction. Furthermore, we show that there exists real number $x$ such that $f(x) \neq t$ (otherwise for all $y$, $t \leqslant yt + t$, which fails when $y < 0$), hence ⑨ hold for this $x$. $\qquad\square$

## Second Day
### 9: 00~13: 30, July 19, 2011

**4** Let $n > 0$ be an integer. We are given a balance and $n$ weights of weight $2^0$, $2^1$, ..., $2^{n-1}$. We are to place each of the $n$ weights on the balance, one after another, in such a way that the right pan is never heavier than the left pan. At each step, we choose one of the weights until all of the weights that have not yet been placed on the balance and place it on either the left pan or the right pan, until all the weights have been placed.

Determine the number of ways in which this can be done.

(posed by Iran)

**Solution.** (Based on the solution by Yao Bowen) The number of ways is $(2n - 1)!! = 1 \times 3 \times 5 \times \cdots \times (2n - 1)$.

We prove the problem by induction on $n$. The case $n = 1$ is clear, put one weight on the left pan. Hence, there is only one way.

Suppose that in the case $n = k$, the number of ways is $(2k - 1)!!$.

Then for the case $n = k + 1$, without affecting the result, multiply both weights by $1/2$, the $k + 1$ weights of weight are $1/2, 1, 2, \ldots, 2^{k-1}$. Since for any positive integer $r$, we have

$$2^r > 2^{r-1} + 2^{r-2} + \cdots + 1 + \frac{1}{2} \geqslant \sum_{i=-1}^{r-1} \pm 2^i.$$

The heavier pan is where the heaviest weight is. Therefore, the heaviest weight on the balance must be on the left pan. In the following, consider the position of the weight $1/2$ in the procession.

(a) If we take weight $1/2$ first, then it must put on the left pan. Then the remaining $k$ weights have $(2k - 1)!!$ ways to do.

(b) If we take weight $1/2$ at step $t = 2, 3, \ldots, k + 1$, we can put it either on the left pan or on the right pan because weight $1/2$ is the lightest one, so the number of ways is $(2k - 1)!!$

To sum up, when $n = k + 1$, there are totally $(2k - 1)!! + k \times (2k - 1)!! = (1 + 2k)(2k - 1)!! = (2k + 1)!!$ ways, which completes the induction. $\qquad\square$

---

**5** Let $f$ be a function from the set of integers to the set of positive integers. Suppose that, for any two integers $m$ and $n$, the difference $f(m) - f(n)$ is divisible by $f(m - n)$. Prove that, for any two integers with $f(m) < f(n)$, the number is divisible by $f(m)$. (posed by Iran)

**Solution.** (Based on the solution by Long Zichao) According to the problem, we have $f(k) \mid f(x + k) - f(x)$, $\forall x \in \mathbf{Z}$ for any integer $k \neq 0$. Consequently, $f(k) \mid f(x + tk) - f(x)$,

$\forall x \in \mathbf{Z}$, $t \in \mathbf{Z}$. So $f(k) \mid f(m) - f(n)$ for any $m$, $n$ such that $k \mid m - n$. Thus,

$$f(n) \mid f(m+n) - f(m), \qquad\qquad ①$$

$$f(m - (-n)) \mid f(m) - f(-n). \qquad\qquad ②$$

We now show that for any integer $n$, $f(n) = f(-n)$. Since otherwise, without loss of generality, if there is an integer $n \neq 0$, such that $f(n) > f(-n) > 0$, then $0 < f(n) - f(-n) < f(n)$, which contradicts $f(n) \mid f(n) - f(-n)$.

Therefore, by ②, we have

$$f(m+n) \mid f(m) - f(n). \qquad\qquad ③$$

If $0 < f(m) < f(n)$, then by ③, we have $0 < f(m+n) < f(n)$. Combining ① and ③, we have $f(m+n) = f(m)$. Then by ③, we get the result $f(m) \mid f(n)$. For $f(m) = f(n)$, the result is obvious. $\qquad\qquad\square$

**⑥** Let $\triangle ABC$ be an acute triangle with circumcircle $\Gamma$. Let $l$ be a tangent line to $\Gamma$, and let $l_a$, $l_b$ and $l_c$ be the lines obtained by reflecting $l$ to the lines $BC$, $CA$ and $AB$, respectively. Show that the circumcircle of the triangle formed by lines $l_a$, $l_b$ and $l_c$ is tangent to the circle $\Gamma$. (posed by Japan)

**Solution.** (Based on the solution by Chen Lin) Let $P$ be the tangent point of $l$ to $\Gamma$. Denote the symmetric points of $P$ to $BC$, $CA$, and $AB$ by $P_a$, $P_b$ and $P_c$, respectively. These points are on circles $\Gamma_a$, $\Gamma_b$ and $\Gamma_c$ symmetric to $\Gamma$ to $BC$, $CA$ and $AB$, respectively. Hence, $\ell_a$, $\ell_b$ and $\ell_c$ are tangent to $\Gamma_a$, $\Gamma_b$ and $\Gamma_c$ at points $P_a$, $P_b$ and $P_c$, respectively.

Denote $l_a \cap l_b = C'$, $l_b \cap l_c = A'$, $l_c \cap l_a = B'$.

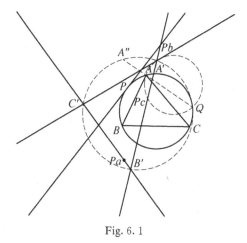

Fig. 6. 1

Define the oriented angle of line $m$ to line $n$ by $\angle(m, n)$. The magnitude of $\angle(m, n)$ is the angle rotated anticlockwise from $m$ to $n$.

We conclude that

(1) Points $P_a$, $P_b$ $P_c$ are collinear.

In fact, the midpoints of $PP_a$, $PP_b$, $PP_c$ are the pedal points of $P$ to $BC$, $CA$ and $AB$, respectively. The pedal points are collinear by Simson's Theorem. So $P_a$, $P_b$ and $P_c$ are collinear.

(2) Denote the circumcircles of $\triangle A'P_bP_c$, $\triangle B'P_cP_a$ and $\triangle C'P_aP_b$ by $\Gamma_1$, $\Gamma_2$ and $\Gamma_3$, respectively. And circumcircles of $\triangle A'B'C'$ by $\Omega$. Then four circles $\Gamma_1$, $\Gamma_2$, $\Gamma_3$ and $\Omega$ are copunctal.

In fact, it is just the Miquel Theorem for complete quadrilateral $A'P_cB'P_aC'P_b$. Denote the copunctal point by $Q$.

(3) Points $A$, $B$ and $C$ are on circles $\Gamma_1$, $\Gamma_2$ and $\Gamma_3$, respectively.

In fact, circles $\Gamma_b$ and $\Gamma_c$ intersect at point $A$, and $\overset{\frown}{AP_c} = \overset{\frown}{AP} = \overset{\frown}{AP_b}$. Rotate an angle of $\angle(P_cA, P_bA)$ about point $A$, then $\Gamma_c \rightarrow \Gamma_b$, $P_c \rightarrow P_b$, and tangent line $l_c \rightarrow l_b$. So, $\angle(l_c, l_b) = \angle(P_cA, P_bA)$ and $\angle(l_c, l_b) = \angle(P_cA', P_bA')$, which means four points are concyclic. That is, point $A$ is on circle $\Gamma_1$. Similarly, points $B$ and $C$ are on circles $\Gamma_2$ and $\Gamma_3$, respectively.

(4) Point $Q$ is on circle $\Gamma$.

By definition of point $Q$,

$$
\begin{aligned}
\measuredangle(AQ, BQ) &= \measuredangle(AQ, P_cQ) + \measuredangle(P_cQ, BQ) \\
&= \measuredangle(AP_b, P_bP_c) + \measuredangle(P_cP_a, BP_a) \\
&= \measuredangle(AP_b, BP_a) \\
&= \measuredangle(AP_b, AC) + \measuredangle(AC, BC) + \measuredangle(BC, BP_a) \\
&= \measuredangle(AC, AP) + \measuredangle(AC, BC) + \measuredangle(BP, BC) \\
&= 2\measuredangle(AC, BC) - \measuredangle(AP, BP) \\
&= \measuredangle(AC, BC).
\end{aligned}
$$

Hence, four points $A$, $B$, $C$ and $Q$ are concyclic, that is, point $Q$ is on circle $\Gamma$.

(5) Circle $\Gamma$ is tangent to circle $\Omega$ at point $Q$.

Let line $QA$ intersect circle $\Omega$ at points $Q$ and $A''$, then we have

$$
\begin{aligned}
\measuredangle(A''B', B'Q) &= \measuredangle(A''B', B'A') + \measuredangle(A'B', B'Q) \\
&= \measuredangle(A''Q, QA') + \measuredangle(A'B', B'Q) \\
&= \measuredangle(AQ, A'Q) + \measuredangle(A'B', B'Q) \\
&= \measuredangle(AP_c, A'P_c) + \measuredangle(A'B', B'Q) \\
&= \measuredangle(AB, BP_c) + \measuredangle(A'B', B'Q) \\
&= \measuredangle(AB, BP_c) + \measuredangle(P_cB', B'Q) \\
&= \measuredangle(AB, BP_c) + \measuredangle(P_cB, BQ) \\
&= \measuredangle(AB, BQ).
\end{aligned}
$$

This shows that the degrees of $\overparen{A''Q}$ and $\overparen{AQ}$ are equal in circle $\Omega$. Note that points $A$, $Q$ and $A''$ are collinear, and point $Q$ is the same point on circles $\Omega$ and $\Gamma$. Therefore, the tangent lines of the two circles at point $Q$ coincide. That is, the circles $\Omega$ and $\Gamma$ are tangent at $Q$. $\qquad\square$

# *2012* (Mar Del Plata, Argentine)

## First Day

9: 00~13: 30, July 10, 2012

**1** Given triangle $ABC$, the point $J$ is the centre of the excircle opposite vertex $A$. This excircle is tangent to the side $BC$ at $M$, and to the lines $AB$ and $AC$ at $K$ and $L$, respectively. The lines $LM$ and $BJ$ meet at $F$, and the lines $KM$ and $CJ$ meet at $G$. Let $S$ be the point of intersection of lines $AF$ and $BC$, and let $T$ be the point of the intersection of lines $AG$ and $BC$.

Prove that $M$ is the midpoint of $ST$. (The excircle of $\triangle ABC$ opposite the vertex $A$ is the circle that is tangent to the line segment $BC$, to the ray $AB$ beyond $B$, and to the ray $AC$ beyond $C$.)

**Solution.** Let $\angle CAB = \alpha$, $\angle ABC = \beta$, $\angle BCA = \gamma$. Since $AJ$ is the bisector of $\angle CAB$, $\angle JAK = \angle JAL = \alpha/2$. Since $\angle AKL = \angle ALJ = 90°$, points $K$ and $L$ are on the circle $\omega$ with diameter $AJ$.

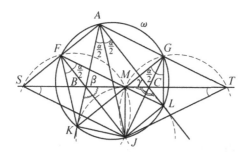

Fig. 1. 1

Since $BJ$ is the bisector of $\angle KBM$, we have $\angle MBJ = 90° -$

$\beta/2$. Similarly, $\angle CML = \gamma/2$ and $\angle MCJ = 90° - \gamma/2$. Consequently, $\angle LFJ = \angle MBJ - \angle BMF = \angle MBJ - \angle CML = 90° - \beta/2 - \gamma/2 = \alpha/2 = \angle JAL$. Hence, point $F$ is on the circle $\omega$. Similarly, point $G$ is also on the circle $\omega$. Since $AJ$ is the diameter of circle $\omega$, we have $\angle AFJ = \angle AGJ = 90°$.

Segments $AB$ and $BC$ are symmetric about the bisector of external angle of $\angle ABC$. And by $AF \perp BF$ and $KM \perp BF$, we see that segments $SM$ and $AK$ are symmetric about $BF$, $SM = AK$. Similarly, $TM = AL$. And since $AK = AL$, we have $SM = TM$, namely, $M$ is the midpoint of $ST$. $\qquad\square$

**2** Let $n \geqslant 3$ be an integer, and let $a_2, a_3, \ldots, a_n$ be positive real numbers such that $a_2 a_3 \cdots a_n = 1$. Prove that $(1 + a_2)^2 (1 + a_3)^3 \cdots (1 + a_n)^n > n^n$.

**Solution.** By the AM – GM inequality, we have

$$(1 + a_k)^k = \left( \frac{1}{k-1} + \frac{1}{k-1} + \cdots + \frac{1}{k-1} + a_k \right)^k$$

$$\geqslant k^k \cdot \left( \frac{1}{k-1} \right)^{k-1} a_k$$

for $k = 2, 3, \ldots, n$.

Thus,

$$(1 + a_2)^2 (1 + a_3)^3 \cdots (1 + a_n)^n$$

$$\geqslant 2^2 a_2 \cdot 3^3 \left( \frac{1}{2} \right)^2 a_3 \cdot 4^4 \left( \frac{1}{3} \right)^3 a_4 \cdot \cdots \cdot n^n \left( \frac{1}{n-1} \right)^{n-1} a_n$$

$$= n^n.$$

The equality holds when $a_k = \dfrac{1}{k-1}$, $k = 2, \ldots, n$, which is not the case since $a_2 a_3 \cdots a_n = 1$.

Hence, $(1 + a_2)^2 (1 + a_3)^3 \cdots (1 + a_n)^n > n^n$. $\qquad\square$

**3** The *liar's guessing game* is a game played between two players $A$ and $B$. The rules of the game depend on two positive integers $k$ and $n$ which are known to both players.

At the start of the game, $A$ chooses integers $x$ and $N$ with $1 \leqslant x \leqslant N$. Player $A$ keeps $x$ secret, and truthfully tells $N$ to player $B$. Now player $B$ tries to obtain information about $x$ by asking player $A$ questions as follows: each question consists of $B$ specifying an arbitrary set $S$ of positive integers (possibly one specified in some previous question), and asking $A$ whether $x$ belongs to $S$. Player $B$ may ask as many such questions as he wishes. After each question, player $A$ must immediately answer it with yes or no, but is allowed to lie as many times as she wants; the only restriction is that, among any $k + 1$ consecutive answers, at least one answer is truthful.

After $B$ has asked as many questions as he wants, he must specify a set of $X$ of at most $n$ positive integers. If $x$ belongs to $X$, then $B$ wins; otherwise, he loses. Prove that:

1. If $n \geqslant 2^k$, then $B$ can guarantee a win;
2. For all sufficiently large $k$, there exists an integer $n > 1.99^k$ such that $B$ cannot guarantee a win.

**Solution.** We rephrase the rule of the game as follows: Given two positive integers $k$ and $n$, player $A$ tells a finite set $T = \{1, 2, \ldots, N\}$ to player $B$ and keeps one element $x$ secret. Player $B$ chooses a finite subset $S$ of $T$ and asks player $A$ whether $x$ belongs to $S$. Player $A$ answers it with yes or no and is allowed to lie at most consecutive $k$ times. After asking finite questions, if player $B$ can specify a subset of $T$ containing $x$ with at most $n$ elements, then player $B$ wins.

(1) If $N > 2^k$, we will show that player $B$ can always

determine some element $y \in T$ such that $y \ne x$. Thus, we can restrict $N \leqslant 2^k$. Consequently, if $n \geqslant 2^k$, then player $B$ wins.

Denote $T = \{1, 2, \ldots, N\}$, player $B$ asks player $A$ same question $k$ times: Is $x = 2^k + 1$? If the answers are all no, then $x \ne 2^k + 1 = y$. Once an answer is yes, then the next first question of player $B$ is: Is $x \in \{t \in \mathbf{Z} \mid 1 \leqslant t \leqslant 2^{k-1}\}$? If $A$ answers yes, then we have to find a $y \in \{t \in \mathbf{Z} \mid 2^{k-1} + 1 \leqslant t \leqslant 2^k\}$ such that $y \ne x$. If $A$ answers no, then we have to find a $y \in \{t \in \mathbf{Z} \mid 1 \leqslant t \leqslant 2^{k-1}\}$ such that $y \ne x$. In this way, by each answer of player $A$, we can reduce half the size of the set containing $y$. When $A$ answers $k$ questions, we can conclude that there is a unique $a$, $1 \leqslant a \leqslant 2^k$. If $a = x$, then $A$ lies consecutive $k + 1$ times. So $y = a \ne x$.

(2) We prove that for any $1 < \lambda < 2$, if $n = [(2 - \lambda)\lambda^{k+1}] - 1$, then player $B$ cannot guarantee a win. Specially, take $\lambda$, $1.99 < \lambda < 2$, for sufficiently large integer $k$, we have

$$n = [(2 - \lambda)\lambda^{k+1}] - 1 > 1.99^k,$$

which is the required conclusion.

Player $A$ chooses $T = \{1, \ldots, n + 1\}$, and chooses any $x \in T$. Denote the maximum number of consecutive lies by $m_i$ when $x = i$. And define $\phi = \sum_{i=1}^{n+1} \lambda^{m_i}$. The strategy of $A$ is to choose to lie or not such that $\phi$ takes the smaller value. We will show that $\phi < \lambda^{k+1}$ at any time with the strategy stated above. Thus $m_i \leqslant k$ for each $i \in T$. So $B$ cannot determine whether $i = x$ or not. Thus, player $B$ cannot guarantee a win.

$$\phi = \min(\phi_1, \phi_2) \leqslant \frac{1}{2}(\phi_1 + \phi_2) = \frac{1}{2}(\lambda\phi + n + 1)$$

$$< \frac{1}{2}(\lambda^{k+2} + (2 - \lambda)\lambda^{k+1}) = \lambda^{k+1}.$$

At the beginning, $m_i = 0$, so $\phi = n + 1 < \lambda^{k+1}$. Suppose that after several answers, $\phi < \lambda^{k+1}$, and $B$ asks: Is $x \in S$? The answer "yes" or "no" corresponding to two numbers of $\phi$, respectively, is as follows: $\phi_1 = \sum_{i \in S} 1 + \sum_{i \notin S} \lambda^{m_i+1}$ and $\phi_2 = \sum_{i \notin S} + \sum_{i \in S} \lambda^{m_i+1}$. By definition,

$$\phi = \min(\phi_1, \phi_2) \leqslant \frac{1}{2}(\phi_1 + \phi_2) = \frac{1}{2}(\lambda\phi + n + 1)$$

$$< \frac{1}{2}(\lambda^{k+2} + (2 - \lambda)\lambda^{k+1}) = \lambda^{k+1}. \qquad \square$$

## Second Day
### 9: 00~13: 30, July 11, 2012

**4** Find all functions $f: \mathbf{Z} \rightarrow \mathbf{Z}$ such that, for all integers $a$, $b$, $c$ that satisfy $a + b + c = 0$, the following equality holds:

$$f^2(a) + f^2(b) + f^2(c)$$
$$= 2f(a)f(b) + 2f(b)f(c) + 2f(c)f(a). \qquad \text{①}$$

(Here $\mathbf{Z}$ denotes the set of integers.)

**Solution.** Let $a = b = c = 0$ yield $3f^2(0) = 6f^2(0)$. Hence,

$$f(0) = 0. \qquad \text{②}$$

Let $b = -a$, $c = 0$ yield $(f(a) - f(-a))^2 = 0$. Hence, $f$ is odd, that is,

$$f(a) = f(-a), \forall a \in \mathbf{Z}. \qquad \text{③}$$

Let $b = a$, $c = -2a$ yield $2f(a)^2 + f(2a)^2 = 2f(a)^2 + 4f(a)f(2a)$. Hence,

$$f(2a) = 0 \text{ or } f(2a) = 4f(a), \forall a \in \mathbf{Z}. \qquad \text{④}$$

If $f(r) = 0$ for some $r \geqslant 1$, then let $b = r$, $c = -a - r$ yield

$(f(a + r) - f(a))^2 = 0$. That is, $f$ is a periodic function of period $r$:

$$f(a + r) = f(a), \ \forall a \in \mathbf{Z}.$$

Especially, $f(a) = 0$, $\forall a \in \mathbf{Z}$, if $f(1) = 0$. And in this case, function $f(a) = 0$, $\forall a \in \mathbf{Z}$, satisfies the condition ①.

Now suppose that $f(1) = k \neq 0$. Then by ③, $f(2) = 0$, or $f(2) = 4k$. If $f(2) = 0$, then $f$ is a periodic function of period 2:

$$f(2n) = 0, \ f(2n + 1) = k, \ \forall n \in \mathbf{Z}.$$

If $f(2) = 4k \neq 0$, then by ④, $f(4) = 0$, or $f(4) = 16k$.

If $f(4) = 0$, then $f$ is a periodic function of period 4, and $f(3) = f(-1) = f(1) = k$. Hence,

$$f(4n) = 0, \ f(4n + 1) = f(4n + 3) = k,$$
$$f(4n + 2) = 4k, \ \forall n \in \mathbf{Z}.$$

If $f(4) = 16k \neq 0$, let $a = 1$, $b = 2$, $c = -3$ yield $f(3)^2 - 10kf(3) + 9k^2 = 0$.

Thus,

$$f(3) \in \{k, 9k\}. \tag{⑤}$$

Let $a = 1$, $b = 3$, $c = -4$ yield $f^2(3) - 34kf(3) + 225k^2 = 0$.

Thus,

$$f(3) \in \{9k, 25k\}. \tag{⑥}$$

Hence, by ⑤ and ⑥, $f(3) = 9k$.

Now we show that $f(x) = kx^2$, $\forall x \in \mathbf{N}$, by induction. For $x = 0, 1, 2, 3, 4$, we have $f(x) = kx^2$. Now suppose that $f(x) = kx^2$ for $x = 0, 1, \ldots, n(n \geq 4)$, then let $a = n - 1$, $b = 2$, $c = -n - 1$ in ① yield

$$f(n+1) \in \{k(n+1)^2, k(n-1)^2\}.$$

And let $a = n - 1$, $b = 2$, $c = -n - 1$ in ① yield

$$f(n+1) \in \{k(n+1)^2, k(n-3)^2\}.$$

Thus, $f(n+1) = k(n+1)^2$ for $n \geqslant 4$. Therefore, $f(x) = kx^2$ for all $x \in \mathbf{N}$. By ③, $f(x) = kx^2$ for all $x \in \mathbf{Z}$.

To sum up all the above cases, we now check the final result:

$$f_1(x) = 0, \quad f_2(x) = kx^2,$$

$$f_3(x) = \begin{cases} 0, & x \equiv 0 \pmod 2, \\ k, & x \equiv 1 \pmod 2, \end{cases}$$

$$f_4(x) = \begin{cases} 0, & x \equiv 0 \pmod 4, \\ k, & x \equiv 1 \pmod 2, \\ 4k, & x \equiv 2 \pmod 4. \end{cases}$$

Obviously, $f_1$, $f_2$ satisfy ①. For $f_3$, if $a$, $b$, $c$ are even, then $f(a) = f(b) = f(c) = 0$ satisfy ①; if there are two odd and one even in $a$, $b$ and $c$, then both sides of ① are equal to $2k^2$.

For $f_4$, since $a + b + c = 0$, $(f(a), f(b), f(c))$ has only four cases: $(0, k, k)$, $(4k, k, k)$, $(0, 0, 0)$ and $(0, 4k, 4k)$, obviously, they all satisfy ①. ☐

**5** Let $\triangle ABC$ be a triangle with $\angle BCA = 90°$, and let $D$ be the foot of the altitude from $C$. Let $X$ be a point in the interior of the segment $CD$. Let $K$ be the point on the segment $AX$ such that $BK = BC$. Similarly, let $L$ be the point on the segment $BX$ such that $AL = AC$. Let $M$ be the point of intersection of $AL$ and $BK$.

Show that $MK = ML$.

**Solution.** Let points $C'$ and $C$ be symmetric over the line $AB$, and $\omega_1$ and $\omega_2$ be circles with centres $A$ and $B$, radius $AL$ and $BK$, respectively. Since $AC' = AC$ = $AL$ and $BC' = BC = BK$, points $C$ and $C'$ are both on circles $\omega_1$ and $\omega_2$. Since $\angle BCA = 90°$, lines $AC$ and $BC$ are tangent to circles $\omega_2$ and $\omega_1$, respectively, at point $C$. Let $K_1$ be another intersection point of line $AX$ and circle $\omega_2$ different to $K$, and $L_1$ be another intersection point of line $BX$ and circle $\omega_1$ different to $L$.

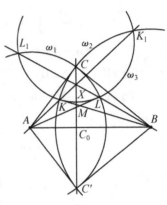

Fig. 5. 1

By the Circle-Power Theorem, we have

$$XK \cdot XK_1 = XC \cdot XC' = XL \cdot XL_1,$$

hence $K_1$, $L$, $K$, $L_1$ are four concyclic points at circle denoted by $\omega_3$.

By applying Circle-Power Theorem to circle $\omega_2$, we have

$$AL^2 = AC^2 = AK \cdot AK_1.$$

This implies the line $AL$ is tangent to circle $\omega_3$ at point $L$. By the similar argument, line $BK$ is tangent to circle $\omega_3$ at point $K$.

Therefore, $MK$ and $ML$ are two tangent lines from point $M$ to circle $\omega_3$, thus $MK = ML$. $\qquad\square$

**6** Find all positive integers $n$ for which there exist non-negative integers $a_1$, $a_2$, $a_3$, $\ldots$, $a_n$, such that

$$\frac{1}{2^{a_1}} + \frac{1}{2^{a_2}} + \cdots + \frac{1}{2^{a_n}} = \frac{1}{3^{a_1}} + \frac{2}{3^{a_2}} + \cdots + \frac{n}{3^{a_n}} = 1. \qquad ①$$

**Solution.** Suppose that $n$ satisfies condition ①, that is, there exist non-negative integers $a_1 \leqslant a_2 \leqslant a_3 \leqslant \cdots \leqslant a_n$, such that

$$\sum_{i=1}^{n} 2^{-a_i} = \sum_{i=1}^{n} i \cdot 3^{-a_i} = 1.$$

Multiplying both sides of the second equality by $a_n$, and then taking mod 2, we have

$$\sum_{i=1}^{n} i = \frac{1}{2}n(n+1) \equiv 1 \pmod 2,$$

hence, $n \equiv 1$, $2 \pmod 4$. In the following, we shall show that this is also a sufficient condition. That is, $n$ are all integers such that $n \equiv 1$, $2 \pmod 4$.

We call a set of finite repeatable positive integers $B = \{b_1, b_2, \ldots, b_n\}$ as "*feasible*", if the elements of $B$ can correspond to non-negative integers $a_1$, $a_2$, $\ldots$, $a_n$ such that

$$\sum_{i=1}^{n} 2^{-a_i} = \sum_{i=1}^{n} b_i 3^{-a_i} = 1.$$

Note an important fact that if $B$ is feasible, replacing any element $b$ of $B$ by two positive integers $u$ and $v$ with $u + v = 3b$, then the resulting set $B'$ is also feasible. In fact, if $b$ corresponds to non-negative integer $a$, we take both $u$ and $v$ corresponding to $a + 1$, keeping others correspondence unchanged. Since

$$2^{-a-1} + 2^{-a-1} = 2^{-a},$$
$$u \cdot 3^{-a-1} + v \cdot 3^{-a-1} = b \cdot 3^{-a},$$

$B'$ is feasible. If $B'$ can be obtained by such finite replacements from $B$, we denote them by $B \rightsquigarrow B'$. Particularly, if $b \in B$, then we can replace $b$ by $b$ and $2b$, thus, $B \rightsquigarrow B \cup \{2b\}$.

We shall show that for each positive integer $n \equiv 1$, 2(mod

4), $B_n = \{1, 2, \ldots, n\}$ is feasible. $B_1$ is feasible, since we can take $a_1 = 0$. $B_2$ is feasible, since $B_1 \rightsquigarrow B_2$. If $B_n$ is feasible for $n \equiv 1 \pmod{4}$, then since $B_n \rightsquigarrow B_{n+1}$, $B_{n+1}$ is also feasible.

Thus, it suffices to show that $B_n$ is feasible for $n \equiv 1 \pmod{4}$. $B_5$ is feasible, since

$$B_2 \rightsquigarrow \{1, 3, 3\} \rightsquigarrow \{1, 3, 4, 5\} \rightsquigarrow B_5.$$

$B_9$ is feasible, since

$$B_5 \rightsquigarrow \{1, 2, 3, 4, 6, 9\} \rightsquigarrow \{1, 2, 3, 5, 6, 7, 9\} \rightsquigarrow B_9 \setminus \{8\} \rightsquigarrow B_9.$$

$B_{13}$ is feasible, since

$$B_9 \rightsquigarrow \{1, 2, 3, 4, 5, 6, 7, 9, 11, 13\} \rightsquigarrow B_{13},$$

The last step is obtained by appending $2b$ several times. Appending 8, 10 and 12 successively, we get that $B_{17}$ is feasible, since

$$\begin{aligned}
B_6 &\rightsquigarrow B_5 \cup \{7, 11\} \rightsquigarrow B_8 \cup \{11\} \rightsquigarrow B_7 \cup \{9, 11, 15\} \\
&\rightsquigarrow B_{12} \cup \{15\} \rightsquigarrow B_{17} \setminus \{10, 14, 16\} \rightsquigarrow B_{17}.
\end{aligned}$$

Lastly, we show that $B_{4k+2} \rightsquigarrow B_{4k+13}$ for any integer $k \geqslant 2$, and complete the proof.

By successively appending $4k + 4$, $4k + 6$, $\ldots$, $4k + 12$, we note that

$$(4k + 12)/2 \leqslant 4k + 2.$$

In the remaining six odd numbers $4k + 3$, $4k + 5$, $\ldots$, $4k + 13$, denote the numbers that are multiples of 3 by $u_1$, $v_1$, the sum of two of which is a multiple of 3 by $u_2$, $v_2$ and $u_3$, $v_3$. Then substitute $b_i = \frac{1}{3}(u_i + v_i)$ by $u_i$, $v_i$ for $i = 1, 2, 3$. Note that $b_i$ is even, so we append $b_i/2$ and get $B_{4k+13}$.  $\square$

# 2013 (Santa Marta, Colombia)

## First Day

9: 00~13: 30, July 23, 2013

**1** Prove that for any pair of positive integers $k$ and $n$, there exist $k$ positive integers $m_1$, $m_2$, ..., $m_k$ (not necessarily different) such that

$$1 + \frac{2^k - 1}{n} = \left(1 + \frac{1}{m_1}\right)\left(1 + \frac{1}{m_2}\right)\cdots\left(1 + \frac{1}{m_k}\right). \qquad ①$$

(posed by Japan)

**Solution.** *Method 1*. By induction on $k$, the case of $k = 1$ is trivial. Suppose that the proposition is true for $k = j - 1$, we show that the case of $k = j$ is also true.

If $n$ is odd, that is, $n = 2t - 1$ for some positive integer $t$, note that

$$\begin{aligned}
1 + \frac{2^j - 1}{2t - 1} &= \frac{2(t + 2^{j-1} - 1)}{2t} \cdot \frac{2t}{2t - 1} \\
&= \left(1 + \frac{2^{j-1} - 1}{t}\right)\left(1 + \frac{1}{2t - 1}\right),
\end{aligned}$$

and by the hypothesis of induction, we can find $m_1$, ..., $m_{j-1}$ such that

$$1 + \frac{2^{j-1} - 1}{t} = \left(1 + \frac{1}{m_1}\right)\left(1 + \frac{1}{m_2}\right)\cdots\left(1 + \frac{1}{m_{j-1}}\right).$$

Thus, we need only to take $m_j = 2t - 1$.

If $n$ is even, that is, $n = 2t$ for some positive integer $t$, note that

$$1 + \frac{2^j - 1}{2t} = \frac{2t + 2^j - 1}{2t + 2^j - 2} \cdot \frac{2t + 2^j - 2}{2t}$$

$$= \left(1 + \frac{1}{2t + 2^j - 2}\right)\left(1 + \frac{2^{j-1} - 1}{t}\right)$$

and $2t + 2^j - 2 > 0$, by the hypothesis of induction, we can find $m_1, \ldots, m_{j-1}$ such that

$$1 + \frac{2^{j-1} - 1}{t} = \left(1 + \frac{1}{m_1}\right)\left(1 + \frac{1}{m_2}\right)\cdots\left(1 + \frac{1}{m_{j-1}}\right).$$

Thus, we need only to take $m_j = 2t + 2^j - 2$.

*Method 2.* Consider the binary expansion of the remainders of mod $2^k$ of $n$ and $-n$:

$$n - 1 \equiv 2^{a_1} + 2^{a_2} + \cdots + 2^{a_r} \pmod{2^k},$$

where $0 \leqslant a_1 < a_2 < \cdots < a_r \leqslant k - 1$, and

$$-n \equiv 2^{b_1} + 2^{b_2} + \cdots + 2^{b_s} \pmod{2^k},$$

where $0 \leqslant b_1 < b_2 < \cdots < b_s \leqslant k - 1$.

Since

$$-1 \equiv 2^0 + 2^1 + \cdots + 2^{k-1} \pmod{2^k},$$

we have

$$\{a_1, a_2, \ldots, a_r\} \cup \{b_1, b_2, \ldots, b_s\}$$
$$= \{0, 1, \ldots, k - 1\}, \text{ and } r + s = k.$$

For $1 \leqslant p \leqslant r$, $1 \leqslant q \leqslant s$, we denote

$$S_p = 2^{a_p} + 2^{a_{p+1}} + \cdots + 2^{a_r},$$
$$T_q = 2^{b_1} + 2^{b_2} + \cdots + 2^{b_q}.$$

And define $S_{r+1} = T_0 = 0$. Note that

$$S_1 + T_s = 2^k - 1$$

and $n + T_s \equiv 0 \pmod{2^k}$. We have

$$1 + \frac{2^k - 1}{n} = \frac{n + S_1 + T_s}{n}$$

$$= \frac{n + S_1 + T_s}{n + T_s} \cdot \frac{n + T_s}{n}$$

$$= \prod_{p=1}^{r} \frac{n + S_p + T_s}{n + S_{p+1} + T_s} \cdot \prod_{q=1}^{s} \frac{n + T_q}{n + T_{q-1}}$$

$$= \prod_{p=1}^{r} \left(1 + \frac{2^{a_p}}{n + S_{p+1} + T_s}\right) \cdot \prod_{q=1}^{s} \left(1 + \frac{2^{b_q}}{n + T_{q-1}}\right).$$

Thus, if for $1 \leqslant p \leqslant r$, $1 \leqslant q \leqslant s$, define $m_p = \frac{n + S_{p+1} + T_s}{2^{a_p}}$ and $m_{r+q} = \frac{n + T_{q-1}}{2^{b_q}}$, then we get the required equality. We need only to prove that all $m_i$ are integers. For $1 \leqslant p \leqslant r$, we know that

$$n + S_{p+1} + T_s \equiv n + T_s \equiv 0 \pmod{2^{a_p}}.$$

And for $1 \leqslant q \leqslant s$, we have

$$n + T_{q-1} \equiv n + T_s \equiv 0 \pmod{2^{b_q}},$$

which obtains the conclusion.

*Method 3*. For any $a (\neq 0, -1)$, we have

$$\left(1 + \frac{1}{a}\right)\left(1 + \frac{1}{a+1}\right) = \left(1 + \frac{1}{\frac{a}{2}}\right). \qquad \textcircled{2}$$

We rewrite the left-hand side of $\textcircled{1}$ in the form of the product of $2^k - 1$ fractions:

$$\frac{n + 2^k - 1}{n} = \left(1 + \frac{1}{n}\right)\left(1 + \frac{1}{n+1}\right)\left(1 + \frac{1}{n+2}\right) \cdots$$
$$\left(1 + \frac{1}{n + 2^k - 3}\right)\left(1 + \frac{1}{n + 2^k - 2}\right). \qquad \textcircled{3}$$

For the even $n$, grouping successively the right-hand side of $\textcircled{3}$ in pairs from the left to the right and by using $\textcircled{2}$ we get the form of product of $2^{k-1}$ fractions:

$$\left(\left(1+\dfrac{1}{\dfrac{n}{2}}\right)\left(1+\dfrac{1}{\dfrac{n}{2}+1}\right)\cdots\left(1+\dfrac{1}{\dfrac{n}{2}+2^{k-1}-2}\right)\right)\left(1+\dfrac{1}{n+2^{k}-2}\right).$$

④

For the even $n$, grouping the right-hand side of ③ in pare from the right to the left and by using ②, we get the form of product of $2^{k-1}$ fractions:

$$\left(1+\dfrac{1}{n}\right)\left(\left(1+\dfrac{1}{\dfrac{n+1}{2}}\right)\left(1+\dfrac{1}{\dfrac{n+1}{2}+1}\right)\cdots\right.$$

⑤

$$\left.\left(1+\dfrac{1}{\dfrac{n+1}{2}+2^{k-1}-3}\right)\left(1+\dfrac{1}{(n+1)/2+2^{k-1}-2}\right)\right).$$

Repeating the above grouping process to the fractions in big parentheses of ④ and ⑤ $k-2$ times, respectively, we will get the form of the right-hand side of ①. □

**2** A configuration of 4027 points in the plane is called *Colombian* if it consists of 2013 red points and 2014 blue points, and no three points of the configuration are collinear. By drawing some lines, the plane is divided into several regions. An arrangement of lines is *good* for a Colombian configuration if the following two conditions are satisfied:

• No line passes through any point of the configuration.
• No region contains points of both colours.

Find the least value of $k$ such that for any Colombian configuration of 4027 points, there is a good arrangement of $k$ lines. (posed by Australia)

**Solution 1.** The answer is $k = 2013$.

Firstly, we show that $k \geqslant 2013$ with an example. We mark 2013 red points and 2013 blue points on a circle alternatively. Then there are 4026 arcs on the circle with different endpoint colours. Mark another point in blue on the plane. If $k$ lines is good, then each arc intersects with some line and any line intersects the circle at most two points, hence there are at least $4026/2 = 2013$ lines.

In the following, we show that there exists a good arrangement of 2013 lines.

First, note that for two points $A$ and $B$ with the same colour, we can separate these two points with others by drawing two sufficient near parallel lines to $AB$ on each side of $AB$.

Let $P$ be the convex hull of all coloured points. Take two adjacent vertices of $P$, say points $A$ and $B$. Then other points are located on one side of line $AB$. If one of them is red, say $A$. Then we can draw one line to separate $A$ with other points. The remaining 2012 red points can be grouped in 1006 pairs; each pair of red points can be separated by two lines. So all together 2013 lines meet the requirement. If $A$ and $B$ are all blue, then they can be separated with other vertices by using a line. The remaining 2012 blue points can be grouped in 1006 pairs; each pair of blue points can be separated by 2 lines. So all together 2013 lines meet the requirements.

**Solution 2.** We have a more general result as follows.

Suppose there are $n$ points in red or blue on a plane, and no three of them are collinear. Then there exist $[n/2]$ lines that meet the requirements.

**Proof.** We prove by induction. The conclusion is obvious if $n \leqslant 2$. For $n \geqslant 3$, consider a line $\ell$ passing points $A$ and $B$, such that the remaining points are on one side of line $\ell$. By the hypotheses of induction, the remaining points have $[n/2] - 1$ lines that meet the requirements.

If $A$ and $B$ have the same colour or in different colours but are separated by a line, we can separate $A$ and $B$ with other points by a line parallel to $\ell$. Thus, $[n/2]$ lines meet the requirements. If $A$ and $B$ are in different colours but both in a region $R$, then the remaining points in the region $R$ do not have a colour, say, blue, we can draw a line to separate the blue point of $A$ or $B$ with other points. Thus, the $[n/2]$ lines meet the requirements and complete the induction.    □

*Remark:* We can ask a general problem that substitutes 2013 and 2014 by any positive integers $m$ and $n$, respectively, $m \leqslant n$. Denoted the solution to the general problem by $f(m, n)$.

We can obtain that $m \leqslant f(m, n) \leqslant m + 1$ by the idea of Solution 1. And if $m$ is even, then $f(m, n) = m$. On the other hand, for the case that $m$ is odd, then there is an $N$, such that for any $m \leqslant n \leqslant N$, we have $f(m, n) = m$, and for any $n > N$, we have $f(m, n) = m + 1$.

③ Suppose that the excircle of $\triangle ABC$ opposite the vertex $A$ touches the side $BC$ at the point $A_1$. Define the points $B_1$ on $CA$ and $C_1$ on $AB$ analogously by using the excircles opposite to $B$ and $C$, respectively. Suppose that the circumcentre of $\triangle A_1 B_1 C_1$ lies on the circumcircle of

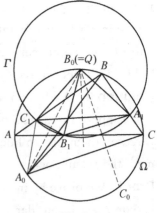

Fig. 3. 1

$\triangle ABC$. Prove that $\triangle ABC$ is right-angled. (The excircle of $\triangle ABC$ opposite the vertex $A$ is the circle that touched the line segment $BC$ to the ray $AB$ beyond $B$ and to the ray $AC$ beyond $C$. The excircles opposite $B$ and $C$ are similarly defined. ) (posed by Russia)

**Solution.** Denote the circumcircles of $\triangle ABC$ and $\triangle A_1 B_1 C_1$ by $\Omega$ and $\Gamma$, respectively. Let $A_0$ be the midpoint of arc $\overset{\frown}{BC}$ on $\Omega$ containing point $A$. Points $B_0$ and $C_0$ are defined analogously. Let $Q$ be the centre of circle $\Gamma$, then $Q$ is on $\Omega$ by the hypothesis of the problem. First, we give the following lemma.

*Lemma.* $A_0 B_1 = A_0 C_1$. Points $A$, $A_0$, $B_1$ and $C_1$ are concyclic.

If points $A_0$ and $A$ coincide, then $\triangle ABC$ is isosceles triangle, thus $AB_1 = AC_1$. Otherwise, $A_0 B = A_0 C$ by the definition of $A_0$. It is evident that

$$BC_1 = CB_1 \left( = \frac{1}{2}(b + c - a) \right),$$

and

$$\angle C_1 BA_0 = \angle ABA_0 = \angle ACA_0 = \angle B_1 CA_0.$$

Thus, $\qquad \triangle A_0 BC_1 \cong \triangle A_0 CB_1.$ $\qquad\qquad$ ①

So, we have $A_0 B_1 = A_0 C_1$.

Also, by ①, we have $\angle A_0 C_1 B = \angle A_0 B_1 C$, hence $\angle A_0 C_1 A = \angle A_0 B_1 A$. Thus, points $A$, $A_0$, $B_1$ and $C_1$ are concyclic.

Obviously, points $A_1$, $B_1$ and $C_1$ are on a semi-arc of $\Gamma$, thus $\triangle A_1 B_1 C_1$ is an obtuse-angled triangle. Without loss of generality, we may suppose that $\angle A_1 B_1 C_1$ is obtuse angle, then points $Q$ and $B_1$ are on different sides of $A_1 C_1$. Obviously, so are points $B$ and $B_1$, hence, points $Q$ and $B$ are on the same side of $A_1 C_1$.

Note that the perpendicular bisector of $A_1 C_1$ intersects with $\Gamma$ at two points which are on different sides of $A_1 C_1$. By the above argument, $B_0$ and $Q$ are among the intersection points, and $B_0$ and $Q$ are on the same side of $A_1 C_1$. So $B_0 = Q$, as shown in the diagram.

By the lemma, lines $QA_0$ and $QC_0$ are perpendicular bisectors of $B_1 C_1$ and $A_1 B_1$, respectively, and $A_0$ and $C_0$ are midpoints of arcs $CB$ and $BA$, respectively. Therefore,

$$\angle C_1 B_0 A_1 = \angle C_1 B_0 B_1 + \angle B_1 B_0 A_1 = 2\angle A_0 B_0 B_1 + 2\angle B_1 B_0 C_0$$
$$= 2\angle A_0 B_0 C_0 = 180° - \angle ABC.$$

On the other hand, by the lemma once more, we have

$$\angle C_1 B_0 A_1 = \angle C_1 B A_1 = \angle ABC.$$

Hence, $\angle ABC = 180° - \angle ABC$, so $\angle ABC = 90°$. This completes the proof. □

# Second Day
### 9: 00~13: 30, July 24, 2013

**4** Let $\triangle ABC$ be an acute-angled triangle with orthocenter $H$, and let $W$ be a point on the side $BC$, lying strictly between $B$ and $C$. The points $M$ and $N$ are the feet of the altitudes from $B$ and $C$, respectively. Denote by $\omega_1$ the circumcircle of $\triangle BWN$, and let $X$ be the point on the $\omega_1$ such that $WX$ is the diameter of $\omega_1$. Analogously, denote by $\omega_2$ the circumcircle of $\triangle CWM$, and $Y$ be the point on $\omega_2$ such that $WY$ is a diameter of $\omega_2$. Prove that $X$, $Y$ and

*H* are collinear. (posed by Thailand)

**Solution.** Let *AL* be the altitude to side *BC*, and *Z* be the intersection point of circles $\omega_1$ and $\omega_2$ other than *W*. We show that points *X*, *Y*, *Z* and *H* are collinear.

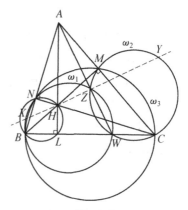

Points *B*, *C*, *M* and *N* are concyclic (denote the circle by $\omega_3$) since $\angle BNC = \angle BMC = 90°$. *WZ*, *BN* and *CM* intersect at a point since they are the radical axis of $\omega_1$ and $\omega_2$, $\omega_1$ and $\omega_3$, $\omega_2$ and $\omega_3$, respectively. And since *BN* and *CM* intersect at *A*, *WZ* passes through *A*.

Fig. 4.1

$\angle WZX = \angle WZY = 90°$ since *WX* and *WY* are the diameters of $\omega_1$ and $\omega_2$, respectively. Thus, points *X* and *Y* are on the line *l* perpendicular from *Z* to *WZ*.

So, points *B*, *L*, *H* and *N* are concyclic since $\angle BNH = \angle BLH = 90°$. By the Circle-Power Theorem,

$$AL \cdot AH = AB \cdot AN = AW \cdot AZ. \qquad ①$$

If point *H* is on the line *AW*, then *H* and *Z* coincide. Otherwise, by ①, we have

$$\frac{AZ}{AH} = \frac{AL}{AW}.$$

Thus, $\triangle AHZ \backsim \triangle AWL$, consequently, $\angle HZA = \angle WLA = 90°$, hence, point *H* is also on line *l*. $\qquad \square$

**5** Let $\mathbf{Q}^+$ be the set of all positive rational numbers. Let *f* : $\mathbf{Q}^+ \to \mathbf{R}$ be a function satisfying the following three conditions:

(i) for all $x$, $y \in \mathbf{Q}^+$, we have $f(x)f(y) \geqslant f(xy)$;

(ii) for all $x$, $y \in \mathbf{Q}^+$, we have $f(x+y) \geqslant f(x)+f(y)$;

(iii) there exists a rational number $a > 1$ such that $f(a) = a$.

    Prove that $f(x) = x$ for all $x \in \mathbf{Q}^+$. (posed by Bulgaria)

**Solution.** Let $x = 1$, $y = a$ in (i), we have

$$f(1) \geqslant 1. \qquad \qquad ①$$

By (ii) and induction on $n$, we have

$$f(nx) \geqslant nf(x), \forall n \in \mathbf{Z}^+ \text{ (the set of all positive integers)},$$
$$\forall x \in \mathbf{Q}^+. \qquad \qquad ②$$

Taking $x = 1$ in ②, we have

$$f(n) \geqslant nf(1) \geqslant n. \ \forall n \in \mathbf{Z}^+. \qquad \qquad ③$$

By (i) once again, we have

$$f\left(\frac{m}{n}\right)f(n) \geqslant f(m), \forall m, n \in \mathbf{Z}^+. \qquad \qquad ④$$

Hence, by ③ and ④, we have

$$f(q) > 0, \forall q \in \mathbf{Q}^+. \qquad \qquad ⑤$$

Then by (ii) and ⑤, $f$ is strictly increasing, and

$$f(x) \geqslant f(\lfloor x \rfloor) \geqslant \lfloor x \rfloor > x - 1, \forall x \in \mathbf{Q}^+, x > 1. \qquad ⑥$$

By (i) and induction on $n$, we have

$$f^n(x) \geqslant f(x^n), \forall n \in \mathbf{Z}^+, \forall x \in \mathbf{Q}^+, \qquad \qquad ⑦$$

Thus, by ⑥ and ⑦, we have

$$f^n(x) \geqslant f(x^n) > x^n - 1, \forall x \in \mathbf{Q}^+, x > 1. \qquad ⑧$$

Then by ⑧, we have

$$f(x) > \sqrt[n]{x^n - 1} \, , \, \forall n \in \mathbf{Z}^+ , \, \forall x \in \mathbf{Q}^+ , \, x > 1. \qquad ⑨$$

Taking the limit as $n$ tends to the infinity in ⑨, we have

$$f(x) \geqslant x \,\, \forall x \in \mathbf{Q}^+ , \, x > 1. \qquad ⑩$$

By (iii), ⑦ and ⑩, we have $a^n = f^n(a) \geqslant f(a^n) \geqslant a^n$. Thus,

$$f(a^n) = a^n. \qquad (11)$$

For all $x \in \mathbf{Q}^+ , \, x > 1$, we may choose $n \in \mathbf{Z}^+$, such that $a^n - x > 1$. By (11), (ii) and ⑩, we have

$$a^n = f(a^n) \geqslant f(x) + f(a^n - x) \geqslant x + (a^n - x) = a^n.$$

Therefore,

$$f(x) = x \text{ for all } x \in \mathbf{Q}^+ , \, x > 1. \qquad (12)$$

Lastly, for all $x \in \mathbf{Q}^+$, and for all $n \in \mathbf{Z}^+$, by (12), (i) and ②, we have

$$nf(x) = f(n)f(x) \geqslant f(nx) \geqslant nf(x),$$
$$\forall n \in \mathbf{Z}^+ , \, n > 1. \,\, \forall x \in \mathbf{Q}^+ , \qquad (13)$$

that is,

$$f(nx) = nf(x), \, \forall n \in \mathbf{Z}^+ , \, n > 1, \, \forall x \in \mathbf{Q}^+ , \qquad (14)$$

Taking $x = m/n$ for all $m \leqslant n \in \mathbf{Z}^+ , \, n > 1$, we have

$$f\left(\frac{m}{n}\right) = \frac{f(m)}{n} = \frac{m}{n}.$$

That is, $f(x) = x$ for all $x \in \mathbf{Q}^+ , \, x \leqslant 1$. $\qquad \square$

**Remark.** The condition $f(a) = a > 1$ is essential. In fact, for $b \geqslant 1$, function $f(x) = bx^2$ satisfies (i) and (ii) for all $x$, $y \in \mathbf{Q}^+$, and $f$ has a unique fixed point $\frac{1}{b} \leqslant 1$.

**6** Suppose that integer $n \geq 3$, and consider a circle with $n + 1$ equally spaced points marked on it. Consider all labellings of these points with the numbers $0, 1, \ldots, n$ such that each number is used exactly once; two such labellings are considered to be the same if one can be obtained from the other by a rotation of the circle. A labelling is called *beautiful* if, for any four labels $a < b < c < d$ with $a + d = b + c$, the chord $(a, d)$ does not intersect the chord $(b, c)$.

Let $M$ be the number of beautiful labellings, and let $N$ be the number of chords $(x, y)$ such that $x < y$, $x + y \leq n$ and gcd $(x, y) = 1$. Prove that $M = N + 1$. (posed by Russia)

**Solution 1.** Note that the distance between marked points does not matter. The intersection of the chords only depends on the order of the points. For a circulation of permutation of $[0, n] = \{0, 1, \ldots, n\}$, we call a chord $(x, y)$ a $k$-chord if $x + y = k$. If $x = y$, then the chord degenerates. We call three disjoint chords are in order if a chord parts other two chords, e. g., in Fig. 6. 1, chords $A$, $B$ and $C$ are in order but chords $B$, $C$ and $D$ not. We call $m \geq 3$ disjoint chords are in order, if any three chords are in order.

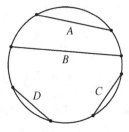

Fig. 6. 1

**Auxiliary Lemma.** In a beautiful labelling, all $k$-chords are in order for any integer $k$.

**Proof.** We prove it by induction on $n$. The lemma is trivial for $n \leq 3$. For $n \geq 4$, suppose that there were a beautiful labelling $S$, such that three $k$-chords $A$, $B$ and $C$ were not in order. If $n$ is not the end point of chords $A$, $B$ and $C$, we can delete the point $n$ and obtain a beautiful labelling $S\backslash\{n\}$ of $[0, n-1]$. By the hypothesis of induction, chords $A$, $B$ and $C$ are in order. Similarly, If $0$ is not the end point of chords $A$, $B$ and $C$, we can delete the point $0$, and minus each labelling by $1$, so we obtain a beautiful labelling $S\backslash\{0\}$. By the hypothesis of induction, chords $A$, $B$ and $C$ are in order, which is a contradiction. Thus, $0$ and $n$ must appear among the end points $n$-chords of $A$, $B$ and $C$. Suppose that chords $(0, x)$ and $(y, n)$ are among $A$, $B$ and $C$. Then $n \geqslant 0 + x = k = n + y \geqslant n$, thus $x = n$ and $y = 0$. That is, $(0, n)$ is one of $n$-chords of $A$, $B$ and $C$. Without loss of generality, suppose that $C = (0, n)$.

Let chord $D = (u, v)$ be adjacent and parallel to chord $C$, see Fig. 6.2, and denote $t = u + v$. If $t = n$, then $n$-chords $A$, $B$ and $D$ are not in order in the beautiful labelling $S\backslash\{0, n\}$, which contradicts the hypothesis of induction. If $t < n$, then $t$-chord $(0, t)$ does not intersect $D$, then chord $C$ is

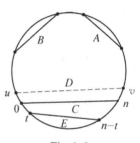

Fig. 6.2

apart point $t$ and chord $D$. And $n$-chord $E = (t, n-t)$ does not intersect chord $C$. So, points $t$ and $n-t$ are on the same side of $C$. But chords $A$, $B$ and $E$ are not in order, which is a contradiction.

Lastly, since the mapping of $x$ to $n - x$ conserves the beautifulness of a circular permutation, a $t$-chord maps to $(2n - t)$-chord, that is, $t > n$ equivalents to $t < n$. Thus, we have proved the auxiliary lemma.

In the following, we prove the problem by induction on $n$. The case of $n = 2$ is trivial. Let $n \geqslant 3$, and $S$ be a beautiful permutation of $[0, n]$. $T$ is obtained by deleting $n$ in $S$. All

$n$-chords in $T$ are in order, their end points include numbers $[0,$ $n-1]$. We call such $T$ is of first kind if $0$ is located between two $n$-chords, otherwise, we call such $T$ is of second kind. We shall show that each first kind of beautiful permutation of $[0, n-1]$ corresponds to exact one beautiful permutation of $[0, n]$. And each second kind of permutation of $[0, n-1]$ corresponds to exactly two beautiful permutations of $[0, n]$.

If $T$ is of the first kind, suppose that $0$ locates on the arc between chords $A$ and $B$. Since chords $A$, $(0, n)$ and chords $B$ in $S$ are in order, $n$ must locate on the other arc between chords $A$ and $B$. Thus, we can retrieve $S$ from $T$ uniquely. On the other hand, for each $T$ of the first kind, we can add $n$ in the above manner to obtain $S$. We shall show that cyclic permutation $S$ must be beautiful.

For $0 < k < n$, the $k$-chord of $S$ is also the $k$-chord of $T$, so $k$-chords are in order.

If $T$ is of the second kind, then the position of $n$ to the corresponding $S$ has two possibilities, that is, $n$ adjacent to $0$ on either side. Similarly, we can check that $S$ is a beautiful permutation of $[0, n]$.

Denote the total number of beautiful permutation of $[0, n]$ by $M_n$, the total number of the second kind of beautiful permutation of $[0, n-1]$ by $L_n$. Then we have

$$M_n = (M_{n-1} - L_{n-1}) + 2L_{n-1}$$
$$= M_{n-1} + L_{n-1}.$$

It suffices to show that $L_{n-1}$ is the number of positive integer pairs $(x, y)$ with the constraints $x + y = n$ and gcd $(x, y) = 1$.

For $n \geqslant 3$, define

$$\varphi(n) = \# \{x : 1 \leqslant x \leqslant n, \gcd(x, n) = 1\}.$$

We shall show that $L_{n-1} = \varphi(n)$. Consider a second kind of beautiful permutation of $[0, n-1]$, mark numbers $0, \ldots, n-1$ clockwise on the circle such that number $0$ coincides with the position $0$. Denote the number of position $i$ as $f(i)$. Suppose that $f(a) = n - 1$.

Since each number of $[1, \ldots, n-1]$ is at an end point of $n$-chord, and all $n$-chords are in order, by the definition of the second kind, $(1, n-1)$ is an $n$-chord. Thus, all $n$-chords are parallel. That is, $f(i) + f(n-i) = n$ for $i = 1, \ldots, n-1$.

Similarly, since all $(n-1)$-chords are in order and each point is an end point of an $(n-1)$-chord, all $(n-1)$-chords are parallel. Thus, $f(i) + f(\mod(a-i, n)) = n - 1$ for $i = 1, \ldots, n-1$.

Therefore, $f(-i) = f(\mod(a-i, n)) + 1$. Consequently, by taking $i = a, 2a, \ldots, (n-1)a$ successively, and $f(0) = 0$, we have

$$f(-ak) = \mod(k, n), \quad k = 0, 1, 2, \ldots, (n-1). \qquad ①$$

Since $f$ is a permutation, we must have $\gcd(a, n) = 1$. That is, $L_{n-1} \leqslant \varphi(n)$. To show that the equality holds for ①, we need only to show that the permutation $f$ given by ① is beautiful. To see this, we consider four numbers $w, x, y$ and $z$ on the circle satisfying $w + y = x + z$. Their corresponding positions on the circle satisfy $(-aw) + (-ay) \equiv (-ax) + (-az) \mod (n)$, that is, the chord $(w, y)$ and chord $(x, z)$ are parallel. Thus, the permutation ① is beautiful, and is of the second kind by construction. $\qquad \square$

## First Day

9: 00~13: 30, July 8, 2014

**1** Let $a_0 < a_1 < a_2 < \cdots$ be an infinite sequence of positive integers. Prove that there exists a unique positive integer $n$ such that

$$a_n < \frac{a_0 + a_1 + \cdots + a_n}{n} \leqslant a_{n+1}.$$

(posted by Austria)

**Solution.** Define $d_n = (a_0 + a_1 + \cdots + a_n) - na_n$, $n = 1, 2, \ldots$.

Note that

$$na_{n+1} - (a_0 + a_1 + \cdots + a_n)$$
$$= (n+1)a_{n+1} - (a_0 + a_1 + \cdots + a_n + a_{n+1})$$
$$= -d_{n+1}.$$

So, the problem is equivalent to prove that there exists a unique positive integer $n$ such that $d_n > 0 \geqslant d_{n+1}$.

We see that $d_1 = (a_0 + a_1) - 1 \cdot a_1 = a_0 > 0$, and

$$d_{n+1} - d_n = ((a_0 + a_1 + \cdots + a_n) - na_{n+1}) -$$
$$((a_0 + a_1 + \cdots + a_n) - na_n)$$
$$= n(a_n - a_{n+1}) < 0,$$

that is, $\{d_n\}$ is a strictly decreasing sequence of integers with the first term being positive. Hence, there exists a unique positive integer $n$, such that $d_n > 0 \geqslant d_{n+1}$. $\square$

**Remark.** It is essentially an intermediate theorem in discrete form.

**2** Let $n \geqslant 2$ be an integer. Consider an $n \times n$ chessboard consisting of $n^2$ unit squares. A configuration of $n$ rooks on this board is *peaceful* if every row and every column contains exactly one rook. Find the greatest positive integer $k$ such that for each peaceful configuration of $n$ rooks, there is an empty $k \times k$ square without any rooks on any of its $k^2$ unit squares. (posed by Croatia)

**Solution.** We shall show that the answer is $k_{max} = \left[\sqrt{n-1}\right]$ by two steps. Let $l$ be a positive integer.

(1) If $n > l^2$, then there exists an empty $l \times l$ square for any peaceful configuration.

(2) If $n \leqslant l^2$, then there exists a peaceful configuration, such that each $l \times l$ square is not empty.

**Proof of (1).** There is a row $R$ with first column has a rook. Take successively $l$ rows containing row $R$, which are denoted by $U$. If $n > l^2$, then $l^2 + 1 \leqslant n$, and from column 2 to column $l^2 + 1$ in $U$, containing $l$ $l \times l$ squares which have at most $l - 1$ rooks. So, at least one $l \times l$ square is empty.

**Proof of (2).** For $n = l^2$, we shall find a peaceful configuration which has no empty $l \times l$ square. We label the rows from bottom to top and the columns from left to right both by $0, 1, \ldots, l^2 - 1$. So denote by $(r, c)$ the unit square at the row $r$ and column $c$.

We put a rook at $(il + j, jl + i)$ for $i, j = 0, 1, 2, \ldots, l - 1$. The figure below shows the case for $l = 3$. Since each number between 0 to $l^2 - 1$ can be written uniquely in the form of $il + j$, $(0 \leqslant i, j \leqslant l - 1)$, such a configuration is peaceful.

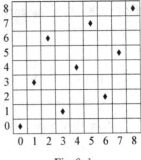

Fig. 2. 1

For any $l \times l$ square $A$, suppose that the lowest row of $A$ is row $pl + q$, $0 \leqslant p, q \leqslant l - 1$ (since $pl + q \leqslant l^2 - l$). There is one rook in $A$ by the configuration, and the column labels of the rook may be $ql + p$, $(q + 1)l + p, \ldots, (l - 1)l + p, p + 1, l + (p + 1), \ldots, (q - 1)l + p + 1$. Rearrange these numbers in increasing order to

$$p + 1, l + (p + 1), \ldots, (q - 1)l + p + 1, ql + p,$$
$$(q + 1)l + p, \ldots, (l - 1)l + p.$$

Then the first number is less than or equal to $l$, the last is greater than or equal to $(l - 1)l$ and the difference between two adjacent numbers is $l$. Therefore, there exists one rook in $A$.

For the case of $n < l^2$, consider the configuration above, but delete $l^2 - n$ columns and rows from the right and from the bottom, respectively, we get an $l \times l$ square. Thus, we obtain an $n \times n$ square, where there is no empty $l \times l$ square. But some rows and columns may be empty. We can put rooks at the cross squares of empty rows and columns to obtain the desired peaceful configuration. $\square$

**Remark.** The answer could also be in the form of $\lceil \sqrt{n} \rceil - 1$.

**3** A convex quadrilateral $ABCD$ has $\angle ABC = \angle CDA = 90°$. Point $H$ is the foot of the perpendicular from $A$ to $BD$. Points $S$ and $T$ lie on sides $AB$ and $AD$, respectively, such that $H$ lies inside $\triangle SCT$ and $\angle CHS - \angle CSB = 90°$,

$\angle THC - \angle DTC = 90°$.

Prove that line $BD$ is tangent to the circumcircle of $\triangle TSH$. (posed by Iran)

**Solution.** Suppose that the line passing $C$ and perpendicular to line $SC$ intersects line $AB$ at point $Q$, see the figure below. Then $\angle SQC = 90° - \angle BSC = 180° - \angle SHC$.

Hence, points $C$, $H$, $S$ and $Q$ are concyclic with $SQ$ its diameter. Therefore, the circumcentre $K$ of $\triangle SHC$ is on line $AB$. In the same manner, the circumcentre $L$ of $\triangle CHT$ is on line $AD$.

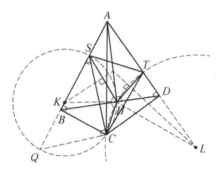

Fig. 3. 1

To show that line $BD$ is tangent to the circumcircle of $\triangle TSH$, it suffices to show that the intersection point of perpendicular bisectors of $HS$ and $HT$ is on line $AH$. But the perpendicular bisectors are just the angle bisectors of $\angle AKH$ and $\angle ALH$, respectively. By the Internal Angle Bisector Theorem, it suffices to show that

$$\frac{AK}{KH} = \frac{AL}{LH}. \qquad \text{①}$$

In the following, we will give two proofs of ①.

**Proof 1.** Let $M$ be the intersection point of lines $KL$ and $HC$, see Fig. 3. 2.

Since $KH = KC$ and $LH = LC$, points $H$ and $C$ are symmetric over line $KL$. Thus, $M$ is the mid-point of $HC$. Let $O$ be the circumcentre of the quadrilateral $ABCD$. Hence, $O$ is the mid-point of $AC$ and consequently $OM \parallel AH$, therefore $OM \perp BD$. Further by $OB = OD$, we see that $OM$ is the perpendicular bisector of $BD$, thus $BM = DM$.

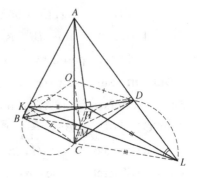

Fig. 3. 2

Since $CM \perp KL$, points $B$, $C$, $M$ and $K$ are concyclic with $KC$, the diameter. Similarly, points $L$, $C$, $M$ and $D$ are concyclic with $LC$, the diameter. So, by the Sine Theorem, we have

$$\frac{AK}{AL} = \frac{\sin \angle ALK}{\sin \angle AKL} = \frac{DM}{CL} \cdot \frac{CK}{BM} = \frac{CK}{CL} = \frac{KH}{LH},$$

that is, ①.

**Proof 2.** If points $A$, $H$ and $C$ are collinear, then $AK = AL$, $KH = LH$, thus ① follows. Else, let $\omega$ be the circle passing $A$, $H$ and $C$. Sine points $A$, $B$, $C$ and $D$ are on a circle

$$\angle BAC = \angle BDC = 90° - \angle ADH = \angle AHD.$$

Let $N \neq A$ be the other intersection point of the circle $\omega$ and the bisector of $\angle CAH$, then $AN$ is also the bisector of $\angle BAD$. Since points $H$ and $C$ are symmetric over line $KL$, and $HN = NC$, we see that point $N$ and the centre of $\omega$ are both on line $KL$. That is, the circle $\omega$ is the Apollonius circle of points $K$ and $L$, thus ① follows. □

**Remark.** Problem 3 has an extension by the problem selection committee as follows.

For a convex quadrilateral $ABCD$, points $S$ and $T$ are on sides $AB$ and $AD$, respectively. And point $H$ is located in the inner part of $\triangle SCT$, satisfying $\angle BAC = \angle DAH$, $\angle CHS - \angle CSB = 90°$ and $\angle THC - \angle DTC = 90°$.

Then the circumcentre of $\triangle TSH$ is on $AH$ and the circumcentre of $\triangle SCT$ is on $AC$. ☐

## Second Day

9: 00~13: 30, July 9, 2014

**4** Points $P$ and $Q$ lie on side $BC$ of acute-angled $\triangle ABC$ so that $\angle PAB = \angle BCA$ and $\angle CAQ = \angle ABC$. Points $M$ and $N$ lie on lines $AP$ and $AQ$, respectively, such that $P$ is the midpoint of $AM$, and $Q$ is the midpoint of $AN$. Prove that lines $BM$ and $CN$ intersect on the circumcircle of $\triangle ABC$. (posed by Georgia)

**Solution.** Let $S$ be the intersection point of lines $BM$ and $CN$, see the figure below. Denote $\beta = \angle QAC = \angle CBA$, $\gamma = \angle PAB = \angle ACB$, then we can see that $\triangle ABP \backsim \triangle CAQ$, thus

$$\frac{BP}{PM} = \frac{BP}{PA} = \frac{AQ}{QC} = \frac{NQ}{QC}.$$

By $\angle BPM = \beta + \gamma = \angle CQN$, $\triangle BPM \backsim \triangle NQC$, thus $\angle BMP = \angle NCQ$. Consequently, $\triangle BPM \backsim \triangle BSC$, thus $\angle CSB = \angle BPM = \beta + \gamma = 180° - \angle BAC$. ☐

**5** For each positive integer $n$, the bank of Cape Town issues coins of denomination $\frac{1}{n}$. Given a finite collection of such coins (of not necessary different denominations) with total value at most $99 + \dfrac{1}{2}$, prove

that it is possible to split this collection into 100 or fewer groups, such that each group has total value at most 1. (posed by Luxembourg)

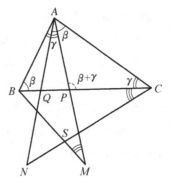

Fig. 5. 1

**Solution.** We shall prove a general conclusion: for any positive integer $N$, given a finite

collection of such coins with a total value at most $N - \dfrac{1}{2}$, prove

that it is possible to split this collection into $N$ or fewer groups, such that each group has a total value of at most 1.

If some coins have total value of $1/k$ ($k$ is a positive integer), we replace these coins by one coin of value $1/k$, which does not affect the problem. In this way, for each even integer $k$, at most one coin has value of $1/k$ (otherwise, two such coins may be replaced by one coin of value $2/k$); for each odd integer $k$, at most $(k-1)$ coins of value $1/k$, (otherwise, $k$ such coins can be replaced by one coin of value 1). So, we may suppose that there is no more replacement can be make for the coins.

First we take each coin of value 1 as a group. Suppose there are $d < N$ such groups. If there are no other coins, then the problem is solved. Otherwise, take coin of value $1/2$ as a group $(d+1)$ if there is any. Let $m = N - d \geqslant 1$. Then for each integer $k$ in $2, \ldots, m$, take coins of value $1/(2k-1)$ and value $1/(2k)$ in group $(d+k)$ if there are any, in which the total value does not exceed $(2k-2)/(2k-1) + 1/(2k) < 1$. For coins of value less than $1/(2m)$, if there are any, we can put them in some group $(d + j)$ such that the total value is less than 1(since if each group $(d + j)$ has value greater than $1 - 1/(2m)$, then the total value will be greater than $d + m(1 - 1/(2m)) = m + d - 1/2 = N - 1/2$). Repeat this procedure finite times, all coins are put in $N$ or fewer groups with each group of value at most one.                                                            $\square$

**6**    A set of lines in the plane is in *general position* if no two of them are parallel and no three of them pass through the same point. A set of lines in general position cuts the plane into regions, some of which have finite area; we call them finite regions. Prove that for all sufficiently large $n$, in any set of $n$ lines in general position, it is possible to colour at least $\sqrt{n}$ of the lines in blue such that none of its finite regions has a completely blue boundary.

(posed by Austria)

**Solution.** Let $B$ be the maximal set of lines such that no finite region has completely blue boundary. Let $|B| = k$, and colour the other $n - k$ lines in red. For each red line $\ell$, there is as least one finite region $A$ whose unique red side lies on $\ell$. We say a point is red if it is the intersection point of red and blue lines. And we say a point is blue if it is the intersection point of two blue lines. Denote the vertices of $A$ clockwise by $(p_1, p_2, \ldots, p_s)$ with point $p_1$ in red and $p_2$ in blue. Then we say the red line $\ell$ corresponds to the red point $p_1$ and blue point $p_2$. Now we are to show that any blue point can be corresponded to at most two red lines. (If this is true, then $n - k \leqslant 2\binom{k}{2} \Leftrightarrow n \leqslant k^2$.)

We prove it by contradiction. If not, suppose that there

are three red lines $\ell_1$, $\ell_2$ and $\ell_3$ correspond to a blue point $b$, and the corresponding red points on red lines $\ell_1$, $\ell_2$ and $\ell_3$ are $r_1$, $r_2$ and $r_3$, respectively. Let $b$ be the intersection point of two blue lines. Without loss of generality, we may suppose that sides $(r_2, b)$ and $(r_3, b)$ are on one of the two blue lines, and $(r_1, b)$ is the side of region $A$ on the other blue line. Since $A$ has only one red side, $A$ must be a triangle $\triangle r_1 b r_2$. But then red lines $\ell_1$ and $\ell_2$ pass $r_2$, and a blue line also passes $r_2$, which is a contradiction. □

Printed in the United States
By Bookmasters